NONLINEAR OPTICAL SYSTEMS

Principles,
Phenomena, and
Advanced Signal
Processing

Optics and Photonics

Series Editor
Le Nguyen Binh
Hua Wei Technologies, European Research Center, Munich, Germany

NONLINEAR OPTICAL SYSTEMS

Principles,

Phenomena, and

Advanced Signal

Processing

Edited by

Le Nguyen Binh

and Dang Van Liet

CRC Press
Taylor & Francis Group
Boca Raton London New York

CRC Press is an imprint of the
Taylor & Francis Group, an **informa** business

CRC Press
Taylor & Francis Group
6000 Broken Sound Parkway NW, Suite 300
Boca Raton, FL 33487-2742

First issued in paperback 2017

© 2012 by Taylor & Francis Group, LLC
CRC Press is an imprint of Taylor & Francis Group, an Informa business

No claim to original U.S. Government works
Version Date: 20120126

ISBN 13: 978-1-138-07276-3 (pbk)
ISBN 13: 978-1-4398-4547-9 (hbk)

Contents

Preface

The rates of competition of limited resources from different subsystems of a generic system would lead into chaotic states. This could happen in natural systems as well as in engineering systems, in particular nonlinear photonic ones. In energy storage systems if there is more than one storage element or subsystem then competition for energy would occur, and when they try to extract from or deposit their energy into each other, chaotic, bifurcation, or both phenomena would occur. The system can thus behave in a nonlinear manner, and hence the term *nonlinear systems*.

The most commonly known energy storage systems are those containing capacitors and inductors as charge and discharge competing elements, and photon storages with positive feedback into the resonance subsystem such as fiber lasers. Thus, higher-order differential equations can be employed to represent the dynamics of the evolution of the amplitude of the current or voltages or field amplitudes of the lightwaves in such systems.

This book selects to treat the nonlinear systems in terms of fundamental principles and associated phenomena as well as their applications in signal processing in contemporary optical systems for communications and laser systems with a touch of mathematical representation of nonlinear equations, which provides some insight into the nonlinear dynamics at different phases.

Therefore, Chapters 1 and 2 give an introduction to nonlinear systems and some mathematical representations, especially the routes to chaos and bifurcation and a brief introduction of nonlinear fiber lightwave lasing systems. Chapters 3, 4, 6, 7, and 8 describe the nonlinear phenomena in fiber lasers including both passive and active energy storage cavities, that is when the cavity is or is not modulated by an external excitation source. Soliton pulses with the lightwaves as the carrier under their envelopes are experimentally and theoretically demonstrated. Sequences of a single soliton as well as multiple numbers of solitons assembled in a group, the multibound solitons, are given and demonstrated. The interactions due to the phase states of the lightwaves under the soliton envelope lead to the binding states of these bound solitons, and their dynamics are treated in detail in Chapter 4. In Chapter 4 we then examine the evolution of these bound solitons when they are transmitted through single-mode optical fibers, which compose a phase variation system.

The evolution of the amplitude and phases of optical pulses through a guided medium can be represented by the well-known nonlinear Schrödinger equation (NLSE). Using this equation we studied different dynamics of solitons in fiber ring lasers and processing of optical systems in advanced optical communications systems. Two advanced applications are the processing of an ultrafast data pulse sequence in the optical domain via the representation of their bispectra. The bispectrum is a modern nonlinear processing technique that has been found to be the most appropriate match with our optical nonlinear systems. Furthermore the nonlinear transfer function of the NLSE can be derived via the use of the Volterra series that allows us to compensate for the distortion of pulses transmitted through the single-mode optical fiber in the digital electronic processing domain. This is described in the last chapter of this book.

The motivation and materials for this book were provided mainly by advanced research conducted in the universities of Osaka Prefecture and the Faculty of Physics of the University of Science of Ho Chi Minh City, Australian Universities, Nanayang Technological University of Singapore and in Advanced Technology Research Laboratories of Siemens, Hua Wei and Nortel Networks where one of the editors, Le N. Binh, has spent several fruitful years. An international workshop was held in 2010 at the University of Science of Ho Chi Min (HCM) City to address some issues treated in this book.

The research endeavors of many other people have contributed significantly to some sections of this book, especially research doctoral graduates of Le Binh over the years, in particular the following scholars, Dr. Lam Quoc Huy, Dr. Wenn Jing Lai, Dr. Nam Quoc Ngo, and Dr. Nguyen Duc Nhan.

We wish to thank contributors of a number of chapters of the book and especially Ashley Gasque of CRC Press for her encouragement and assistance in the formation of this book.

Le Nguyen Binh, Munich, Spring 2011
Dang Van Liet, HCM City, 2011

Contributors

Le Nguyen Binh
Hua Wei Technologies
European Research Center
Munich, Germany

Li Liangchuan
Hua Wei Technologies
Optical Networking Research
    Department
Shen Zhen, China

Dang Van Liet
Faculty of Physics
VNU University of Science
Ho Chi Minh City, Vietnam

Liu Ling
Hua Wei Technologies
Optical Networking Research
    Department
Shen Zhen, China

Nguyen Duc Nhan
Institute of Technology for Posts
    and Telecommunications
Hanoi, Vietnam

Tang Ding Yuan
School of Electrical and Electronic
    Engineering
Nanyang Technological University
Singapore

L. M. Zhao
School of Electrical and Electronic
    Engineering
Nanyang Technological University
Singapore

1

Introduction

Le Nguyen Binh

Hua Wei Technologies, European Research Center, Munich, Germany

CONTENTS

1.1 Overview

Nonlinear systems have attracted significant interests over the last century and in the first decade of the 21st century. The subjects continue to create extensive problems in the transmission of lightwaves over very long distance optical fiber links under extremely high bit rates as well as systems in nature and all branches of engineering. This chapter gives an overview of the nonlinear systems and those that attract our detailed treatment in this book.

1.2 Introduction

Over the last few decades, a large number of published materials, books, and research papers on nonlinear dynamics in natural systems have appeared and currently emerged as one of a number of significant phenomena with critical applications in engineering and science. The description of several shapes and structures in nature, the physics and mathematics can explain intricate patterns formed from simple shapes and the repeated playing roles of dynamic procedures. The fundamental rules underlying various physical and mathematical shapes in nature have led to the searching and defining of patterns in systematic scientific relationships such as algebraic equations, ordinary differential equations, partial differential equations, difference equations, and so forth.

Contemporary nonlinear dynamics was fathered by Henry Poincaré (1854 – 1912), and the dynamics studies have focused on analytic solutions of the dynamic equations. In particular, he emphasized the importance of obtaining the global dynamics in nature. Here, nonlinear dynamics is represented by the nonlinear phenomena prescribed by a set of nonlinear equations. The nonlinearity is hereby defined to be the phenomenon such that the commonly defined linear superposition principles do not hold. Furthermore, the fundamental theoretical problem arises from a reduced description of the dynamics that can be approximately described by a system of nonlinear equations. Some simple examples of such systems of equations occur in the fields of mechanics and electric circuits. Thus, nonlinear phenomena concern processes involving physical variables governed by a set of nonlinear equations. In the 1940s, primitive digital computing systems were available to scientific communities. It was also commonly known that analytic solutions to nonlinear equations give the most useful information.

For instance, in the case of solitons, the generation and observation of chaos and fractals are fascinating phenomena of nonlinear science in contemporary engineering and sciences in which carriers can either be included or not. Thus, some fundamental aspects and essential mathematical representations of nonlinear systems are given in Chapter 2. These fundamentals would then be translated to optical systems in which lightwaves act as carriers in the system and the envelope carrying these lightwaves behaves as in normal systems that must satisfy conditions and phase states of normal nonlinear physical systems.

Depending on the medium and material property, nonlinearity depends on the spectrum of the waves that would induce the nonlinear effects, then the carrier and its wave envelope would suffer the nonlinearity. One can use these nonlinearities for the generation of new types of devices, such as optical solitons, and so forth, or equalization to minimize such effects. Optical nonlinearity in communications systems, especially in long haul and ultrahigh capacity, ultrahigh speed transmission of information over a single-mode fiber, also limits the ultimate transmission capacity.

1.2.1 Nonlinear Solitary W56aves

1.2.1.1 Noncarrier Solitary Waves

In 1834, J. S. Russell conducted experiments on a canal to measure the relationship between the speed of a canal boat and its propelling force, in order to explore the design parameters for conversion from horsepower to steam. During the experiments, the boat suddenly stopped, and he discovered a large wave without the change of form and speed in the canal, the solitary wave, preserving its original figure for some 30 feet. Then over a period of a decade, he continued to study the solitary wave in water tanks and canals, finding it to be an independent dynamic wave moving with constant shape and speed [1]. In 1895, D. Korteweg and H. de Vries [2] published a theory of shallow water waves and established the K-dV equation, a nonlinear partial differential equation (PDE), of the form

$$\frac{\partial u}{\partial t} + au\frac{\partial u}{\partial x} + b\frac{\partial^3 u}{\partial x^3} = 0, \tag{1.1}$$

where $u(x, t)$ is the wave amplitude, a is the nonlinear parameter, and b is the dispersive parameter [2]. The second term in (1.1) is nonlinear, and the exact solution of this equation is a traveling wave given as

$$u(x,t) = h \sec h^2(k(x - vt)), \tag{1.2}$$

with $k \propto \sqrt{h}$ implying that higher amplitude waves had a width more narrow with a balance between the dispersion and the nonlinearity. Then, the solitary wave can be considered to be an independent dynamic entity.

On the other hand, Fermi, Pasta, and Ulam (FPU) [3] suggested one of the first scientific problems by using a computing machine. Their system was a chain of equal mass particles connected by weakly nonlinear springs. As a surprising result, when all the energy is originally in the mode of lowest frequency, it returns almost entirely to the initial mode after a period of interaction among a few other lower-frequency modes [3]. Following the FPU recurrence, Zabusky and Kruskal introduced the nonlinear spring-mass system by the K − dV equation (1.3) in 1965, and observed numerically that the K − dV solitary waves pass through each other with no change in shape or speed, and this was termed *soliton* with a particle-like property [4]. The initial condition generates a family of solitons with different speeds, moving apart in the *xt*-plane.

Many other nonlinear PDEs in the 1960s were known to have soliton solutions, such as the modified K – dV equation without any carrier waves, given by

$$\frac{\partial u}{\partial t} + au^2 \frac{\partial u}{\partial x} + b \frac{\partial^3 u}{\partial x^3} = 0, \tag{1.3}$$

The Sine–Gordon equation given as

$$\frac{\partial^2 u}{\partial t^2} - \frac{\partial^2 u}{\partial x^2} + m^2 \sin u = 0, \tag{1.4}$$

The Kadomstev–Petviashvile equation, a high-order two-dimensional (2-D) differential form can be written as

$$\frac{\partial}{\partial x}\left(\frac{\partial u}{\partial t} + 6u \frac{\partial u}{\partial x} + \frac{\partial^3 u}{\partial x^3} \right) + a \frac{\partial^2 u}{\partial y^2} = 0, \tag{1.5}$$

The Toda lattice equation [5] is given as

$$\frac{d^2 Q_n}{dt^2} = \exp(Q_{n-1} - Q_n) - \exp(Q_n - Q_{n+1}) \tag{1.6}$$

which is a nonlinear differential-difference equation with Q_n denoting the displacement of the nth lattice, and the nonlinear Schrödinger (NLS) equation [6]

$$j \frac{\partial u}{\partial t} + \frac{\partial^2 u}{\partial t^2} + 2|u|^2 u = 0 \tag{1.7}$$

This provides a third example of a nonlinear partial differential equation (NLPDE) with N-soliton solutions obtainable by the inverse scattering method. The soliton solution produces a bright soliton as illustrated in Figure 1.1. We note that the carrier is not included in (1.7) but the envelope of the solitary wave is included.

In addition, other integrable PDEs have been derived from eigenvalue problems, and they have been mapped to each other through transformations. As an example, the Harry Dym equation

$$\frac{\partial r}{\partial t} = r^3 \frac{\partial^3 r}{\partial x^3}, \tag{1.8}$$

is integrable but does not possess the Painleve property. Also, (1.8) can be transformed to the modified K – dV Equation (1.3) by the hodograph transformation, a special nonlinear technique determined by

$$u(\tau) = \int^x \frac{1}{r(x,t)} dx, \tau = t, \tag{1.9}$$

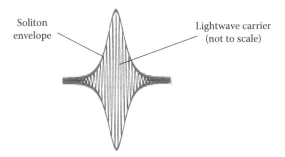

FIGURE 1.1
Bright soliton solution with its envelope and the lightwave carrier.

These integrable-soliton equations are well known to be reducible to the Painleve equations with fixed critical points [7,8]. The first Painleve equation is a second-order nonlinear differential equation [9,10] given by

$$\frac{d^2y}{dx^2} = 6y^2 + x, \tag{1.10}$$

For example, some of the applications of the Sine–Gordon equation include the study of propagation of crystal defects, domain walls in ferromagnetic and ferroelectric materials, self-induced transparency of short optical pulses, and propagation of quantum units of magnetic flux on Josephson supercon-ducting transmission lines.

1.2.1.2 Optical Solitary Waves

The nonlinear Schroedinger (NLS) equation can represent the evolution of the envelope of the nonlinear pulses of optical solitons in the optical trans-mission fiber for communication system depicted in Figure 1.1 [11]. The NLS equation of such solitary waves in a single-mode optical fiber can be derived from the Maxwell equations subject to the nonlinear effects of such guided optical medium as follows (see Appendix A for the detailed derivation and definitions of parameters):

$$\frac{\partial A(z,t)}{\partial z} + \frac{\alpha}{2} A(z,t) - j \sum_{n=1}^{\infty} \frac{j^n \beta_n}{n!} \frac{\partial^n A(z,t)}{\partial t^n} = j\gamma \left(1 + \frac{j}{\omega_0} \frac{\partial}{\partial t}\right)$$

$$\times A(z,t) \int_{-\infty}^{\infty} R(t')|A(z,t-t')|^2 \, dt' \tag{1.11}$$

The nonlinear effects that contributed the variation of the phase of the optical carrier are included in the right-hand side (RHS) of the equation.

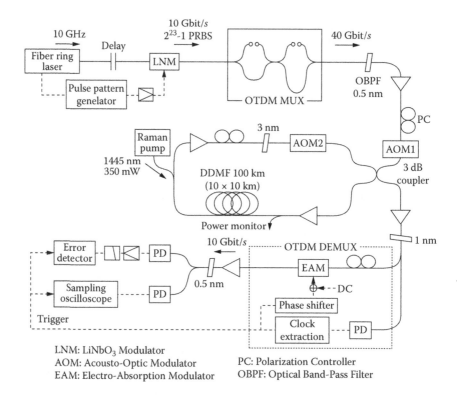

FIGURE 1.2
Typical setup for soliton generation and transmission over optical fibers.

Other nonlinear effects such as stimulated Raman scattering (SRS) and four wave mixing (FWM) can be integrated without much difficulty.

Solitons have been employed in optical transmission systems over single-mode optical fibers in which the linear dispersion of the guided mode, the linearly polarized mode, is balanced by the nonlinear dispersion effects via the nonlinear Kerr's effect. Furthermore the fiber attenuation of the guided modes is equalized by optical amplifications of the optical amplifiers EDFA (Er:doped fiber amplifiers) placed at the end of each fiber transmission length. Figure 1.2 depicts a typical experimental setup for generation and propagation of optical solitons over several thousands of single-mode optical fibers.

In the optical soliton transmission experiment, short pulse sequences are generated by a fiber ring laser, a mode-locked type (for details see Chapter 3), under which the lightwaves are phase modulated by a LiNbO$_3$ modulator (LNM). The pulse shape would follow that of a sech^2 profile of an ideal soliton envelope. They are then time multiplexed by a number of asymmetric Mach Zehnder interferometers (OTDM Mux) in which the two lightwave paths are delayed by half the pulse period so as to double the pulse rate. Thus, a 40 Gb/s soliton data sequence can be generated and launched into and switched out of the fiber transmission link embedded in an optical loop by acoustic modulators (AOM) AOM1 and AOM2. The propagation loss of these solitons is compensated by the distributed amplification along the transmitting fiber using a Raman pump source launched at the input of the fiber link (see loop in Figure 1.2). Likewise another Raman source can be used to pump in a contrapropagation direction from the receiving end. At the output of the fiber link the received solitons are optically preamplified by a lumped amplifier (Er:doped fiber amplifier) and then recovered and time demultiplexed by an electroabsorption (EA) device, and then converted to the electronic domain.

Nonlinear phenomena in the generation of solitons, the transmission of a soliton pulse sequence over the guided optical medium, optical fibers, are examined and investigated in Chapters 3 through 10. The next two sections give a brief introduction of the chaotic patterns and fractals that can also be generated and observed in fiber optic systems that are described in Chapters 4 and 5.

1.2.2 Chaos

1.2.2.1 Mathematical Equation

It has been well known that the simplest first-order nonlinear difference equation possesses a rich spectrum of dynamical behavior, the "chaotic" features. The main features of these nonlinear phenomena have been discovered independently, and some interesting mathematical equations have not been fully resolved [12]. The magnitude of the population in generation $\{n+1, X_{n+1}\}$ can be expressed by that in the preceding generation $\{n, X_n\}$ as a general first-order nonlinear difference equation:

$$X_{n+1} = F(X_n), \tag{1.12}$$

where the function $F(X_n)$ is a density dependent in biology, and X_n is the population density of the nth generation.

The well-known logistic equation is given as

$$\frac{dP}{dt} = P(1-P) \tag{1.13}$$

The exact solution of (1.13) can be written as

$$P(t) = 1/(1+\exp(-t)) \tag{1.14}$$

with a population P. By introducing a differential term

$$dP/dt \approx (x_{n+1} - x_n)/\Delta t \tag{1.15}$$

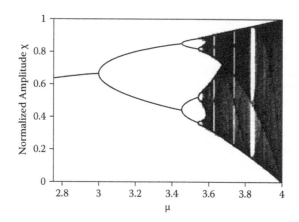

FIGURE 1.3
Representation of Equation 1.16 and bifurcation, normalized amplitude *x* versus m.

we can arrive at a special solution following the simplest nonlinear difference equation, given as

$$x_{n+1} = \mu x_n (1 - x_n),\tag{1.16}$$

This equation is termed the *logistic map* for $0 \leq \mu \leq 4$ as shown in Figure 1.3
which shows the phase space in response to infinitesimal changes of the
parameters of (1.16). Such changes are called *bifurcations* in fiber lasers' operation in a nonlinear regime, which will be illustrated by Chapters 3, 6, and 8.

If a nonlinear system is operating in a chaotic dynamic state, then it is
natural to ask how the complexity evolves due to the variation of parameters.
In Equation (1.16), when $\mu = 0.5$, there is a fixed point at $x_n = 0$ that attracts
all solutions subject to initial values within the range $0 < x_0 < 1$. If $\mu = 4$, the
map has a pure chaos solution

$$x_n = \sin^2 (c2^n),\tag{1.17}$$

where c is a real number [13]. Furthermore, the Lyapunov exponent of the
map can be introduced to estimate the dynamic nonlinear behavior [14].
Feigenbaum has also proposed the period-doubling cascade as one of the
universal properties [15].

1.2.2.2 Chua's Circuits

After several attempts to produce chaos in an electrical analog circuit,
Chua realized that chaos could be produced in a piecewise-linear circuit if
it possessed at least two unstable equilibrium points. Thus, the third-order

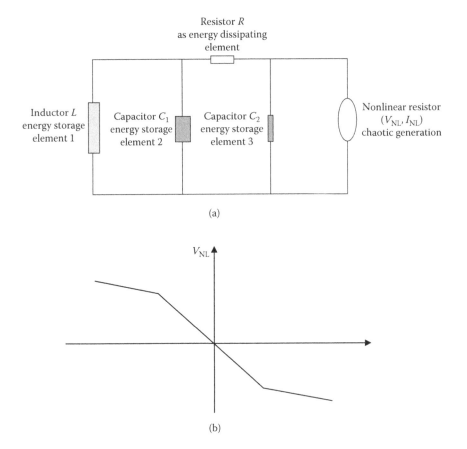

FIGURE 1.4
(a) Chua's circuit with RLC components and a nonlinear resistor; and (b) V-I characteristic of the nonlinear resistor.

circuit, the Chua's circuit (see also Chapter 2) consisting of a single voltage-controlled nonlinear resistor, can be identified, and the active element (e.g., a diode N_r) can be chosen to produce a piecewise nonlinear region of different resistance characteristic or slope of the V-I curve as shown in Figures 1.4a and 1.4b. This circuit has been accepted widely as a powerful paradigm for learning, understanding, and teaching about nonlinear dynamics and chaos. It is noted that three energy storage elements, inductors and capacitors, act as the competitors for the energy available in the system. When the competition occurs in the same mode (i.e., charge or discharge) then chaos or bistability would occur.

On the other hand, the iteration of the function $f(x) = 4x(1-x)$ can generate pseudorandom numbers in deterministic processes [16], and then various chaotic sequences can be proposed for the generation of pseudorandom numbers. In particular, it was shown that the self-synchronization appears to be structurally stable in chaotic systems, and it may offer important applications such as a unique signature to secure communication systems [17]. Several authors proposed the possibility that chaos could be used for secure communications. However, it was pointed out that the message masked by chaos can be easily extracted from the transmitted signal [18]. Also, a high-dimensional chaotic wave form for secure communications has been generated by driving an erbium-doped fiber ring laser [19]. Recently, a simple algorithm for exact long-time chaotic series and the pseudorandom numbers have been proposed, and that could be applied to cryptosystems [20–22]. However, this is outside the scope of this book.

1.2.3 Remarks

In this section, we emphasized that nonlinear dynamics in nature have been considered for the description of intricate patterns formed from simple shapes and the repeated application of dynamic procedures by introducing physics and mathematics. We can assert that "Nature is nonlinear," and the fundamental rules underlying the shapes and the structures have led to dynamic study in the fields of physics, mathematics, engineering, and so forth. Especially, modern nonlinear dynamics have focused on analytic solutions of the dynamic equations, and the nonlinearity is determined based on the linear superposition. However, fundamental theoretical problems arose in physics and mathematics, as well as in engineering systems that can be solved with the assistance of the digital computing techniques.

In Chapters 3 to 9 the solitons and their chaotic dynamics and potential applications to optical systems are introduced and described in the field of nonlinear optics in a number of chapters of this book. The chapters following this chapter will treat soliton and soliton lasers in extensive detail, especially their generation, dynamic behavior, and then transmission over guided optical media.

Surprisingly, simple nonlinear systems are found to have chaotic solutions, which remain within a bounded region. In other words, the nonlinearity has been positively considered, and the result has been applied to the analysis and the design in engineering and technology. Thus, nonlinear dynamics in nature have played a key role in physics, mathematics, and engineering, and would be fundamental tools for many branches of future research. The next section gives briefly the principal functions in nonlinear optical systems and introduces several aspects of such systems, especially in fiber lasers in both passive and active operating regimes. The nonlinear behavior and transitions to chaotic states in stable chaotic, bifurcation, attractors, and repellers are also analyzed in these optical guided wave systems.

1.3 Nonlinear Optical Systems

1.3.1 Ultrashort Optical Pulse Sources as Nonlinear Systems

Ultrashort pulse lasers have reached the stage of rapid evolution. They are the subjects of ultrafast optics research over a number of decades since the invention of the laser in the early 1960s. Optical pulses generated from ultrashort pulse lasers can be in the range from femtoseconds to picoseconds and have been used in several applications. Due to their availability in the market and their improved performance, applications of these ultrashort pulse sources can range from testing and measurement applications [23–36] to advanced technology applications such as electronics and biomedicine [37–39]. Ultrashort pulse lasers have increasingly expanded their applications because of useful characteristics of ultrashort pulses such as high peak power, broad bandwidth, and high temporal and spatial resolution. Because various laser systems generate ultrashort pulses of different characteristics, each system can be suitably used in one or some applications. For example, the ultrashort pulse laser with high peak power is preferably used in nonlinear microscopy [40] and micromachining and marking [41,42], which are very useful in many industrial areas such as electronics, medicine, the automotive industry, and many others. Frequency metrology and optical spectroscopy require the laser systems with ultrashort pulses and long-term coherence [43,44]. Strict requirements such as high repetition rate, low noise, and low jitter are set by telecommunication systems, therefore most research activities on ultrashort pulse lasers have been accomplished to target telecommunication applications. On the other hand, optical fiber communications has driven the development of ultrashort pulse sources.

In the telecommunication area, the ultrashort pulse lasers have became a key component in various subsystems employed in optical fiber communication systems. The progress in telecommunications systems has led to the increasing use of ultrashort pulse lasers. First, the ultrashort pulse lasers with the capability of transform-limited hyperbolic secant pulses at rates from 10 Gbits/s to 40 Gbits/s can be employed in ultralong haul transmission systems based on soliton transmission technology. The soliton transmission distances up to 10,000 km at 80 Gbits/s and 70,000 km at 40 Gbits/s have been proven in the laboratory [45,46]. With the pulse width of 200 fs at 10 GHz, the transmission at ultrahigh speed of 1.28 Tbits/s and 5.1 Tbits/s has been demonstrated in optical time-division multiplexing (OTDM) systems [47,48]. Moreover, the ultrashort pulse laser also proves its importance in wavelength-division multiplexing (WDM) systems in which the broad bandwidth of ultranarrow pulses is sliced into many bands or wavelength channels. For those systems, a supercontinuum source can be built by a chirped short-pulse laser that is passed through a special fiber such as dispersion decreasing fiber or photonic crystal fiber. A flat optical spectrum with a bandwidth exceeding 150 nm can be obtained by using this technique [49]; thus, it is possible to produce thousands of wavelength channels using spectral slicing techniques. The broadband coherence of the ultrashort pulse laser that is a specific and well-defined frequency-dependent phase relation across the spectrum can be used in encoding various types of optical data in optical code-division multiple access (O-CDMA) [50]. Beside the employment in optical data generators, the ultrashort pulse laser has become an indispensable component for optical signal processing. Many processing functions such as OTDM demultiplexing, optical sampling, or optical analog-to-digital converting require a high-performance ultrashort pulse laser. Classification of an ultrashort pulse laser commonly depends on the gain medium that includes solid-state, semiconductor, and fiber lasers. Each category has its own advantages and disadvantages. They have different performance parameters to be able to adapt to a specific application. In optical telecommunications, the semiconductor and fiber lasers are the most commonly used for the generation of ultrashort pulses using some techniques such as mode locking, gain switching, or externally modulating. Although the semiconductor lasers offer a good engineering solution, the mode-locked fiber lasers have been dramatically developed for optical communication systems. Reasons that the mode-locked fiber lasers have remained attractive in research are due to these lasers based on fiber that offer many advantages such as simple doping procedures, low loss, and the possibility of pumping with compact, efficient diodes. The availability of various fiber components allows minimization of the need for bulk optics and mechanical alignment that are very important to reduce the complexity as well as the size and the cost of the laser system for practical use. Moreover, the research interests focused on the mode-locked fiber lasers come from the fact that they can directly produce hyperbolic secant pulses or optical solitons. Soliton pulses are desirable in ultralong haul transmission systems because they remain stable against perturbations of the transmitting medium and keep the pulse shape undistorted during propagation. There has been a significant advance in development of the mode-locked fiber lasers to reduce the size and cost as well as to improve their performance. Although techniques in mode-locked fiber lasers are mature in practical applications today, they have been still a rich research field. Research efforts are being made to explore new properties in various operation schemes, especially in highly nonlinear regimes, of generated short pulses for future potential applications.

1.3.2 Nonlinear Dynamics in Active and Passive Mode-Locked Fiber Lasers

Mode locking is the most popular technique to generate the ultrafast pulses in fiber systems. When hundreds or thousands of longitudinal oscillation modes in a fiber cavity are forced in phase, the short pulses are formed. The pulse-width depends on the number of phase-locked longitudinal modes or is inversely proportional to the spectral bandwidth of the generated pulses. Like any mode-locked lasers, the mode-locked fiber lasers can be classified into two basic categories: active and passive sources.

Passive mode-locked fiber lasers perform mode locking based on the exploitation of some optical effects in an active fiber cavity without use of any external driving signal. On the other hand, the short pulses are generated by the loss modulation of the passive elements that also determine the physical mechanism of passive mode locking in the fiber cavity. There are several typical configurations of passive mode-locked fiber lasers that are based on mechanisms such as saturable absorption, nonlinear amplification loop mirror, and nonlinear polarization evolution [51–53]. Pulse shaping in all passive mode-locking techniques is commonly based on the intensity-dependent discrimination where the peak of optical pulse acquires the lowest loss while the wings of the pulse experience a higher loss per round-trip in the fiber cavity. After many round-trips, the pulse is shortened until its bandwidth is comparable to the gain bandwidth of the medium. The reduced gain in spectral wings then provides a broadening mechanism that stabilizes the pulse width to a specific value. Because operation of the nonlinear amplification loop mirror (NOLM) and nonlinear polarization rotation (NPR) is considered as a fast saturable absorption process, the width of generated pulses based on these mechanisms is often very narrow and can be the order of sub-100 fs. Furthermore, it has been theoretically and experimentally shown that a ring configuration of mode-locked fiber laser initiates more easily at a lower threshold [54,55].

Although the passively mode-locked fiber lasers can easily generate very short pulses with high peak power, the repetition rate of generated pulse train is limited only at scale of the order of 100 MHz. The reason for the limited repetition rate is due to the length of the fiber cavity, which determines the fundamental frequency, being of the order of meters. This limitation prevents the passively mode-locked fiber lasers from high-speed transmission applications. To increase the repetition rate in these fiber lasers, a harmonic mode locking where multiple pulses coexist inside the fiber cavity can be employed. However the generated pulses in this scheme experience amplitude and timing fluctuations, they are unable to meet strict requirements in communication systems. Therefore, active mode locking is an alternative technique to generate the mode-locked pulses for communication-oriented applications.

Unlike passive mode locking, active mode locking requires an active element driven by an external electronic signal to modulate the loss inside the fiber cavity. Although a bulk modulator can be used for mode locking, most active mode-locked fiber lasers use the integrated electro-optic $LiNbO_3$ modulators for mode locking. There are two types of modulator—amplitude and phase—that correspond to two active mode-locking mechanisms—amplitude modulation (AM) and frequency modulation (FM). A typical configuration of an active mode-locked fiber laser using an amplitude modulator or a phase modulator for shaping a pulse in the cavity is shown in Figure 1.5a.

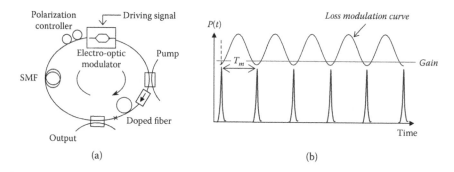

FIGURE 1.5
Typical configurations of actively mode-locked fiber laser: (a) a fiber ring configuration, (b) a description of pulse formation in active mode locking.

To provide a high-speed pulse train, the active modulator is often driven at harmonics of the fundamental mode spacing. In AM mode locking, the loss of cavity is periodically modulated and the pulse is built up at the positions having minimum loss in the fiber cavity. FM mode locking is carried out by repeated up or down frequency chirping. After many round-trips the pulse is built up and shortened because the chirped light is swept out of the gain bandwidth of the laser. From the distinct mode-locking mechanism between AM and FM, there are differences in characteristics between these two types of active mode-locked fiber lasers. Recent studies have shown the superiority of the active FM mode-locked fiber laser compared to the AM mode-locked fiber laser to become a valuable candidate for ultrashort pulse generation at very high speeds [123]. First, a phase modulator is more attractive than an amplitude modulator because it has no DC bias that can avoid the DC bias drift of the modulator. Second, the pulses obtained from the FM mode-locked fiber laser are generally shorter than those obtained from the AM mode-locked fiber laser due to the pulse compression induced by group velocity dispersion (GVD) and chirping from phase modulation [56]. The performance of mode-locked pulses from the FM mode-locked fiber laser can be improved further in optimal combination between GVD, phase modulation, and nonlinear effects in design of the fiber cavity [57]. Furthermore, when the FM mode-locked fiber lasers use the rational harmonic mode locking technique to increase repetition rate, the output pulses do not experience the problem of unequal amplitudes due to the unique property of the phase modulation [58]. Some other advantages such as the reduction of timing jitter or smaller quantum limited jitter also make the FM mode-locked fiber laser more attractive in optical communication systems at very high speed [59].

However, a challenge to realizing ultrahigh speed optical communication is that of generation of a stable short pulse train in the long term. Because an active mode-locked fiber laser often operates in a harmonic mode locking regime, the stability of the pulse train is always a concern in design of these fiber lasers. Hence, if the modulation frequency is not equal to a harmonic of the fundamental frequency, the phases of the cavity modes are not locked, and consequentially, unstable pulsing occurs. The polarization states, the cavity length, or the fundamental frequency change is sensitive to the temperature variations and mechanical vibrations, and this causes a mismatch between the modulation frequency and the cavity mode. On the other hand, it is difficult to maintain an optimum operation condition in the long term. A solution for this problem is to utilize a regenerative mode-locking technique that uses a feedback loop to adjust the phase between the pulse and the clock signal driving the modulator [60,61]. To stabilize the repetition rate at a fixed frequency, the phase locking can be carried out by controlling the fiber cavity length with a fiber stretcher [62–64]. These stabilization techniques really have made the active FM mode-locked fiber laser more attractive and more feasible in practical optical fiber transmission applications.

Although the active mode-locked fiber laser can produce short pulses at speeds of 10 to 40 GHz, the pulse width is only of the order of picosecond. The generation of femtosecond pulses in the GHz region is especially important in ultrahigh-speed transmission systems using OTDM technique. Therefore, it would be desirable to produce a femtosecond pulse train at a repetition rate of 10 to 40 GHz with a conventional high-speed modulator. There are some techniques to compress the mode-locked pulses from the mode-locked fiber laser. A promising technique is to use hybrid mode locking that combines passive and active mode locking in a fiber laser using phase modulator [65]. In this hybrid scheme, the mechanism of passive mode locking such as NPR is used to further shorten the pulse width. The asynchronously driven phase modulator is treated as a polarizing element, which is responsible for both timing of the pulse and shortening the pulse through NPR. This configuration has been used to produce pulse train with 816 fs pulse width at a 10 GHz repetition rate [66] and a shorter pulse width of 400 fs with a repetition rate from 2.5 GHz to 12 GHz recently reported when additionally using a dispersion-shifted fiber (DSF) to compensate the chirp of the output pulses [67].

Another promising technique is the adiabatic soliton pulse compression through dispersion decreasing fibers. The mechanism of the compression is based on the principle of keeping the soliton energy constant. Thus, when a soliton propagates through a fiber with decreasing the GVD, the pulse width of the soliton automatically shortens corresponding to the ratio between the decrease in the dispersion and the original dispersion. By using this technique, a generation of 54 fs pulse train at 10 GHz has been successfully demonstrated [68], and it has been used to realize the first 1.28 Tbit/s OTDM transmission experiment [47]. However, when the pulse width is shortened to femtosecond scale, more effects such as higher-order dispersion, polarization mode dispersion, and the finite nonlinear response time of the fiber are impossibly negligible. They limit the compression ratio as well as the performance of shortened pulses.

1.3.3 Nonlinearity in Fiber Optics

Nonlinearity is a fundamental property of optical waveguides including optical fibers. Origin of the nonlinear response is related to the high-order susceptibility of material under the influence of an applied field. For silica glasses, which is the material of optical fibers and some nonlinear waveguides, only the third-order susceptibility is responsible for nonlinear effects through the contribution to the total polarization $\vec{P}$ which is given by [69] $\vec{P} = \varepsilon_0(\chi^{(1)} \cdot \vec{E} + \chi^{(3)} : \vec{E}\vec{E}\vec{E} + ...)$, where ε_0 is the vacuum permittivity and $\chi^{(j)}$ is jth order susceptibility. In general, the nonlinear effects in optical fiber can be classified into two classes. One class refers to the energy transfer from the optical field to the propagation medium that is the result of stimulated inelastic scattering such as stimulated Raman scattering (SRS) and stimulated Brillouin scattering (SBS). Another class is governed by the third-order susceptibility $\chi^{(3)}$ and responsible for most important nonlinear effects called *Kerr effects*.

The Kerr effect is also known as the nonlinear refractive index because of the intensity dependence of the refractive index. In a high-intensity regime, the refractive index can be given by $n = n_0 + n_2 |E|^2$, where n_0 is the linear part of the refractive index which determines the material dispersion of a propagation medium, n_2 is the nonlinear refractive index that relates to $\chi^{(3)}$, and $|E|^2$ is the intensity of the optical field. Important effects that originate from the nonlinear refractive index consist of third-harmonic generation (THG), four-wave mixing (FWM), self-phase modulation (SPM), and cross-phase modulation (XPM). First, SPM refers to the self-induced phase shift experienced by an optical field during its propagation. The phase of an optical field can be represented by $\phi_{NL} = n_2 k_0 L |E|^2$, , where $k_0 = 2\pi/\lambda$ and L is the length of fiber. Because the phase change is proportional to the intensity, an optical pulse modulates its own phase with its intensity. The alteration of optical phase leads to frequency chirping according to the pulse intensity profile. Therefore, SPM is responsible for spectral broadening and formation of optical soliton in the balance with anomalous dispersion effect. Contrast to SPM, XPM, and FWM effects require the participation of more than one optical field having different wavelength, direction, or polarization state. The main difference between XPM and FWM is that the occurrence of FWM requires a specific phase-matching condition. For XPM, an optical field with high intensity imposes a nonlinear phase shift on another field, and this results in a spectral broadening similar to SPM. The interaction between optical fields in FWM creates new optical waves according to conservation rules of energy-momentum. SPM may be a major concern in a single channel system, while XPM and FWM are the main concerns in WDM systems, especially at the zero-dispersion wavelength. However, these Kerr effects are significant in photonic signal processing systems due to the instantaneous response of the third-order nonlinearity.

In fact, nonlinear effects play an important role in design of fiber systems including the mode-locked fiber lasers. Under strong pump power, the mode-locked pulses in the fiber cavity attain high intensity to cause a sufficiently nonlinear phase shift that affects pulse formation in the cavity. In the nonlinear effects, the Kerr effects that are caused by the intensity-dependent refractive index relate mainly to the performance of the output. In particular, SPM and XPM are the most important effects that play a dominant role in the operation of most passive mode-locked fiber lasers. When the optical power level in the fiber laser is high enough, the nonlinear phase shift produced by SPM and XPM can change the power-dependent transmission of the nonlinear fiber loop mirror as in figure-eight laser type or the polarization states as in nonlinear polarization rotator fiber laser type [52,53]. Owing to the ultrafast response of fiber nonlinearity, passive mode-locked fiber lasers can easily produce ultrashort soliton pulses of femtosecond scale. Together with dispersion, the nonlinear phase shift plays an important role in the evolution of mode-locked pulses as well as in shaping the output pulses. Particularly, in a normal dispersion scheme, the parabolic pulses with high energy are normally formed rather than the Gaussian pulses [70,71], otherwise the soliton pulses are generated in anomalous dispersion regime.

For active mode-locked fiber lasers, mode locking is based on harmonic modulation of active elements as mentioned in the previous section. Although nonlinear effects do not totally determine mode locking, they influence significantly the shapes and the performance of output pulses. Some reports that analyzed the active mode-locked fiber laser taking the nonlinear effects into account indicated the influence of SPM on ultrashort pulse generation. Depending on the optical power levels in the cavity, the active mode-locked fiber laser operates in various regimes that correspond to a certain performance of output pulses [72–75]. In general, the important influence of the SPM in the active mode-locked fiber can exhibit in two following aspects. The first influence is pulse compression based on the soliton effect. When the power in the fiber cavity increases, the SPM effect becomes important and shortens the pulse based on the soliton effect in the anomalous dispersion regime, but a large amount of dropouts may occur. The stable operating regime is only attained when the power exceeds a certain value [73]. Another important impact of SPM in active mode locking is suppression of supermode noise that causes an amplitude fluctuation induced by amplification stimulated emission (ASE) noise. The peak intensity of a high-power pulse with a short pulse width is eventually limited to within the filter bandwidth, so the energy at the pulse center shifts to the wings due to SPM and is eliminated by the filter. Hence, the pulse amplitude is clipped to a certain intensity level at which the pulse energy is stabilized when the pulse spectral width reaches a certain ratio of the filter band width. When the SPM is strong enough, stabilization occurs through the filtering effect to suppress the supermode noise efficiently [76].

Furthermore, for active FM mode-locked fiber laser in linear operating regime it is well known that the pulses of FM mode-locked fiber lasers tend to switch back and forth between the two phase states: up-chirp and down-chirp [76]. In fact such an unstable operation does not occur due to the presence of dispersion in the cavity and the stable pulse state with the pulses locked at up-chirped or down-chirped positions depending on the sign of average cavity dispersion [37,77]. When the optical power increases, the presence of SPM in the cavity greatly modifies the up- and down-chirp caused by the FM modulation. It should be noted that the FM modulation interacts with the SPM in a direct manner. There are two ways of applying an imbalanced FM modulation. One is where the up-chirped modulation produced by the phase modulator is in phase with the chirp caused by the SPM. In this case, the total chirp is increased and the pulse in the cavity is further compressed by anomalous GVD. Thus, the soliton effect helps to generate a shorter pulse. The other way is where the phase modulator produces a down-chirp, and here the chirping is out of phase with the chirp caused by the SPM. In this case, the total down-chirp is reduced and this weakens both the pulse formation via phase modulation and filtering, and the soliton effect. Furthermore, in an asynchronous mode-locking scheme, the SPM effect and filtering also suppress effectively the relaxation oscillation that causes the intensity fluctuations of the output pulse sequence [78].

1.3.4 Signal Processing in Fiber Optics

Optical fiber communication systems can now support Tb/s information capacity over thousands of kilometers [47,79,80]. They form the core technology for high-capacity telecommunication networks. Recently, broadband applications such as high-speed Internet access, multibroadcast systems have been dramatically growing, which require huge transport capacity in core communication networks, especially at 100, 160, or even 320 Gbit/s operating speed. The intense demand of increasing system reach and transport capacity as well as reduction of the cost of the system are a driving factor for research to develop advanced photonic technologies, especially high-speed signal processing. In addition to the matured technologies such as WDM and optical fiber amplification, new transmission techniques such as OTDM, and advanced modulation formats have set new issues to signal processing in future optical networks that are expected to operate at tens of Tbit/s.

OTDM technology seems to be a good option for implementing an Ethernet transmission channel at the rate of Tbit/s. However, the realization of this technology requires all signal processing functions in the system such as pulse generator, switch, and buffer operating at ultrahigh speed. At rates higher than 40 GHz to hundreds of GHz, the electronic domain is technologically and economically limited to perform signal processing functions. Thus, the limited operating bandwidth of functional blocks in optical networks is a major challenge. Furthermore, several technological advancements such as advanced modulation formats have recently drawn a lot attraction in order to extend the reach and capacity [81]. By multilevel coding data bits into the phase of optical carrier in advanced modulation formats such as Mary-phase shifted keying (M-PSK) or quadrature amplitude modulation (M-QAM), the capacity can increase several times while the bandwidth of the optical signal remains unchanged, which is very important in high-capacity communication networks. A high spectral efficiency 1.6-b/s/Hz transmission has been demonstrated by using differential quadrature phase shifting key (DQPSK) modulation and polarization multiplexing where 40 Gbit/s channels can be allocated into the WDM system with a wavelength spacing of only 25 GHz [82]. For a single channel system, a transmission rate of 5 Tbit/s has recently been demonstrated by using the OTDM technique and 8-PSK and 16-QAM modulation formats [83]. The introduction of advanced modulation formats such as differential phase shift keying (DPSK), DQPSK, and quadrature amplitude modulation (QAM) in optical networks results in new challenges that require signal processing techniques transparent to different modulation formats and transmission rates.

In order to overcome challenges in future optical networks, ultrafast photonic processing or all-optical signal processing is expected. By shifting signal processing from the electronic domain to the photonic domain, the limitation of operation bandwidth is easily eliminated by a reduction of optical-to-electrical (O/E) conversions. Ultrafast photonic signal processing can be implemented by exploiting nonlinearity in optical guided systems such as optical fibers and nonlinear waveguides. Kerr effects with ultrafast response can be applied in photonic signal processing to overcome the bandwidth limitation. In addition to the formation of solitons and pulse compression as well as pulse-shaping mechanisms in the mode-locked fiber system, many signal processing functions such as signal regeneration and supercontinuum generation can be implemented by exploiting SPM to high-speed transmission systems [84,85]. Similarly, wavelength conversion, signal regeneration, and high- speed optical switching can be implemented by a NOLM based on the XPM effect. These functions are essential to high-capacity communication networks using OTDM/WDM technologies. However, FWM have recently attracted a lot of attention in photonic signal processing. A broad range of signal processing functions from wavelength conversion, ultrahigh speed switching to distortion compensation can be implemented by exploiting FWM processes with high performance. Another attractive feature of FWM-based signal processing is format independence because FWM can preserve both amplitude and phase information. Therefore, it is advantageous for future high-speed transmission systems using advanced modulation formats.

Rapid progress of nonlinear signal processing is due to the development of nonlinear propagation devices. The dispersion shifted (DS) fiber has been used as a highly nonlinear fiber (HNLF) because of modifying the fiber geometry [86]. The value of the nonlinear coefficient of this fiber is normally ten times higher than the standard single-mode fiber (SSMF) that is mainly achieved by reducing the fiber core. Consequently, the DS-HNLF length of hundreds of meters is required in some nonlinear signal processing applications. Recent advances in the development of HNLFs have focused on reduction of the propagation length and operation power in practical applications. Through modifying the fiber structures as well as the glass composition, many novel HNLFs such as photonic crystal fibers [87], microstructure fibers [88], and Bismuth-based or chalcogenide glass fibers [89,90] have been developed with the achieved nonlinear coefficient beyond hundreds to thousands times higher than SSMF. Another option is based on nonlinear waveguides to enable photonic signal processing in compact devices, which has recently attracted a lot of attention in research. The benefit of using nonlinear waveguides is the feasibility of the development of photonic integrated circuits (PICs) for all-optical signal processing. It recently has been shown that a dispersion engineered planar chalcogenide waveguide is an attractive candidate for photonic signal processing devices in future optical networks. With the nonlinear coefficient of about 10,000 $W^{-1}km^{-1}$, which is ten thousand times greater than that of the SSMF, multifunction processing of optical signals at ultrahigh speed has been demonstrated in compact devices [91,92].

1.4 Objectives and Organization of the Book

The principal objectives of this book and its chapters are to address the principles of nonlinear systems and applications, especially in a number of fiber optic lasers. We focus mainly on nonlinear optical systems after the chapter on fundamental equations of nonlinearity as an introduction to the phase space and nonlinear dynamics. In particular the principles and theoretical as well as experimental principles on several nonlinear fiber lasers in active and passive structures are described.

Active FM mode-locked fiber lasers offer several advantages such as the generation of stable ultrashort pulse sequence at ultrahigh transmission rate, ease of synchronization with electronic components of communication systems, low timing jitter, and so forth. Moreover, active phase modulation has important contributions to shortening and stability of mode-locked pulses, especially in the presence of dispersion and nonlinear effects. Although the active mode-locked fiber lasers have been available for the commercial market, they are still the subject for fundamental research of pulse dynamics in various operating conditions. On the other hand, exploration of new states and their characteristics in the active FM mode-locked fiber laser is of importance for explicit interpretation of the mode-locking and generation mechanisms inside the fiber cavity.

In general the mode-locked fiber ring lasers are ideal systems in the nonlinear fiber optics area. The attention of research has been not only to generate stable solitons with ultrashort width but also to investigate the dynamic characteristics of solitons in these systems. In periodic systems like mode-locked fiber lasers, solitons that can be considered as dissipative solitons experience periodically loss and gain effects, and they behave in various interesting manners when circulating inside the fiber cavity. When the fiber ring cavity operates in a strongly nonlinear regime, soliton pulses can exhibit several exciting characteristics such as bifurcation, period doubling, chaos, and specially bound states observed recently [93–97]. These complicated nonlinear dynamics have attracted a lot of researchers because the knowledge of solitons is necessary to study the mode-locked fiber system as well as to explore new unknown applications of the soliton. However, most research efforts have focused on the passive mode-locked fiber laser system [98–102], it is only recently that attention has been paid to the active mode-locked fiber laser system. On the other hand, the dynamic aspects of the active soliton mode-locked fiber lasers remain very challenging. Moreover, the dynamic behaviors of soliton in the active mode-locked fiber laser are different from that in the passive mode-locked fiber laser due to the presence of an active photonic component (e.g., an electro-optic phase modulator).

Existence of the bound solitons in the passive mode-locked fiber lasers has shown complicated behavior of the fiber cavity through various interactions. The competition with the continuous wave (CW) component and interaction between bound soliton pairs can cause the position of bound solitons to not be fixed and vary from time to time [94]. To generate periodic bound solitons with a fixed time separation between pulses, the requirements in design and tuning parameters of passive mode-locked fiber lasers are very strict. The stability of the bound soliton train is very sensitive to the change of operating conditions in the fiber cavity. Furthermore, the limitation of repetition rate may prevent passive mode-locked fiber lasers from potential applications in telecommunication area. New dynamic phenomena of solitons such as bound states can also present in the active mode-locked fiber lasers. However, not much attention has been paid to bound solitons in such lasing systems. When bound solitons are generated by active mode locking, a high repetition rate of the bound soliton train is easily achieved and synchronized with other electronic components that can create a new soliton source for potential applications in optical communication systems.

The relative phase difference of the lightwaves between adjacent pulses in a generated pulse sequence is the critical parameter that contributes to the stability of multibound solitons in the mode-locking mechanism.

Similar to phase modulation formats in communications such as the DQPSK or QAM in advanced optical communication networks, the phase information becomes significant not only for monitoring but also for signal processing, especially signal recovery. For conventional transmission systems in which information bits modulate the intensity of an optical carrier only, a power spectrum or a Fourier transform of autocorrelation is a popular technique used to characterize the modulated signal. However, this is not sufficient for phase-modulated signals because the phase information is hidden in the commonly known power spectrum analysis. Instead of power spectrum characterization, multidimensional spectra that are known as Fourier transforms of high-order correlation functions can provide us not only the magnitude information but also its associate phase. One technique of the multidimensional spectrum analysis is the bispectrum that can be formed by the Fourier transform of the triple correlation of the signals. It is especially useful in characterizing the non-Gaussian or nonlinear processes and applicable in many various fields such as signal processing, biomedicine, and image reconstruction [99]. However, not much attention has been paid to applying this technique in the fiber systems. Therefore, it is really useful in using this technique to characterize the short pulse generated from the mode-locked fiber lasers, especially in the mode-locked fiber laser operating at high power where remarkably nonlinear effects such as doubling and multipulsing are exhibited. Due to the phase information obtained from bispectra, it is a useful tool to analyze the behaviors of signals generated from these systems such as multibound solitons.

Nonlinear optics and nonlinear fiber optics have been extensively treated by Yeh [124] and Agrawal [125] over the last two decades. In these nonlinear optical systems the generation of ultrashort pulse sequence for application in optical communications and biophysical systems attract much interest and accelerate several experimental works.

In this book, we focus on two issues that investigate the nonlinearity in the guided wave systems including the active mode-locked fiber ring and nonlinear optical waveguides for signal processing. The first issue is the generation of multibound solitons where the actively FM mode-locked fiber ring laser is the subject of this book. The mechanism of multibound soliton formation in an active fiber ring under a highly nonlinear scheme is investigated. In addition, the phase relationship between bound solitons and the influence of active phase modulation on the formation and the stability of multibound solitons are also discussed. Dynamic behaviors of multibound solitons in the fiber ring cavity as well as transmission in fibers are investigated. Second, we also focus on aspects of nonlinearity in photonic signal processing, especially exploiting parametric processes such as FWM to implement some signal processing functions such as parametric amplification, high-speed demultiplexing and bispectrum estimation. The bispectrum technique is proposed to characterize the phase-modulated signals that include multibound solitons and other phase-modulated signals.

This book is thus divided into three principal parts in which the first part (Chapter 2) is devoted to some fundamental mathematical equations of basic nonlinear systems without using any carriers. The systems just operate on their own and reach the nonlinear chaotic or bifurcation state.

The second part (Chapters 3 to 6) is devoted to experimentally and numerically investigate the nonlinear fiber lasers and solitons as well as multibound solitons in the active FM mode-locked fiber ring lasers. Both passive and active fiber lasers as nonlinear systems are described. Transmission of multibound solitons over the optical guided medium, the single-mode optical fibers, is also given to understand the evolution of such a bounded soliton group and the phases of the lightwaves in such phase-sensitive medium due to the group delay.

The third part (Chapters 7 through 10) focuses on numerical investigations on the third-order nonlinearity parametric process as well as the bispectrum technique for photonic signal processing.

Therefore, the chapters involving optical nonlinear systems are organized as below.

Chapter 3 gives an introduction of soliton generation in fiber lasers. A number of nonlinear fiber lasers are described and examined. Fundamentals of pulse propagation in nonlinear dispersive medium are introduced through important nonlinear Schrödinger equations (NSEs) and their analytical solutions. These equations also provide the theoretical basis for building the numerical model of the FM mode-locked fiber laser, the main subject of this book. It is more important to give a foundation of soliton generation in the nonlinear fiber ring resonators based on both passive and active mode-locking techniques. Experimental works on the active FM mode-locked fiber lasers are conducted to investigate soliton generation in a single pulse scheme. Numerical simulations are also used to investigate the pulse formation inside the fiber cavity as well as to verify the experimental results. Besides, detuning effects in the FM mode-locked fiber laser are experimentally and numerically investigated. All results obtained in this chapter are the foundation for the rest of the chapters of the book.

Chapter 4 investigates multibound solitons in nonlinear fiber ring lasers. First, the review of bound solitons generated by passive mode-locked fiber lasers is given and the mechanisms of multibound soliton formation and dynamics in the fiber ring cavity are introduced. Then the conditions of multibound solitons in an active fiber ring laser are determined and discussed with the participation of the active phase modulation in the balance of soliton interactions inside the fiber ring cavity. Experimental and simulation works on the generation of multibound solitons in the active FM mode-locked fiber ring are conducted and described in this chapter. For the first time, the existence of multibound solitons up to the fourth order has been demonstrated [126–128]. States of multibound solitons at higher order up to the sixth order have been published [129,130].

In an active FM mode-locked fiber ring laser, the electro-optic (EO) phase modulator is a key component for not only mode locking but also multibound soliton generation. Therefore, Chapter 4 is devoted to the influence of the EO phase modulator on pulse shaping and multibounding solitons. Important characteristic measurements of the phase modulator, such as dynamic response and half-wave voltage, are implemented to show the difference between two typical types of the EO phase modulators: lumped and traveling wave. The influence of the EO phase modulators on the operating states of active FM mode-locked fiber ring laser exhibits in two aspects: the artificial comb-filtering generated from the modal birefringence of the phase modulator and the chirping caused by the radio-frequency (RF) driving signal. Obtained results have shown these effects on wavelength tunability of the fiber laser as well as stability and formation of multibound solitons in our system. Some obtained results in this chapter were published [69,103–109,131,132].

Chapter 5 then gives a detailed study of passive soliton fiber lasers in which nonlinear evolution of the nonlinear and linear interaction of polarized guided modes occurs, and exchanges of optical energy from one polarization to the other and vice versa happen chaotically, as well as stable states are achieved for the solitons generated inside the fiber ring [110–122].

An advantage of the active mode-locked fiber ring is the generation of a stable periodic train of multibound solitons. Therefore, we experimentally investigate the propagation of multibound solitons in optical fibers in Chapter 6. A theoretical basis of soliton transmission in optical fibers is also reviewed. Numerical simulations are also implemented to investigate dynamic behaviors of the different multibound solitons during propagation along the fibers. The results of this chapter were presented in the literature [133].

In Chapter 6 most of the emphasis is on photonic signal processing. Although an approach of temporal imaging–based signal processing using an EO phase modulator was described in the literature [134], most of the content of this chapter focuses on photonic signal processing exploiting the parametric process in the third-order nonlinearity. Hence, a fundamental theory of FWM is reviewed, and then important applications based on FWM, such as parametric amplification and OTDM demultiplexing using nonlinear waveguides, are investigated via simulation. A part of these works was presented in the literature [135]. In another approach to characterize the phase-modulated signal, the bispectrum analysis, one of high-order spectra, is proposed. Therefore, a basis of bispectrum is introduced before this technique is applied to the analysis of multibound solitons as well as other phase-modulated signals. Furthermore, we consider a possibility of bispectrum estimation based particularly on the FWM process. Limitation of this estimation as well as application in optical receivers of the transmission systems is also discussed and presented in the literature [136].

Chapter 7 presents nonlinear fiber lasers. Different types on such nonlinear fiber lasers are described, and their experimental setup and generated pulse sequences are given.

Chapter 8 presents higher-order spectral analyses for examining the phase states of the lightwave carrier underneath the envelope of multibound solitons. This is typically the nonlinear signal processing applied to the study of nonlinear systems, especially the phase states of the bounded soliton pulse group.

Chapter 9 illustrates the nonlinear switching phenomena in planar-guided optical medium to add another dimension of nonlinear optical systems in channel-guided structures rather than circular-guided optical medium.

Chapter 10 presents the generic transfer function of single-mode optical fibers operating under nonlinear effects, which is very important in current advanced optical transmission systems. Volterra series are employed to present the nonlinear transfer section of the fiber propagation equation given in Chapter 2. Furthermore, the inverse of such a Volterra transfer function is employed to equalize distorted pulse sequences transmitted over the single fiber under nonlinear effects. This is implemented in electronic digital signal processing at the receiving ends. The effectiveness of this equalization technique is demonstrated.

A number of appendices related to a number of chapters are also included.

References

1. J. R. Russell, Report on Waves, *14th Meeting of the British Association for the Advancement of Science*, London, pp. 311–339, 1844.
2. D. J. Korteweg and H. de Vries, On the Change of Form of Long Waves Advancing in a Rectangular Canal, and on a New Type of Long Stationary Waves, *Phil. Mag.*, 39, 422–443, 1895.
3. E. Fermi, J. Pasta, and S. Ulam, Studies of Nonlinear Problems, Los Alamos Scientific Laboratory Report, no. LA-1940, 1955.
4. N. J. Zabusky and M. D. Kruskal, Interactions of Solitons in a Collisionless Plasma and the Recurrence of Initial States, *Phys. Rev. Lett.*, 15, 240–243, 1965.
5. M. Toda, Vibration of a Chain with Nonlinear Interactions, *J. Phys. Soc. Jpn.*, 22, 431–436, 1967.
6. V. E. Zakharov and A. B. Shbat, Exact Theory of Two-Dimensional Self-Focusing and One-Dimensional Self-Modulation of Waves in Nonlinear Media, *Soviet Phys., JETP*, 34, 62–69, 1970.
7. S. Kawamoto, An Exact Transformation from the Harry Dym Equation to the mKdV Equation, *J. Phys. Soc. Jpn.*, 54(5), 2055–2056, 1985.
8. D. J. Kaup and S. Kawamoto, A New Integrable System Including the Oscillating Two-Stream Instability Equation, *Proceedings of the IVth Workshop on Nonlinear Evolution Equations and Dynamical Systems*, World Scientific, pp. 181–189, 1987.
9. E. L. Ince, *Ordinary Differential Equations*, Dover, New York, 1944.
10. S. Kawamoto, Derivation of Nonlinear Partial Differential Equations Reducible to the Painleve Equations, *J. Phys. Soc. Jpn.*, 52(12), 4059–4065, 1983.
11. H. Toda and T. Yoshikawa, 40 Gbit/s Soliton Transmission Experiment in a Dense Dispersion Managed Fiber with 100 km Repeater Spacing, *Proceedings of OECC/COIN2004*, no. 14C3, Yokohama, 2004.
12. R. M. May, Simple Mathematical Models with Very Complicated Dynamics, *Nature*, 261, 459–467, 1976.
13. S. Kawamoto and T. Tsubata, Integrable Chaos Maps, *J. Phys. Soc. Jpn.*, 65(9), 5501–5502, 1996.
14. P. Coullet and C. Tresser, Critical Transition to Stochasticity for Some Dynamical Systems, *J. Physique Lett.*, 41, L255–L258, 1980.

15. M. J. Feigenbaum, Quantitative Universality for a Class of Nonlinear Transformations, *J. Stat. Phys.*, 19, 25–52, 1978.
16. S. L. Ulam and J. von Neumann, On Combination of Stochastic and Deterministic Processes, *Bull. Am. Math. Soc.*, 53, 1120, 1947.
17. L. M. Pecora and T. L. Carroll, Synchronization in Chaos Systems, *Phys. Rev. Lett.*, 64, 821–824, 1990.
18. G. Perez and H. A. Cerdeira, Extracting Message Masked by Chaos, *Phys. Rev. Lett.*, 74, 1970–1973, 1995.
19. G. D. V. Wiggeren and R. Roy, Optical Communication with Chaotic Waveforms, *Phys. Rev. Lett.*, 81, 3547–3550, 1998.
20. S. Kawamoto and T. Horiuchi, Algorithm for Exact Long Time Chaotic Series and Its Application to Cryptosystems, *Intl. J. Bifurcation and Chaos*, 14(10), 3607–3611, 2004.
21. J. Toyama and S. Kawamoto, Generation of Pseudo-Random Numbers by Chaos-Type Function and Its Application to Cryptosystems, *Elec. Eng. Jpn.*, 163(2), 67–74, 2008.
22. S. Kawamoto, Chaos and Its Application to Cryptosystems, *Proceedings of the Sixth Scientific Conference*, no. II-O-1.1, Ho Chi Minh City, Vietnam, 2008.
23. K. Minoshima and H. Matsumoto, High-Accuracy Measurement of 240-m Distance in an Optical Tunnel by Use of a Compact Femtosecond Laser, *Appl. Opt.*, 39, 5512–5517, 2000.
24. Y. Ding et al., Femtosecond Pulse Shaping by Dynamic Holograms in Photorefractive Multiple Quantum Wells, *Opt. Lett.*, 22, 718–720, 1997.
25. M. E. Zevallos et al., Picosecond Electronic Time-Gated Imaging of Bones in Tissues, *IEEE J. Selected Topics in Quantum Electronics*, 5, 916–922, 1999.
26. R. Cubeddu et al., Fluorescence Lifetime Imaging: An Application to the Detection of Skin Tumors, *IEEE J. Selected Topics in Quantum Electronics*, 5, 923–929, 1999.
27. H. N. Paulsen et al., Coherent Anti-Stokes Raman Scattering Microscopy with a Photonic Crystal Fiber Based Light Source, *Opt. Lett.*, 28, 1123–1125, 2003.
28. A. Tünnermann et al., Ultrashort Pulse Fiber Lasers and Amplifiers, in *Femtosecond Technology for Technical and Medical Applications*, vol. 96, F. Dausinger et al., Eds., Springer, Berlin/Heidelberg, 2004, pp. 35–54.
29. D. Breitling et al., Drilling of Metals, in *Femtosecond Technology for Technical and Medical Applications*, vol. 96, F. Dausinger et al., Eds., Springer, Berlin/Heidelberg, 2004, pp. 131–156.
30. T. Udem et al., Optical Frequency Metrology, *Nature*, 416, 233–237, 2002.
31. I. Hartl et al., Ultrahigh-Resolution Optical Coherence Tomography Using Continuum Generation in an Air/Silica Microstructure Optical Fiber, *Opt. Lett.*, 26, 608–610, 2001.
32. K. Suzuki et al., 40 Gbit/s Single Channel Optical Soliton Transmission over 70000 km Using In-Line Synchronous Modulation and Optical Filtering, *Electron. Lett.*, 34, 98–100, 1998.
33. M. Nakazawa et al., Single-Channel 80 Gbit/s Soliton Transmission over 10000 km Using In-Line Synchronous Modulation, *Electron. Lett.*, 35, 162–164, 1999.
34. M. Nakazawa et al., 1.28 Tbit/s-70 km OTDM Transmission Using Third- and Fourth-Order Simultaneous Dispersion Compensation with a Phase Modulator, *Electron. Lett.*, 36, 2027–2029, 2000.
35. H. G. Weber et al., Ultrahigh-Speed OTDM-Transmission Technology, *IEEE J. Lightwave Technol.*, 24, 4616–4627, 2006.

36. K. Mori et al., Flatly Broadened Supercontinuum Spectrum Generated in a Dispersion Decreasing Fiber with Convex Dispersion Profile, *Electron. Lett.*, 33, 1806–1808, 1997.
37. C. C. Chang et al., Code-Division Multiple-Access Encoding and Decoding of Femtosecond Optical Pulses over a 2.5-km Fiber Link, *IEEE Photonics Technol. Lett.*, 10, 171–173, 1998.
38. M. Zirngibl et al., 1.2 ps Pulses from Passively Mode-Locked Laser Diode Pumped Er-Doped Fiber Ring Laser, *Electron. Lett.*, 27, 1734–1735, 1991.
39. I. N. Duling, III, Subpicosecond All-Fiber Erbium Laser, *Electron. Lett.*, 27, 544–545, 1991.
40. K. Tamura et al., Self-Starting Additive Pulse Mode-Locked Erbium Fiber Ring Laser, *Electron. Lett.*, 28, 2226–2228, 1992.
41. K. Tamura et al., Unidirectional Ring Resonators for Self-Starting Passively Mode-Locked Lasers, *Opt. Lett.*, 18, 220–222, 1993.
42. F. Krausz and T. Brabec, Passive Mode Locking in Standing-Wave Laser Resonators, *Opt. Lett.*, 18, 888–890, 1993.
43. D. Kuizenga and A. Siegman, FM and AM Mode Locking of the Homogeneous Laser—Part I: Theory, *IEEE J. Quantum Electron.*, 6, 694–708, 1970.
44. K. Tamura and M. Nakazawa, Pulse Energy Equalization in Harmonically FM Mode-Locked Lasers with Slow Gain, *Opt. Lett.*, 21, 1930–1932, 1996.
45. Y. Shiquan and B. Xiaoyi, "Rational Harmonic Mode-Locking" in a Phase-Modulated Fiber Laser, *Photonics Technol. Lett., IEEE*, 18, 1332–1334, 2006.
46. M. E. Grein et al., Quantum-Limited Timing Jitter in Actively Modelocked Lasers, *IEEE J. Quantum Electron.*, 40, 1458–1470, 2004.
47. H. Takara et al., Stabilisation of a Modelocked Er-Doped Fiber Laser by Suppressing the Relaxation Oscillation Frequency Component, *Electron. Lett.*, 31, 292–293, 1995.
48. M. Nakazawa and E. Yoshida, A 40-GHz 850-fs Regeneratively FM Mode-Locked Polarization-Maintaining Erbium Fiber Ring Laser, *Photonics Technol. Lett., IEEE*, 12, 1613–1615, 2000.
49. M. Nakazawa et al., Ideal Phase-Locked-Loop (PLL) Operation of a 10 GHz Erbium-Doped Fiber Laser Using Regenerative Modelocking as an Optical Voltage Controlled Oscillator, *Electron. Lett.*, 33, 1318–1320, 1997.
50. W. -W. Hsiang et al., Long-Term Stabilization of a 10 GHz 0.8 ps Asynchronously Mode-Locked Er-Fiber Soliton Laser by Deviation-Frequency Locking, *Opt. Express*, 14, 1822–1828, 2006.
51. Y. Shiquan et al., Stabilized Phase-Modulated Rational Harmonic Mode-Locking Soliton Fiber Laser, *Photonics Technol. Lett., IEEE*, 19, 393–395, 2007.
52. C. R. Doerr, et al., Asynchronous Soliton Mode Locking, *Opt. Lett.*, 19, 1958–1960, 1994.
53. W. -W. Hsiang et al., Direct Generation of a 10 GHz 816 fs Pulse Train from an Erbium-Fiber Soliton Laser with Asynchronous Phase Modulation, *Opt. Lett.*, 30, 2493–2495, 2005.
54. S. B. Eduardo et al., 396 fs, 2.5–12 GHz Asynchronous Mode-Locking Erbium Fiber Laser, Conference on Lasers and Electro Optics CLEO 2007, 1–2, Baltimore, MD, Paper: CMCZ.
55. K. R. Tamura and M. Nakazawa, 54-fs, 10-GHz Soliton Generation from a Polarization-Maintaining Dispersion-Flattened Dispersion-Decreasing Fiber Pulse Compressor, *Opt. Lett.*, 26, 762–764, 2001.

56. G. P. Agrawal, *Nonlinear Fiber Optics*, 3rd ed., Academic Press, San Diego, 2001.
57. F. Ilday et al., Self-Similar Evolution of Parabolic Pulses in a Laser, *Phys. Rev. Lett.*, 92, 213902, 2004.
58. A. Chong et al., Properties of Normal-Dispersion Femtosecond Fiber Lasers, *J. Opt. Soc. Am. B*, 25, 140–148, 2008.
59. F. X. Kärtner et al., Solitary-Pulse Stabilization and Shortening in Actively Mode-Locked Lasers, *J. Opt. Soc. Am. B*, 12, 486–496, 1995.
60. T. F. Carruthers and I. N. Duling Iii, 10-GHz, 1.3-ps Erbium Fiber Laser Employing Soliton Pulse Shortening, *Opt. Lett.*, 21, 1927–1929, 1996.
61. M. Horowitz et al., Theoretical and Experimental Study of Harmonically Modelocked Fiber Lasers for Optical Communication Systems, *J. Lightwave Technol.*, 18, 1565–1574, 2000.
62. A. Zeitouny et al., Stable Operating Region in a Harmonically Actively Mode-Locked Fiber Laser, *IEEE J. Quantum Electron.*, 41, 1380–1387, 2005.
63. M. Nakazawa et al., Supermode Noise Suppression in a Harmonically Modelocked Fiber Laser by Selfphase Modulation and spe, *Electron. Lett.*, 32, 461, 1996.
64. N. G. Usechak et al., FM Mode-Locked Fiber Lasers Operating in the Autosoliton Regime, *IEEE J. Quantum Electron.*, 41, 753–761, 2005.
65. S. Yang et al., Experimental Study on Relaxation Oscillation in a Detuned FM Harmonic Mode-Locked Er-Doped Fiber Laser, *Opt. Commun.*, 245, 371–376, 2005.
66. Y. Frignac et al., Transmission of 256 Wavelength-Division and Polarization-Division-Multiplexed Channels at 42.7Gb/s (10.2Tb/s Capacity) over 3 × 100km of TeraLight™ Trade/Fiber, in *Optical Fiber Communication Conference and Exhibit, 2002. OFC 2002*, 2002, pp. FC5-1–FC5-3.
67. A. H. Gnauck et al., 25.6-Tb/s WDM Transmission of Polarization-Multiplexed RZ-DQPSK Signals, *J. Lightwave Technol.*, 26, 79–84, 2008.
68. P. J. Winzer and R.-J. Essiambre, Advanced Modulation Formats for High-Capacity Optical Transport Networks, *J. Lightwave Technol.*, 24, 4711–4728, 2006.
69. C. Wree et al., High Spectral Efficiency 1.6-b/s/Hz Transmission (8 × 40 Gb/s with a 25-GHz grid) over 200-km SSMF using RZ-DQPSK and polarization multiplexing, *Photonics Technol. Lett., IEEE*, 15, 1303–1305, 2003.
70. C. Schmidt-Langhorst et al., Generation and Coherent Time-Division Demultiplexing of up to 5.1 Tb/s Single-Channel 8-PSK and 16-QAM Signals, in *Optical Fiber Communication (incudes post deadline papers), 2009. Conference on OFC 2009*, 2009, pp. 1–3.
71. A. G. Striegler and B. Schmauss, Analysis and Optimization of SPM-Based 2R Signal Regeneration at 40 Gb/s, *J. Lightwave Technol.*, 24, 2835, 2006.
72. H. Murai et al., Regenerative SPM-Based Wavelength Conversion and Field Demonstration of 160-Gb/s All-Optical 3R Operation, *J. Lightwave Technol.*, 28, 910–921, 2010.
73. M. Takahashi et al., Low-Loss and Low-Dispersion-Slope Highly Nonlinear Fibers, *J. Lightwave Technol.*, 23, 3615–3624, 2005.
74. J. C. Knight et al., Anomalous Dispersion in Photonic Crystal Fiber, *Photonics Technol. Lett., IEEE*, 12, 807–809, 2000.
75. J. K. Ranka et al., Optical Properties of High-Delta Air Silica Microstructure Optical Fibers, *Opt. Lett.*, 25, 796–798, 2000.
76. S. Naoki et al., Bismuth-Based Optical Fiber with Nonlinear Coefficient of 1360 $W^{-1}km^{-1}$, Optical Fiber Communication Conference, 2004. OFC 2004 23–27, Feb. 2004, Paper: PDP26, Los Angeles, CA, USA.

77. R. E. Slusher et al., Large Raman Gain and Nonlinear Phase Shifts in High-Purity As2Se3 Chalcogenide Fibers, *J. Opt. Soc. Am. B*, 21, 1146–1155, 2004.

78. M. D. Pelusi et al., Wavelength Conversion of High-Speed Phase and Intensity Modulated Signals Using a Highly Nonlinear Chalcogenide Glass Chip, *Photonics Technol. Lett., IEEE*, 22, 3–5, 2010.

79. D. V. Trung et al., Photonic Chip Based 1.28 Tbaud Transmitter Optimization and Receiver OTDM Demultiplexing, Optical Fiber Conference 2010, California, USA, 2010, p. PDPC5.

80. D. Y. Tang et al., Observation of Bound States of Solitons in a Passively Mode-Locked Fiber Laser, *Phys. Rev. A*, 64, 033814, 2001.

81. D. Y. Tang et al., Bound-Soliton Fiber Laser, *Phys. Rev. A*, 66, 033806, 2002.

82. P. Grelu et al., Phase-Locked Soliton Pairs in a Stretched-Pulse Fiber Laser, *Opt. Lett.*, 27, 966–968, 2002.

83. L. M. Zhao et al., Period-Doubling and Quadrupling of Bound Solitons in a Passively Mode-Locked Fiber Laser, *Opt. Commun.*, 252, 167–172, 2005.

84. A. Haboucha et al., Coherent Soliton Pattern Formation in a Fiber Laser, *Opt. Lett.*, 33, 524–526, 2008.

85. S. T. Cundiff, Soliton Dynamics in Mode-Locked Lasers, in *Dissipative Solitons*. vol. 661, N. Akhmediev and A. Ankiewicz, Eds., Springer, Berlin/Heidelberg, 2005, pp. 183–206.

86. C. L. Nikias and J. M. Mendel, Signal Processing with Higher-Order Spectra, *Signal Processing Magazine, IEEE*, 10, 10–37, 1993.

87. Y. Shiquan et al., Stabilized Phase-Modulated Rational Harmonic Mode-Locking Soliton Fiber Laser, *Photonics Technol. Lett., IEEE*, 19, 393–395, 2007.

88. C. R. Doerr et al., Asynchronous Soliton Mode Locking, *Opt. Lett.*, 19, 1958–1960, 1994.

89. W.-W. Hsiang et al., Direct Generation of a 10 GHz 816 fs Pulse Train from an Erbium-Fiber Soliton Laser with Asynchronous Phase Modulation, *Opt. Lett.*, 30, 2493–2495, 2005.

90. S. B. Eduardo et al., 396 fs, 2.5-12 GHz Asynchronous Mode-Locking Erbium Fiber Laser, Conference CLEO, Baltimore, MD, USA, 2007, p. CMC2.

91. K. R. Tamura and M. Nakazawa, 54-fs, 10-GHz Soliton Generation from a Polarization-Maintaining Dispersion-Flattened Dispersion-Decreasing Fiber Pulse Compressor, *Opt. Lett.*, 26, 762–764, 2001.

92. G. P. Agrawal, *Nonlinear Fiber Optics* 3rd ed., Academic Press, San Diego, 2001.

93. F. Ilday et al., Self-Similar Evolution of Parabolic Pulses in a Laser, *Phys. Rev. Lett.*, 92, 213902, 2004.

94. A, Chong, W.H. Renninger, and F.W. Wise, Properties of Normal-Dispersion Femtosecond Fiber Lasers, *J. Opt. Soc. Am. B*, 25, 140–148, 2008.

95. F. X. Kärtner et al., Solitary-Pulse Stabilization and Shortening in Actively Mode-Locked Lasers, *J. Opt. Soc. Am. B*, 12, 486–496, 1995.

96. T. F. Carruthers and I. N. Duling Iii, 10-GHz, 1.3-ps Erbium Fiber Laser Employing Soliton Pulse Shortening, *Opt. Lett.*, 21, 1927–1929, 1996.

97. M. Horowitz et al., Theoretical and Experimental Study of Harmonically Modelocked Fiber Lasers for Optical Communication Systems, *J. Lightwave Technol.*, 18, 1565–1574, 2000.

98. A. Zeitouny, Y.N. Parkhomenko, and M. Horowitz, Stable Operating Region in a Harmonically Actively Mode-Locked Fiber Laser, *IEEE J. Quantum Electronics*, 41, 1380–1387, 2005.

99. M. Nakazawa et al., Supermode Noise Suppression in a Harmonically Modelocked Fibre Laser by Selfphase Modulation and spe, *Electron. Lett.*, 32, 461, 1996.

100. N. G. Usechak et al., FM Mode-Locked Fiber Lasers Operating in the Autosoliton Regime, *IEEE J. Quantum Electron.*, 41, 753–761, 2005.

101. S. Yang et al., Experimental Study on Relaxation Oscillation in a Detuned FM Harmonic Mode-Locked Er-Doped Fiber Laser, *Opt. Commun.*, 245, 371–376, 2005.

102. Y. Frignac et al., Transmission of 256 Wavelength-Division and Polarization-Division-Multiplexed Channels at 42.7Gb/s (10.2Tb/s Capacity) over 3 × 100km of TeraLight™ fiber, in *Optical Fiber Communication Conference and Exhibit, 2002. OFC 2002*, 2002, pp. FC5-1–FC5-3.

103. H. Gnauck et al., 25.6-Tb/s WDM Transmission of Polarization-Multiplexed RZ-DQPSK Signals, *J. Lightwave Technol.*, 26, 79–84, 2008.

104. P. J. Winzer and R.-J. Essiambre, Advanced Modulation Formats for High-Capacity Optical Transport Networks, *J. Lightwave Technol.*, 24, 4711–4728, 2006.

105. C. Wree et al., High Spectral Efficiency 1.6-b/s/Hz Transmission (8 × 40 Gb/s with a 25-GHz Grid) over 200-km SSMF Using RZ-DQPSK and Polarization Multiplexing, *Photonics Technol. Lett., IEEE*, 15, 1303–1305, 2003.

106. C. Schmidt-Langhorst et al., Generation and Coherent Time-Division Demultiplexing of up to 5.1 Tb/s Single-Channel 8-PSK and 16-QAM Signals, in *Optical Fiber Communication (incudes postdeadline papers), 2009. Conference on OFC 2009*, 2009, pp. 1–3.

107. G. Striegler and B. Schmauss, Analysis and Optimization of SPM-Based 2R Signal Regeneration at 40 Gb/s, *J. Lightwave Technol.*, 24, 2835, 2006.

108. H. Murai et al., Regenerative SPM-Based Wavelength Conversion and Field Demonstration of 160-Gb/s All-Optical 3R Operation, *J. Lightwave Technol.*, 28, 910–921, 2010.

109. M. Takahashi et al., Low-Loss and Low-Dispersion-Slope Highly Nonlinear Fibers, *J. Lightwave Technol.*, 23, 3615–3624, 2005.

110. J. C. Knight et al., Anomalous Dispersion in Photonic Crystal Fiber, *Photonics Technol. Lett., IEEE*, 12, 807–809, 2000.

111. J. K. Ranka et al., Optical Properties of High-Delta Air Silica Microstructure Optical Fibers, *Opt. Lett.*, 25, 796–798, 2000.

112. S. Naoki et al., Bismuth-Based Optical Fiber with Nonlinear Coefficient of 1360 $W^{-1}km^{-1}$, Optical Fiber Conference, 2004, 23–27 Feb. 2004, Los Angeles, CA, USA, 2004, p. PD26.

113. R. E. Slusher et al., Large Raman Gain and Nonlinear Phase Shifts in High-Purity As2Se3 Chalcogenide Fibers, *J. Opt. Soc. Am. B*, 21, 1146–1155, 2004.

114. M. D. Pelusi et al., Wavelength Conversion of High-Speed Phase and Intensity Modulated Signals Using a Highly Nonlinear Chalcogenide Glass Chip, *Photonics Technol. Lett., IEEE*, 22, 3–5, 2010.

115. V. Trung et al., Photonic Chip Based 1.28 Tbaud Transmitter Optimization and Receiver OTDM Demultiplexing, National Fiber Optics Engineers Conference, San Diego, CA, USA, March 21, 2010, p. PDPC5.

116. Y. Tang et al., Observation of Bound States of Solitons in a Passively Mode-Locked Fiber Laser, *Phys. Rev. A*, 64, 033814, 2001.

117. D. Y. Tang et al., Bound-Soliton Fiber Laser, *Phys. Rev. A*, 66, 033806, 2002.

118. P. Grelu et al., Phase-Locked Soliton Pairs in a Stretched-Pulse Fiber Laser, *Opt. Lett.*, 27, 966–968, 2002.

119. L. M. Zhao et al., Period-Doubling and Quadrupling of Bound Solitons in a Passively Mode-Locked Fiber Laser, *Opt. Commun.*, 252, 167–172, 2005.

120. A. Haboucha et al., Coherent Soliton Pattern Formation in a Fiber Laser, *Opt. Lett.*, 33, 524–526, 2008.

121. S. T. Cundiff, Soliton Dynamics in Mode-Locked Lasers, in *Dissipative Solitons*, vol. 661, N. Akhmediev and A. Ankiewicz, Eds., Springer, Berlin/Heidelberg, 2005, pp. 183–206.

122. L. Nikias and J. M. Mendel, Signal Processing with Higher-Order Spectra, *Signal Processing Magazine, IEEE*, 10, 10–37, 1993.

123. L. N. Binh and N. Q. Ngo, *Ultra-Fast Fiber Lasers: Principles and Applications with MATLAB Models (Optics and Photonics)*, CRC Press, Boca Raton, FL, 2010.

124. P. Yeh, *Introduction by Photorefractive Nonlinear Optics*, Wiley-Interscience, New York, 1993.

125. G. P Agrawal, *Nonlinear Fiber Optics*, 4th edition, Academic Press, New York, 2002.

126. N. D. Nguyen and L. N. Binh, Generation of Bound Solitons in Actively Phase Modulation Mode-Locked Fiber Ring Resonators, *Opt. Commun.*, 281, 2012–2022, 2008.

127. L. N. Binh, N. D. Nguyen, and W. J. Lai, Nonlinear Photonic Fibre Ring Lasers: Stability, Harmonic Detuning, Temporal Diffraction and Bound States, in *Progress in Nonlinear Optics Research*, M. Takahashi and H. Goto, Eds., Nova Science, New York, 2008, pp. 1–62.

128. L.N. Binh, N.D. Nguyen, and T. L. Huynh, Multi-Bound Solitons in a FM Mode-Locked Fiber Laser, in *Optical Fiber Communication/National Fiber Optic Engineers Conference, 2008. OFC/NFOEC 2008 Conference on*, 2008, pp. 1–3.

129. N. D. Nguyen and L. N. Binh, Generation of High Order Multi-bound Solitons and Propagation in Optical Fibers, *Opt. Commun.*, 282, 2394–2406, 2009.

130. L. N. Binh and N. D. Nguyen, Active Multi-bound Soliton Lasers: Generation of Dual to Sextuple States, in *Optical Fiber Communication (incudes postdeadline papers), 2009. OFC 2009. Conference on*, 2009, pp. 1–3.

131. N. D. Nguyen, L. N. Binh, and T. L. Huynh, Bound-Soliton States under a Periodic Phase Modulation, in *Lasers and Electro-Optics, 2008 and 2008 Conference on Quantum Electronics and Laser Science. CLEO/QELS 2008. Conference on*, 2008, pp. 1–2.

132. N. D. Nguyen and L. N. Binh, Solitonic Interactions in Actively Multi-bound Soliton Fiber Lasers, in *Lasers and Electro-Optics, 2009 and 2009 Conference on Quantum Electronics and Laser Science Conference. CLEO/QELS 2009. Conference on*, 2009, pp. 1–2.

133. N. D. Nguyen, L. N. Binh, and K. K. Pang, Propagation of Multi-bound Soliton States in Optical Fibers, in *Optical Fiber Communication (incudes postdeadline papers), 2009. OFC 2009. Conference on*, 2009, pp. 1–3.

134. N. D. Nguyen, L. N. Binh, K. K. Pang, T. Vo, and T. L. Huynh, Temporal Imaging and Optical Repetition Multiplication via Quadratic Phase Modulation, in *Information, Communications and Signal Processing, 2007 Sixth International Conference on*, 2007, pp. 1–5.

135. N. Nguyen and L. N. Binh, Demultiplexing Techniques of 320 Gb/s OTDM-DQPSK Signals: A Comparison by Simulation, presented at the *IEEE International Conference on Communication Systems*, Singapore, 2010.

136. L. N. Binh, N. Nguyen, Martin Firus, M. Steve, and L. Dang, Nonlinear Photonic Pre-Processing Bi-Spectrum Optical Receivers for Long Haul Optically Amplified Transmission Systems, presented at the *IEEE International Conference on Communication Systems*, Singapore, 2010.

2

Nonlinear Systems and Mathematical Representations

Dang Van Liet

Faculty of Physics, VNU University of Science, Ho Chi Minh City, Vietnam

Le Nguyen Binh

Hua Wei Technologies, European Research Center, Munich, Germany

CONTENTS

2.1 Introduction

Nonlinear dynamics have attracted significant research interest and was introduced briefly in Chapter 1. This chapter gives further mathematical details of generic nonlinear dynamic equations and simplified solutions whereby one can generate these nonlinear phenomena in practice.

It is appreciable that nonlinear dynamics can happen due to the nonlinear variations of parameters of physical systems or subsystems in which partial feedback of the outputs of the system reaches an indeterminate state on multistable states at a particular instant. For example in a system consisting of energy storage elements (e.g., inductors and capacitors), the charge and discharge of electrons or currents through or across these elements determine the states of dynamics and stability of the systems, so when both energy storage elements compete for the charges available in the system then chaotic or bistability may occur depending on the rates of storage of energy.

The fundamental rules underlying the shapes and the structures have led to the dynamic study of the nonlinear systems. Especially, modern nonlinear dynamics have focused on analytic solutions of the dynamic equations, and the nonlinearity is determined based on the principles of linear superposition under some approximations such as perturbation techniques. However, fundamental theoretical problems that arise in physics and mathematics as well as in engineering systems can be solved with the assistance of the digital computing techniques.

Solitons and chaos are briefly given, and their potential applications to engineering are given and described in the field of nonlinear optics in the rest of the chapters of this book. In subsequent chapters we will treat soliton and soliton lasers in detail, especially their generation, dynamic behavior, and then transmission over guided optical media.

Surprisingly, simple nonlinear systems are found to have chaotic solutions that remain within a bounded region. In other words, the nonlinearity has been positively considered, and the result has been applied to the analysis and the design in engineering and technology. Thus, nonlinear dynamics in nature have played a key role in physics, mathematics, and engineering, and would be fundamental tools for many branches of future research. This chapter gives a number of mathematical relationships of the principal functions in a nonlinear system and states of the systems in stable or chaotic dynamics bifurcation, as attractors and repellers in such systems. The dynamics of nonlinear systems are also treated in this chapter.

2.2 Nonlinear Systems, Phase Spaces, and Dynamical States

Nonlinear dynamical systems and chaotic phenomena are used widely in physics and engineering. This section describes some mathematical tools for the analysis of dynamical systems, especially the main pathways to chaos.

2.2.1 Phase Space

Consider a dynamical system represented by a differential equation subject to some initial conditions, for example, a well-known damped oscillation of a mass-spring system, as a mechanical system is governed by the differential equation and initial conditions at t_0 [1] given as

$$x'' = f(x, x') = -\frac{k}{m}x - \frac{c}{v}x'$$

$$x(t_0) = x_0 \quad \text{and} \quad x'(t_0) = v_0$$

(2.1)

where m is the mass, k is the stiffness of the spring, and c is the damping coefficient. Equation (2.1) is called *autonomous* because its right-hand side (RHS) does not depend on the time variable. The prime indicates the order of differentiation. In general one can reduce an nth order *autonomous* ordinary differential equation (ODE) to a system of n first-order *DEs* then applying the fourth-order Runge–Kutta method to obtain the final solutions of the system.

Let a sinusoidal driving force F exciting on the system, then the DE (2.1) can be rewritten as

$$x'' = g(t, x, x') = -\frac{k}{m}x - \frac{c}{v}x' + F\cos\omega t$$

(2.2)

the RHS of (2.2) depends explicitly on time. Equation (2.2) is *nonautonomous* and can be reduced to a system of $(n + 1)$ *first-order DE* by replacing the variable t by another dummy variable z. For example, (2.1) can be solved numerically by transforming it into two first-order DEs given by

$$x' = v; \quad x(t_0) = x_0$$

$$v' = -\frac{k}{m}x - \frac{c}{v}v; \quad v(t_0) = v_0$$

(2.3)

where $x(t)$ and $v(t)$ represent the displacement and velocity, respectively, and the prime indicates differentiation. In order to examine the behavior of the mass-spring system, the variation of the velocity versus displacement, $v(x)$, is used instead of the time-dependent displacement, $x(t)$. The x-v coordinate system is thus called a phase space, or a two-dimensional (2-D) phase state, of the nonlinear system. Each coordinate represents a variable of the system. The system can thus be represented by an nth-order autonomous differential equation so its phase space has n dimensions. The solutions of the system of (2.3) are subject to a given set of initial conditions under which $x(t)$ and $v(t)$ are determined by a point P in the phase space. When the time t changes then the locus of P follows a curve in the phase space, the *phase space trajectory* or *phase curve*. Furthermore, a set of trajectories forms a *phase portrait* that illustrates the dynamical behavior of the system.

In general, there are three kinds of trajectories in the phase space: fixed points, closed trajectories, and nonclosed trajectories. Figure 2.1 depicts a trajectory of the damped mass-spring system in the phase space operating under the following parameters: $m = 1$ kg, $k = 1.1$ Nm^{-1}, $c = 0.1$ kg^{-1}, $x_0 = 1$, and $v_0 = 0$, $t = [0, 20\pi]$. The energy system dissipates its energy in air so the trajectory is a spiral curve toward to the origin O(0,0); the point (0,0) is thus a stable attractor. Figure 2.2 shows the phase portrait of a periodic mass-spring

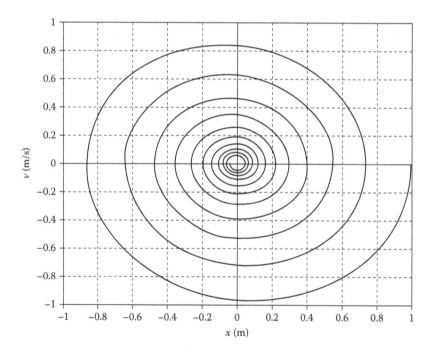

FIGURE 2.1
Trajectory of a damped mass-spring in phase space with an attractor at (0,0).

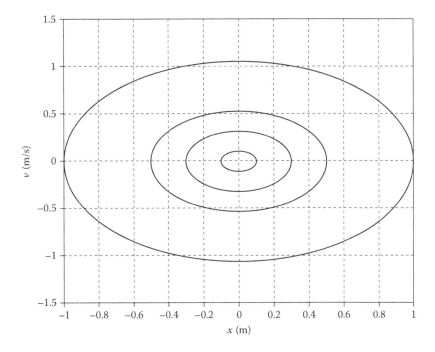

FIGURE 2.2
Phase portrait of harmonic mass-spring in phase space.

system in the phase space under different initial conditions given by $m = 1$ kg, $k = 1.1$ Nm–1, $c = 0$, $x_0 = [1, 0.5, 0.3, 0.1]$, *and* $v_0 = 0$, $t = [0, 20\pi]$.

Thus, one can deduce the no-intersecting theorem as follows:

Two distinct state space trajectories can neither intersect (in a finite period of time), nor can a single trajectory cross itself at a phase space at a later time. ∎

2.2.2 Critical Points

The critical point or fixed point in the phase space is the point that corresponds to the equilibrium of the dynamical system. The fixed point can also be the equilibrium point that is an important position, because a trajectory starts at a fixed point and then it stays at this point and the characteristic of the fixed point shows the behavior of trajectories at both its sides.

There are three kinds of fixed points: the attractors, the repellers, and the saddle points. Attractors or nodes or sinks are stable fixed points that attract neighboring trajectories. Repellers or sources are unstable fixed points that repel neighboring trajectories. Saddle points are the semistable fixed points that attract neighboring trajectories on one side and repel those on the other side. Fixed points can be identified without much difficulty. Let's consider the fixed points in one-dimensional (1-D) phase space which is just a line, we can then extend to those in the 2-D phase space.

2.2.2.1 Fixed Points in One-Dimensional Phase Space

The dynamic equation in 1-D phase space can be written as

$$x' = f(x) \tag{2.4}$$

where the prime indicates the differentiation with respect to the position x with respect to time. The fixed points of (2.4) are the locations in the phase space with their value $x^{*\prime}$, which must satisfy the condition

$$x^{*\prime} = f(x^*) = 0 \tag{2.5}$$

Thus, to find the fixed points, one can solve the equation $f(x) = 0$, and the solutions x^* are the positions of the fixed points. Let's consider the point x^r on the right side and nearby the fixed point x^*, if $f(x^r) > 0$ so that $x' > 0$, the trajectory starts at x^r and moves away from the fixed point; if $f(x^r) < 0$ so that $x' < 0$, the trajectory starts at x^r moving toward the fixed point, and vice versa for the point x^l on the left side and in the neighborhood of the fixed point. If the trajectories at both sides of the fixed point move toward the fixed point, then it can now be called the *attractor*. On the other hand, if the trajectories at two sides of the fixed point move away from the fixed point then it is termed the *repeller*. If the trajectory at one side of the fixed point moves toward the fixed point and vice versa for the trajectory at the other side, then the fixed point is classified as the *saddle point*.

Alternatively, we can use an eigenvalue, λ a characteristic value, to distinguish different types of fixed points that can be obtained by setting

$$\lambda = \left. \frac{df(x)}{dx} \right|_{x^*} \tag{2.6}$$

Then we can approximate this by using the Taylor series to distinguish the kind of fixed point. Let $\zeta = x - x^*$ and keeping the first two terms of the Taylor series expansion [2,8] in the neighborhood of x^*, we have

$$f(\zeta + x^*) = f(x^*) + \zeta f'(x^*) + \Theta(\zeta^2) \tag{2.7}$$

On the RHS of this equation the term $f(x^*) = 0$, because x^* is defined as the fixed point. So

$$\zeta' = f'(x^*)\zeta \tag{2.8}$$

The solution of (2.8) is

$$\zeta(t) = Ae^{\lambda t} \tag{2.9}$$

ζ is the quantity that measures the distance between a point of the trajectory and the fixed point. The exponential coefficient λ influences the trajectory as follows: (1) If $\lambda < 0$, the trajectory moves toward the fixed point exponentially so the fixed point is the attractor; (2) If $\lambda > 0$, the trajectory moves away from the fixed point exponentially so the fixed point acts as the repeller; (3) If $\lambda = 0$, the fixed point may be the saddle point or the attractor or the repeller. In this case we must compute the second derivative of $f(x)$. If $f''(x)$ has the same sign at both sides of the fixed point, then that is the saddle point; if $f''(x) > 0$ is on the left side of the fixed point and $f'(x) < 0$ is on the right side of the fixed point, then that is the attractor; otherwise the fixed point is the repeller.

As an example of the attractor and the repeller, consider the equation

$$x' = f(x) = x^2 - 1 \tag{2.10}$$

$x^* = 1$ and $x^* = -1$ are the fixed points because $f(x^*) = 0$; the eigenvalue at $x^* = 1$ is $\lambda = 2 > 0$ so the fixed point at $x = 1$ is the repeller and its eigenvalue located at $x^* = -1$ with $\lambda = -2 < 0$; hence, the fixed point at $x = -1$ is the attractor (Figure 2.3). The saddle point of the equation $x' = f(x) = x^2$ is depicted in Figure 2.4 in which the saddle point is located at the origin $O(0,0)$.

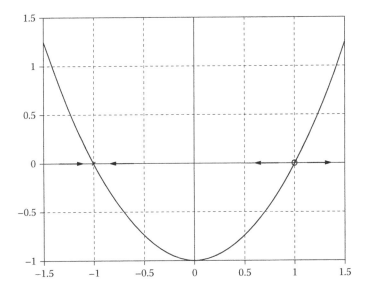

FIGURE 2.3
Attractor at (–1,0). Repeller at (1,0).

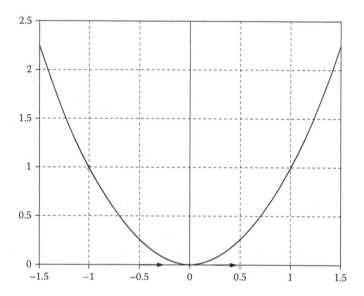

FIGURE 2.4
Saddle point at O(0,0).

2.2.2.2 Fixed Points in Two-Dimensional Phase Space

We can now extend the fixed points in 1-D phase space to 2-D phase space by considering the system of equations

$$x' = f(x,y)$$
$$y' = g(x,y)$$

(2.11)

The phase space is the (x-y) plane, and the behavior of the system is represented by the trajectories in the phase space. Let's (x_c, y_c) be the coordinate of the fixed points satisfying the equations

$$f(x_c, y_c) = 0$$
$$g(x_c, y_c) = 0$$

(2.12)

The first step is to find the positions of fixed points by solving the system of equations (2.12). Then determine the type of fixed points by finding the characteristic values of the fixed points. The functions f and g depend on the two variables, so the characteristic values depend on their partial derivatives. Similar to the case of 1-D, we are concerned only with the characteristic values in the x and y directions by setting

$$\lambda_x = \frac{\delta f}{\delta x}\bigg|_{(x_c, y_c)}$$

$$\lambda_y = \frac{\delta g}{\delta y}\bigg|_{(x_c, y_c)}$$

(2.13)

Thus, we can extend the nature of the eigenvalues in 1-D to classify the types of fixed points in 2-D phase space. We thus consider two distinct cases as follows:

2.2.2.2.1 λ_x and λ_y as Real Numbers

$\lambda_x < 0$ and $\lambda_y < 0$: fixed points are attractors
   Considering the equations

$$x' = -x$$
$$y' = -2y$$

(2.14)

whose solutions can be found as

$$x = C_1 e^{-t} \quad \text{and} \quad y = C_2 e^{-2t}$$

(2.15)

We can eliminate the variable t in both x and y to give

$$y = K x^2, \quad K = \frac{C_2}{C_1^2}$$

(2.16)

   The fixed point at the origin O(0,0) and the $\lambda_x = -1 < 0$ and $\lambda_y = -2 < 0$, so O(0,0) is the attractor. We can check this property by inspecting the trajectories. The phase space trajectories are a set of parabolas. For $K = 0$ the trajectory is the x-axis, when $t \to 0$, $x \to 0$ and $y \to 0$. So all trajectories move toward the origin, and the attractor as indicated in Figure 2.5.

$\lambda_x > 0$ and $\lambda_y > 0$: fixed points are repellers
   Considering the equation

$$x' = x$$
$$y' = 2y$$

(2.17)

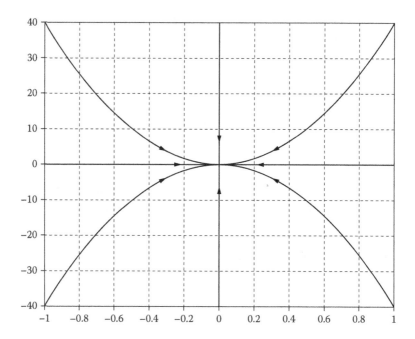

FIGURE 2.5
Attractor in two-dimensional phase space.

The fixed point is also at the origin O(0,0) and the $\lambda_x = 1 > 0$ and $\lambda_y = 2 > 0$, so O(0,0) is the repeller. The phase space trajectories are the same as the trajectories shown in Figure 2.5, but moving away from the origin, the *repeller* (Figure 2.6).

$\lambda_x < 0$ *and* $\lambda_y > 0$. Fixed points as saddle points
In the case that the fixed point on the x-axis is an attractor ($\lambda_x < 0$) and the fixed point on the y-axis is a repller ($\lambda_y > 0$), then the fixed point is the saddle point in the 2-D phase space.

When $\lambda_x > 0$ and $\lambda_y < 0$, we have the opposite situation and the fixed point is now the saddle point.

For example, considering the set of equations

$$x' = x$$
$$y' = -y$$
(2.18)

The variable t can be eliminated in the solution of (2.18), and we have

$$y = \frac{K}{x}$$
(2.19)

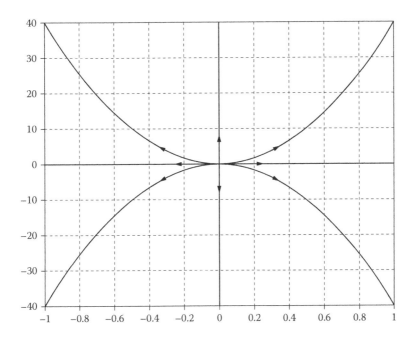

FIGURE 2.6
Repeller in two-dimensional phase space.

The phase space trajectories shown in Figure 2.7, are a set of hyperbolae. When $K = 0$ the trajectory is the x-axis, when $t \to 0$, $x \to \infty$ and $y \to 0$, so the origin is thus the saddle point.

2.2.2.2.2 λ_x and λ_y as Complex Conjugates
Suppose that

$$\lambda_{x,y} = \alpha + i\beta \tag{2.20}$$

Then the solutions are

$$x(t) = e^{\alpha t} \cos \beta t \quad and \quad y(t) = e^{\alpha t} \sin \beta t \tag{2.21}$$

- If $\alpha = 0$: the solution $x(t) = cos\beta t$ and $y(t) = sin\beta t$; so the trajectories in the space phase are circles, the fixed point at the origin is called a *center* (Figure 2.8).
- If $\alpha \neq 0$ and $\beta \neq 0$: the fixed point is at the origin that is called a *focal point*; the solutions are shown by (2.21), and the phase space trajectories oscillate with an angular frequency β. Its amplitudes vary exponentially, so the trajectories move spirally to the origin (attractor) ($\alpha < 0$) or out of the origin (repeller) ($\alpha > 0$) (Figure 2.9).

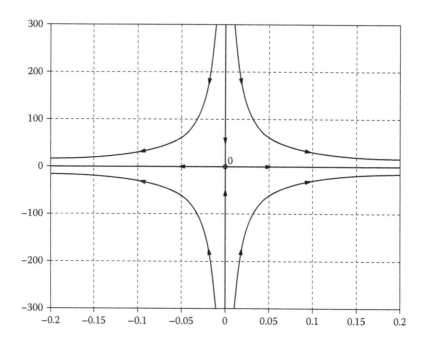

FIGURE 2.7
Saddle point in two-dimensional phase space.

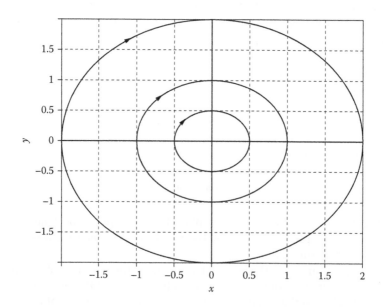

FIGURE 2.8
Center.

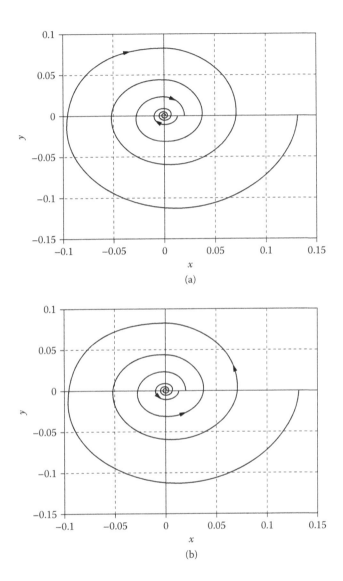

FIGURE 2.9
Focus. (a) Attractor. (b) Repeller.

2.2.3 Limit Cycles

In 2-D phase space or higher dimensions, consider a new fixed point that is a limit cycle defined as an isolated closed-loop trajectory in phase space. The neighboring trajectories are opened loops moving spirally in or out of the limit cycle when $t \rightarrow \infty$ or $t \rightarrow -\infty$, respectively. If the neighboring trajectories at both sides move toward the limit cycle, that is a *stable limit cycle*; on the

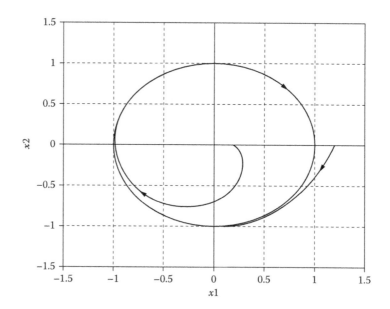

FIGURE 2.10
Limit cycle.

contrary, the limit cycle is an *unstable limit cycle*. If the neighboring trajectories at either side of the limit cycle move toward the limit cycle and the others move away, then the limit cycle is a *semistable limit cycle*. The limit cycle only appears in the nonlinear dynamical system. There are also closed loops in the linear system but they are not isolated, so these closed loops are not limit cycles as shown in Figure 2.2.

Figure 2.10 represents the table limit cycle in phase space $(x_1 - x_2)$, x_1 and x_2 are solutions of the equations

$$x_1' = \mu x_1 + x_2 - x_1\left(x_1^2 + x_2^2\right)$$
$$x_2' = -x_1 + \mu x_2 - x_1\left(x_1^2 + x_2^2\right)$$

(2.22)

where μ is a parameter and the system has a stable limit cycle for $\mu > 0$ like Figure 2.10.

This section can thus be concluded by stating the *Poincaré–Bendixson* theorem:

Supposing that there is a bounded region that contains no fixed point, if a trajectory is confined in this region then only either of the following two possible cases can be true: (1) The trajectory is a closed orbit as $t \to \infty$; (2) The trajectory spirals towards a close orbit as $t \to \infty$. ∎

The Poincaré–Bendixson theorem implies that if a trajectory is confined in a closed region that does not contain any fixed point, then at last this trajectory spirals to a closed orbit. For a higher-order dimensional system ($n \geq 3$) the trajectories may wander forever in a bounded region. The trajectories are attracted to a complex geometrical object, the strange attractor. This is the characteristic of a chaotic phenomenon. The chaotic trajectories cannot appear in 2-D phase spaces.

2.3 Bifurcation

As described above, a dynamical system can be represented by a differential equation. If the equation depends on only one or more parameters then the fixed points of the system can be altered accordingly. These qualitative variations of the dynamical system are termed *bifurcation*. In this section, we briefly present the bifurcation in 1-D and 2-D phase spaces. Further, we restrict our discussion to the most common types: the pitchfork, the saddle-node, the transcritical, and the Hopf bifurcations [1,2,8].

2.3.1 Pitchfork Bifurcation

This bifurcation occurs most commonly in symmetrical systems. As an example of pitchfork bifurcation consider the equation

$$x' = f(x,\mu) = \mu x - x^3 \tag{2.23}$$

where μ is a real parameter. This equation is invariant when x is replaced by $-x$. To study the bifurcation of this system, we have to find the fixed points and their characteristics.

- The positions of fixed points are determined by

$$x' = f(x,\mu) = \mu x - x^3 = 0 \tag{2.24}$$

so there are three fixed points at $x = 0$, $x = \sqrt{\mu}$, and $x = -\sqrt{\mu}$.

- The characteristics of the fixed points are determined by

$$\lambda = \frac{df}{dx} = \mu - 3x^2 \tag{2.25}$$

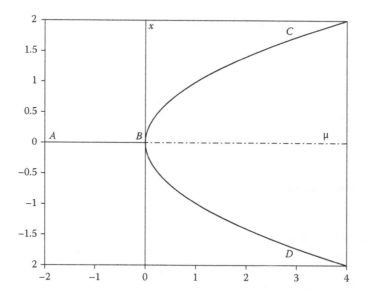

FIGURE 2.11
Pitchfork bifurcation diagram. *Solid lines:* stable state equilibrium. *Dashed line:* unstable state equilibrium.

> When $\lambda < 0$, the fixed point is stable, the *attractor*. However, if $\lambda > 0$, then the fixed point is unstable, the *repeller*.
>
> - If $\mu < 0$, there is only one fixed point at the origin ($x_1 = 0$) and it is stable ($\lambda = \mu < 0$).
> - If $\mu > 0$, there are three cases: (1) The fixed point at $x_1 = 0$ is unstable ($\lambda = \mu > 0$); (2) the fixed point at $x_2 = \sqrt{\mu}$ is stable ($\lambda = -2\mu < 0$); and (3) the fixed point at $x_3 = -\sqrt{\mu}$ is stable ($\lambda = -2\mu < 0$).

We can thus solve the system of equations (2.24) and (2.25) to obtain $x = 0$ and $\mu = 0$. The origin ($x = 0$, $\mu = 0$) is the *bifurcation* point.

The diagram of the pitchfork bifurcation is shown in Figure 2.11, the solid lines indicate the positions of stable fixed points, and the dashed line indicates the positions of unstable fixed points. This case is the *supercritical pitchfork bifurcation* because the bifurcation branches are stable. The other case is called the *subcritical pitchfork bifurcation* when the bifurcation branches are unstable.

2.3.2 Saddle-Node Bifurcation

This bifurcation is the basic mechanism. In this case the fixed points are created or destroyed as one or more parameters of the system change. As an example of the saddle-node bifurcation, consider

$$x' = f(x,\mu) = \mu + x^2 \tag{2.26}$$

with μ a parameter that can take positive, negative, or zero values. The fixed point and its characteristics are determined by

$$x' = f(x, \mu) = \mu + x^2 = 0 \tag{2.27}$$

and

$$\lambda = \frac{df}{dx} = 2x \tag{2.28}$$

- If $\mu < 0$ the system has two fixed points, the stable fixed point A at $x_1 = -\sqrt{-\mu}$ $(1 < 0)$ and the unstable fixed point B at $x_2 = \sqrt{-\mu}$ $(1 > 0)$ (Figure 2.12a).
- If $\mu \to 0$, the parabola moves up and two fixed points A and B move together. When $\mu = 0$ two fixed points A and B add together to be a half-stable fixed point at the origin O $(x = 0)$ (Figure 2.12b).
- If $\mu > 0$ the system does not have any fixed point (Figure 2.12c).

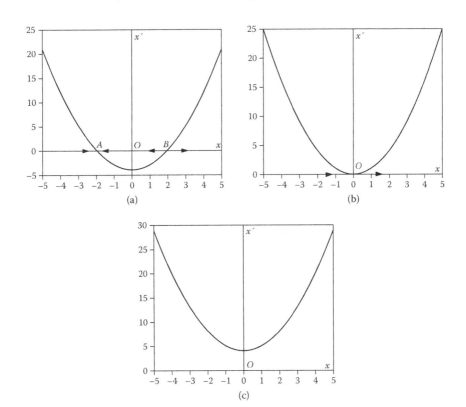

FIGURE 2.12
Saddle-node bifurcation process in the phase space: (a) m < 0; (b) m = 0; (c) m > 0.

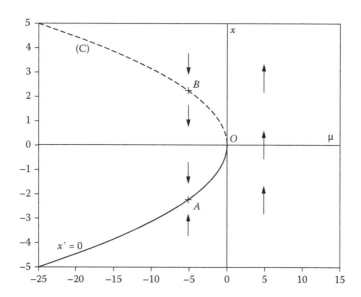

FIGURE 2.13
Saddle-node bifurcation diagram.

   The system of equations (2.27) and (2.28) can be solved to give solutions $x = 0$ and $\mu = 0$, so the bifurcation point can be located at the origin ($x = 0$, $\mu = 0$).
   We can now depict the curves of $x' = 0$ in the space (μ, x); the curves (C) of function $\mu = -x^2$ represent equilibrium positions of x on μ (Figure 2.13). Figure 2.13 shows the saddle-node bifurcation diagram. This is the common way to illustrate the saddle-node bifurcation, which can also be called a *fold bifurcation* or a *turning-point bifurcation* because the point of $x = 0$ and $\mu = 0$ is a turning point.

2.3.3 Transcritical Bifurcation

In this bifurcation the fixed point of the system always exists but its characteristics change as one or more parameters change. As an example of the transcritical bifurcation consider the equation

$$x' = f(x,\mu) = x(\mu - x) \tag{2.29}$$

where μ is a parameter that can take values of positive, negative, or zero. The fixed point and its characteristics can be determined by

$$x' = x(\mu - x) = 0 \tag{2.30}$$

and

$$\lambda = \frac{df}{dx} = \mu - 2x \tag{2.31}$$

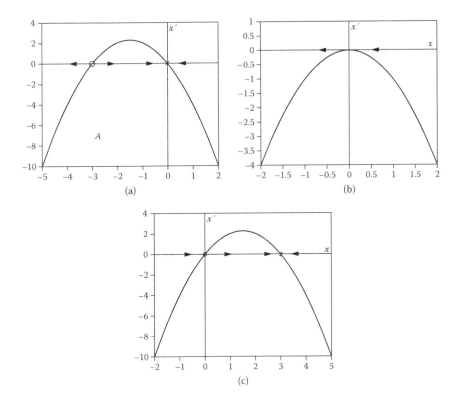

FIGURE 2.14
Transcritical bifurcation process in the phase space: (a) $\mu < 0$; (b) $\mu = 0$; (c) $\mu > 0$.

- If $\mu < 0$ the system has two fixed points, the stable fixed point at $x_1 = 0$ ($\lambda < 0$) and the unstable fixed point $x_2 = -\mu)$ ($\lambda > 0$) (Figure 2.14a).
- If $\mu = 0$ the system has a half-stable fixed point at the origin O ($x = 0$) ($\lambda = 0$) (Figure 2.14b).
- If $\mu > 0$ then the system has solutions of two fixed points, the unstable fixed point at $x_1 = 0$ ($\lambda > 0$) and stable fixed point $x_2 = \mu$ ($\lambda < 0$) (Figure 2.14c).

Combining three cases shown in Figures 2.14a, 2.14b, and 2.14c, we see that the fixed point located at the origin ($x = 0$) always exists but its stability changes as m varies. So this is the case of transcritical bifurcation.

The solution of system of equations (2.30) and (2.31) gives the bifurcation point at the origin ($x = 0$ and $\mu = 0$). Thus we also use the space (μ, x) to represent the transcritical bifurcation diagram as shown in Figure 2.15.

2.3.4 Hopf Bifurcation

Hopf bifurcation occurs when a limit cycle is taken from a fixed point as one or more parameters of the system vary, and it takes place only in the phase

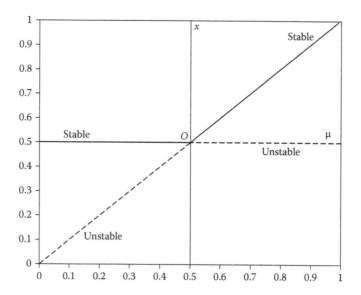

FIGURE 2.15
Transcritical bifurcation diagram.

space of higher-order dimensions (≥2). In this section we present the Hopf bifurcation in 2-D phase space. As an illustration of the Hopf bifurcation, consider the system of equations

$$x_1' = \mu x_1 + x_2 - x_1\left(x_1^2 + x_2^2\right)$$

$$x_2' = -x_1 + \mu x_2 - x_1\left(x_1^2 + x_2^2\right)$$

(2.32)

where μ is a real parameter. This system is simpler when transformed into polar coordinates (r,θ) by setting $x_1 = r\cos\theta$ and $x_2 = r\sin\theta$ with $r^2 = x_1^2 + x_2^2$ (2.32), we thus have

$$r' = r(\mu - r^2)$$

$$\theta' = -1$$

(2.33)

Equation (2.33) shows that there are two fixed points at $r_1 = 0, r_2 = \sqrt{\mu}$ because the solution $r_2 = -\sqrt{\mu}$ can be eliminated by setting $r > 0$ and the eigenvalues $\lambda = \mu \pm i$.

If $\mu < 0$ there is only a stable fixed point at the origin that is the focus; the trajectories spiral the origin (Figure 2.16a). If $\mu = 0$, the origin, that is a

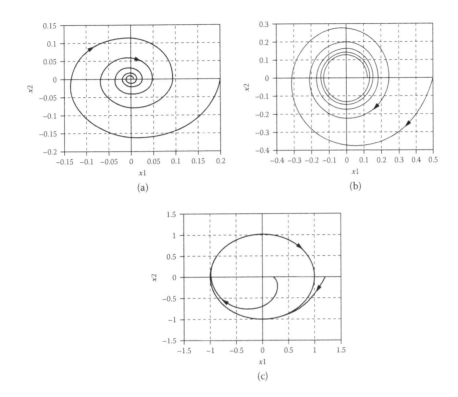

FIGURE 2.16
The birth of the limit cycle of Hopf bifurcation: (a) m = –0.1 < 0; (b) m = 0; (c) m = 0.3 > 0.

center, is also a stable fixed point, the trajectories also spiral the origin slowly (Figure 2.16b). If $\mu > 0$ the fixed point at the origin ($r = 0$) is unstable and the fixed point at $r = \sqrt{\mu}$ is stable, the trajectories become a stable limit cycle (Figure 2.16c).

We can see that when $\mu < 0$ the origin is stable and when $\mu > 0$ the origin is unstable and appears to be a stable limit cycle with $r = \sqrt{\mu}$. So the Hopf bifurcation occurs at $\mu = 0$, and it is called the *supercritical Hopf bifurcation* because the limit cycle is stable. The diagram of the Hopf bifurcation is depicted in Figure 2.17. We also have the subcritical Hopf bifurcation when the limit cycle is unstable. Hopf bifurcation is also called the *oscillatory bifurcation*.

Bifurcations play important roles, because in some dynamical systems, when the control parameter increases for a long time, the bifurcations may appear several times. This is then followed by the wandering unpredictably of the system's trajectories in the phase space. This is the chaotic phenomenon.

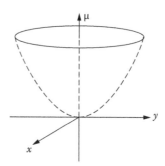

FIGURE 2.17
Hopf bifurcation diagram.

2.4 Chaos

2.4.1 Definition

There is no exact definition of chaos, but the commonly accepted definition [8] is: "Chaos is aperiodic long term behavior in a deterministic system which is strongly dependent on its initial conditions."

Under this definition, a chaotic phenomenon appears only when the system evolves for a long time, during which times there are no intersections of different aperiodic trajectories produced by the system represented by a DE associated with some certain deterministic initial conditions. The system is sensitive to its initial conditions meaning that there are two neighboring trajectories beginning at the distance d_0, and the time-dependent distance is given by $d(t) = d_0 e^{\lambda t}$ ($\lambda > 0$), with λ is known as the Lyapunov exponent, so two neighboring trajectories can move away with respect to each other exponentially.

E. N. Lorenz (1963) conducted the first experiment on chaos [3]. His model is a simplified model of the fundamental Navier-Stokes equation of the dynamics of fluids given by

$$x' = p(y - x)$$
$$y' = rx - y - xz \qquad (2.34)$$
$$z' = xy - bz$$

where p is the Prandtl number, r is the Rayleigh number, and $b > 0$ is an unknown parameter. He solved (2.34) numerically and realized that the system was sensitive to its initial conditions, and the system trajectories neither settle at any point nor repeat again but always follow a spiral, and no

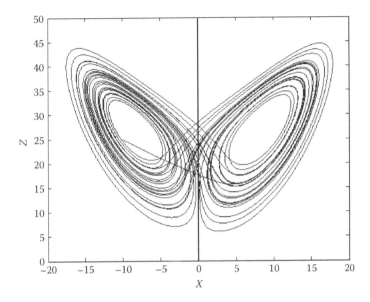

FIGURE 2.18
Variation of z with respect to x. Strange attractor.

intersection of trajectories would occur. He then termed this image the *Lorenz attractor* (Figure 2.18) ($p = 10$, $r = 28$, $b = 8/3$). This is a chaotic phenomenon. Currently this is called the *attractor of a chaotic system,* which is a strange attractor or chaotic attractor or fractal attractor as shown in Figure 2.18.

2.4.2 Routes to Chaos

The trajectories of nonlinear systems may move from a regular to a chaotic route pattern when a control parameter changes. These are the transitional

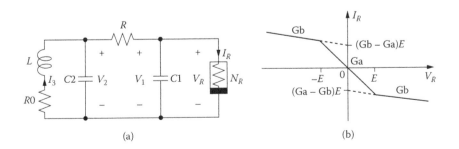

FIGURE 2.19
(a) Chua's circuit. (b) The driving-point characteristic of the nonlinear resistor N_R in Chua's circuit. (Chua's equations are as given in M. P. Kennedy (1994), "ABC (Adventures in Bifurcations and Chaos): A Program for Studying Chaos," *Journal of the Franklin Institute,* 331B(6), 631–658.)

routes to chaos. There are several types of routes to chaos [4]. In this section we review only some main routes to chaos.

2.4.2.1 Period Doubling

As described in the above sections the limit cycle was born from the bifurcation that relates to a fixed point. As a control parameter changes, the limit cycle would be unstable and onset to periodic-doubling limit cycles. The control parameter continues to change so that the period doubling becomes unstable and a period-four cycle starts to appear. The process will continue to infinite-period cycle, and then trajectories follow the chaotic behavior.

As an illustration of period doubling, consider Chua's circuit, a simple nonlinear circuit that leads to chaotic behavior:

$$\frac{dI_3}{dt} = -\frac{R_0}{L}I_3 - \frac{1}{L}V_2$$

$$\frac{dV_2}{dt} = \frac{1}{C_2}I_3 - \frac{G}{C_2}(V_2 - V_1)$$

$$\frac{dV_1}{dt} = \frac{G}{C_1}(V_2 - V_1) - \frac{1}{C_1}f(V_1)$$

$$= \begin{cases} \dfrac{G}{C_1}V_2 - \dfrac{G'_b}{C_1}V_1 - \left(\dfrac{G_b - G_a}{C_1}\right)E & \text{if } V_1 < -E \\[2ex] \dfrac{G}{C_1}V_2 - \dfrac{G'_a}{C_1}V_1 & \text{if } -E \le V_1 \le E \\[2ex] \dfrac{G}{C_1}V_2 - \dfrac{G'_b}{C_1}V_1 - \left(\dfrac{G_a - G_b}{C_1}\right)E & \text{if } V_1 > E \end{cases} \qquad (2.35)$$

in which the conductance of the elements is given as $G = 1/R$, $G'_a = G + G_a$ and $G'_b = G + G_b$.

Using the Kennedy's values [5], $L = 18$ mH, $R_0 = 12.5\ \Omega$, $C_2 = 100$ nF, $G_a = -757.576\ \mu S$, $G_b = -409.09\ \mu S$, $E = 1$ V, $C_1 = 10$ nF. By changing the control parameter G we have the phase portraits in Figure 2.20. Figure 2.20a shows a single-period limit cycle, $G = 530\ \mu S$. Figure 2.20b shows a period-doubling limit cycle with $G = 537\ \mu S$, and Figure 2.20c is a quadrupling period limit cycle, $G = 539\ \mu S$. Figure 2.20d depicts a spiral Chua's chaotic attractor with $G = 541\ \mu S$. Figure 2.20 thus shows the sequence of dynamics leading to the doubling period and thence the route to chaos.

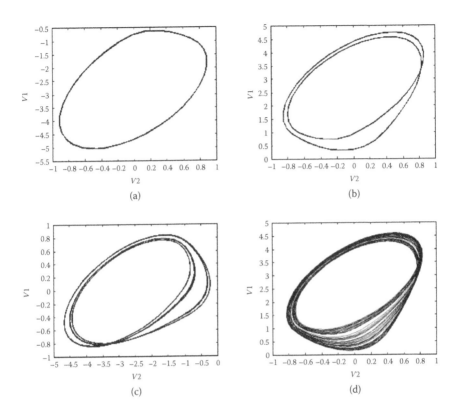

FIGURE 2.20
Phase portraits. (a) T period solution, $G = 530$ mS. (b) 2T period solution; $G = 537$ mS; (c) 4T period solution; $G = 539$ mS; and (d) chaotic solution, $G = 541$ mS.

2.4.2.2 *Quasi-periodicity*

The motion of a nonlinear dynamic system that associates with two frequencies is called the *quasi-periodic motion,* and it appears when the ratio of two frequencies is not a ratio of integers. So the motion does not repeat itself exactly but is not chaotic.

The quasi-periodic route to chaos is connected to the Hopf bifurcation. In Section 2.2 we considered the Hopf bifurcation that associates with the onset and the start of the limit cycle, a periodic motion, from a fixed point. In some nonlinear dynamical systems, the first and then the second Hopf bifurcation appears when the control parameter changes. If the ratio of frequencies of the second and the first motion is not rational, then the motion of the system is quasi-periodic. If the control parameter continues to change, then a chaotic motion would appear. This route is called the Rouelle–Takens route.

Figure 2.21 represented a quasi-periodic signal that is the voltage of the second capacitor in Chua's circuit using an inductor gyrator (see Figure 2.23).

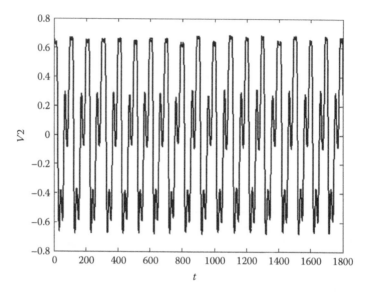

FIGURE 2.21
Quasi-periodic signal.

2.4.2.3 Intermittency

The intermittency is the motion that is nearly periodic with irregular bursts appearing from time to time. The phenomenon appears pseudo-randomly and not purely random because the system is represented by a deterministic equation. There can be two approaches to the intermittency: (a) the motion of system changes between the periodic motion and the chaotic motion and (b) the motion of system changes between the periodic motion and the quasi-periodic motion.

In some nonlinear dynamical systems, the control parameters of the system change further and the irregular bursts appear frequently so that the intermittency behaviors become chaotic behaviors. We will observe the intermittency motion in the next section.

2.4.3 Chaotic Nonlinear Circuit

The Chua's circuit using the inductor gyrator was investigated by Bharathwaj Muthuswamy et al. [7]. The schematic of the inductor gyrator is given in Figure 2.22, and its impedance is given:

$$Z_{in} = R_L + j\omega R_L R_g C \tag{2.36}$$

In 1992, K. Murali and M. Lakshamanan [6] examined the effect of the external periodic excitation on Chua's circuit. We can now consider a Chua's

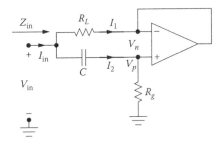

FIGURE 2.22
Circuit of inductor gyrator. (From B. Muthuswamy, T. Blain, and K. Sundqvist (2009), "A Synthetic Inductor Implementation of Chua's Circuit," *Technical Report No. UCB/EECS-2009-20.* Access date: June 2011, http://www.eecs.berkeley.edu/Pubs/TechRpts/2009/EECS-2009-20. html. With permission.)

circuit with external driving sinusoidal excitation, and this circuit used the inductor gyrator instead of an inductor because the inductor is cumbersome for the integrated circuit to be used in an experiment.

Figure 2.23 shows the schematic of Chua's circuit using an inductor gyrator with a sinusoidal excitation. The state equations for Chua's circuit (See Figure 2.19a) are given:

$$C_1 \frac{dv_1}{dt} = \frac{v_2 - v_1}{R} - i_R$$

$$C_2 \frac{dv_2}{dt} = \frac{v_2 - v_1}{R} - i_L - \frac{v_2 - i_L R_L - V_f}{R_g} \tag{2.37}$$

$$C \frac{di_L}{dt} = \frac{v_2 - i_L R_L - V_f}{R_L R_g}$$

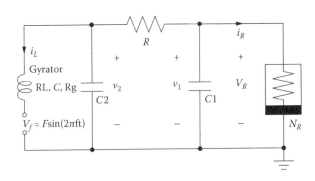

FIGURE 2.23
Chua's circuit using inductor gyrator and sinusoidal excitation.

where $i_R = g(v_R) = g(v_1)$ is a piecewise-linear function given by

$$g(v_R) = G_b v_R + \frac{1}{2}(G_a - G_b)(|v_R + E| - |v_R - E|) \qquad (2.38)$$

G_a, G_b, and E were shown in Figure 2.19b, $V_f = Fsin(2\pi ft - F)$ and f is the frequency of the external sinusoidal excitable source connected to the induction gyrator.

From the schematics of Figure 2.22 and Figure 2.23 we can design an electronic circuit as shown in Figure 2.24. Figure 2.24a,b,c shows the circuit diagram, the realization of the electronic Chua's circuit using the inductor gyrator, and the experimental measurement setup, respectively. A signal generator excites a sinusoidal voltage to the circuit, and an oscilloscope is used to capture the dynamic nonlinear behavior of the output point of the circuit versus the input excitation source.

We fixed the values of all elements of Chua's circuit and consider F and f as the control parameters. If the amplitude F and frequency f are varied by changing the external excitable voltage, the dynamical behavior of Chua's circuit can be observed and examined.

2.4.3.1 Simulation Results

Using equations (2.27) and (2.38), setting the values $R_L = 10\ \Omega$, $R_g = 100\ k\Omega$, $C = 16\ nF$, $C_1 = 9.8\ nF$, $C_2 = 100\ nF$, $G_a = -0.756\ mS$, $G_b = -0.409\ mS$, $E = 1.08\ V$, $R = 1770\ \Omega$. The amplitude of excitation voltage $F = 275\ mV$ and its frequency can be employed as the control parameters, and we can observe the following cases:

- *Case 1:* $V_f = 0$, this is the motion of the Chua's circuit using the inductor gyrator, and we get the typical Chua's attractors as in Figure 2.25.
- *Case 2:* $F = 275\ mV$, f *varies from* 25 Hz *to* 500 Hz, we also get the typical double scroll of Chua's attractor and look like a combination of two single attractors. The motion behaviors are more complex (Figure 2.26).
- *Case 3:* $F = 275\ mV$ and f *varies from* 516 Hz *to* 1545 Hz, we get a cascade of period adding when decreasing the frequency of sinusoidal excitation (Figure 2.27).
- *Case 4:* Keeping $F = 275\ mV$ and f *increasing from* 1800 Hz, we can realize some different cases. At $f = 1929$ Hz, the dynamical system is represented by a point-attractor (Figure 2.28a) and after that the system will be chaotic. At $f = 1967$ Hz we get two single scroll attractors

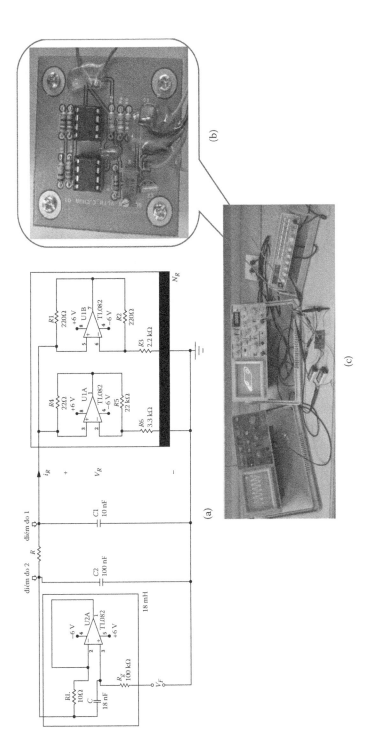

FIGURE 2.24
Chua's circuit experiment: (a) principle diagram with the details of elements, (b) electronic Chua's circuit using inductor gyrator, and (c) experimental system.

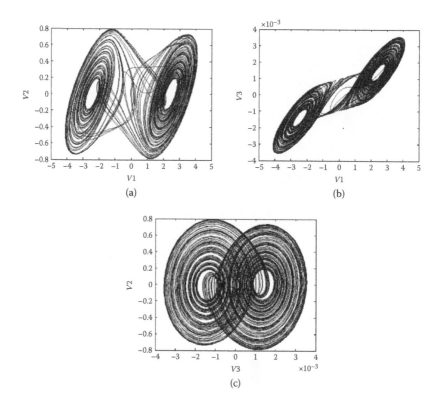

FIGURE 2.25
Typical Chua's attractor: (a) $(v_1\text{-}v_2)$ phase space, (b) $(v_1\text{-}v_3)$ phase space, (c) $(v_3\text{-}v_2)$ phase space. (In this figure $v_3 = i_L$.)

and it also exits limit cycles (Figure 2.28b). At $f = 3333$ Hz, two single scroll attractors extend their sizes and meet together to be a double scroll attractor with limit cycles at the outside. (Figure 2.28c) At $f = 4287$ Hz, there is a typical double scroll attractor (Figure 2.28d).

2.4.3.2 Experimental Results

We can set $R = 1720\ \Omega$ so that the circuit behaves chaotically by itself without any sinusoidal excitation or self-excitation in order to study the dynamical behavior of the circuit when it is excited by an external sinusoidal force.

- Bifurcation diagram

    We change the value of driving source amplitude F from 25 mV to 400 mV with the step of 25 mV; at every amplitude we decrease

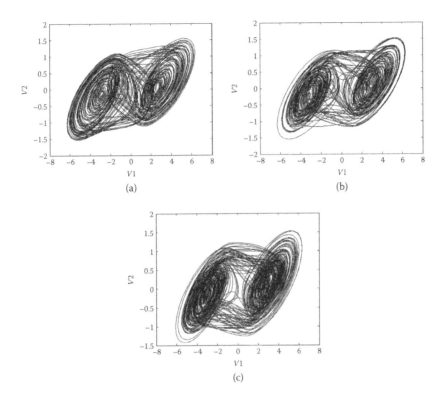

FIGURE 2.26
Attractors at $F = 275$ mV, (a) $f = 27$ Hz, (b) $f = 275$ Hz, (c) $f = 400$ Hz.

the driving source frequency f from 9 KHz to 0 KHz with step varying from 5 Hz to 20 Hz. We monitored $(v_1 - v_2)$ (v_1 and v_2 are voltages dropped across the capacitors C_1 and C_2, respectively), the phase space on the oscilloscope depicts the bifurcation diagram in the $(F\text{-}f)$ plane. Figure 2.29 is the bifurcation diagram with f changes from 200 Hz to 4000 Hz; the numbers denote the periods (adding periods) of attractors, and the shaded regions denote chaos.

In the region of the driving source frequencies greater than 4000 Hz, the bifurcation only changes from single-scroll to double-scroll attractors and vice versa as illustrated in Figure 2.30.

- Period doubling

A cascade of period doubling bifurcations appears in the middle frequencies from 3000 Hz to 4000 Hz (Figure 2.29). At $F = 275$ mV, typically one period appears at driving source frequency

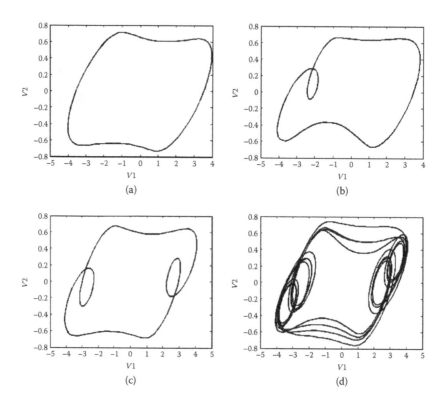

FIGURE 2.27
Period adding at $F = 275$ mV. (a) Period 1, $f = 1545$ Hz. (b) Period 2, $f = 1017$ Hz. (c) Period 3, $f = 798$ Hz. (d) Period 4, $f = 624$ Hz.

$f = 3582$ Hz (Figure 2.31a). We then decrease f to 3333 Hz to obtain a double-scroll attractor (Figure 2.31b). At $f = 3040$ Hz a Hopf bifurcation appears with an outer limit cycle (Figure 2.31c). However we do not observe any period doubling, period quadrupling, or period octupling trajectories in this cascade structure. This may be due to the fact that the state of the system has reached the limit of the bandwidth of the oscilloscope.

- Period adding

  In Figure 2.29, the region of excitable frequencies of less than 1200 Hz is called the *region of periodic windows*. In this region chaotic behaviors and the period doubling appear sequentially. Figure 2.32 represents a cascade of period doubling at the excitable amplitude $F = 275$ mV, and the order of period increases when the excitable frequency decreases.

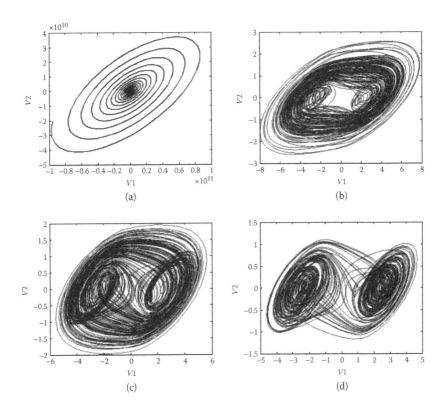

FIGURE 2.28
Attractors at $F = 275$ mV. (a) $f = 1929$ Hz. (b) $f = 1967$ Hz. (c) $f = 3333$ Hz. (d) $f = 4287$ Hz.

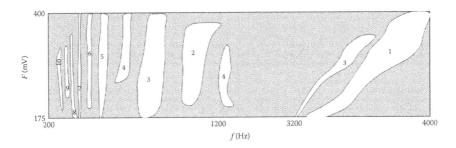

FIGURE 2.29
Bifurcation diagram in $(F\text{-}f)$ plane. The numbers are the periods, and the shaded regions are chaos.

FIGURE 2.30
Relation of single-scroll attractors and double-scroll attractors in $(v_1\text{-}v_2)$ phase space. (a) $f = 8.29$ KHz. (b) $f = 6.8$ KHz. (c) $f = 6$ KHz. (d) $f = 4.4$ KHz (*horizontal axis: 1 V/div and vertical axis: 0.5 V/div*).

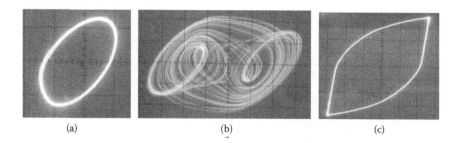

FIGURE 2.31
Period doubling at $F = 275$ mV. (b) Two-scroll attractor, $f = 3333$ Hz. (a) and (b) h axis: 1 V/div and v. axis: 0.5 V/div. (c) Limit cycle, $f = 3040$ Hz, h. axis: 5 V/div, v. axis: 2 V/div.

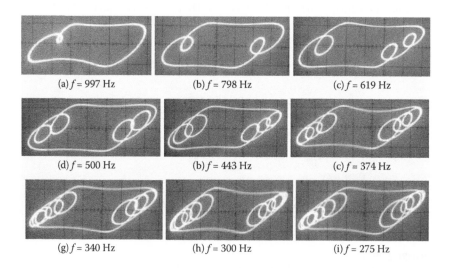

FIGURE 2.32
A cascade of period doubling in $(v_1\text{-}v_2)$ phase space at $F = 275$ mV. (a) through (i) from Period 1 to Period 10. (*Horizontal axis: 1 V/div and vertical axis: 0.5 V/div.*)

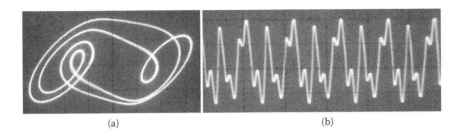

(a)   (b)

FIGURE 2.33
Quasi-periodicity at $F = 100$ mV and $f = 2954$ Hz. (a) Trajectories in $(v_1\text{-}v_2)$ phase space. (*Horizontal axis: 1 V/div and vertical axis: 0.5 V/div.*) (b) v_1 voltage.

- Quasi-periodicity

  With quasi-periodicity, the system has two different frequencies that associate together so the motion of the system is called *quasi-periodic motion* and it repeats itself inexactly. In our experiment, the quasi-periodic behaviors at the excitable frequencies are less than 3000 Hz according to the excitable amplitude appropriately. Figure 2.33 depicts the trajectories in the phase space $(v_1\text{-}v_2)$ and its signal v_1. When the excitable frequency is changed slowly, the quasi-periodic motion alters to chaotic motion.

- Intermittency

  In this experiment, the intermittent behaviors appear in the excitable region with frequency greater than 1000 Hz. Figure 2.34 depicts the trajectories in the phase space $(v_1 - v_2)$ and its voltage v_1 to illustrate the intermittent behavior. We realized that the periodic oscillations are interrupted by the intermittent bursts. Figure 2.35 depicts the intermittent behavior near the boundary crisis region.

(a)   (b)

FIGURE 2.34
Intermittency at $F = 275$ mV and $f = 1731$ Hz. (a) Trajectories in $(v_1\text{-}v_2)$ phase space. (*Horizontal axis: 1 V/div and vertical axis: 0.5 V/div.*) (b) v_1 voltage with intermittent bursts.

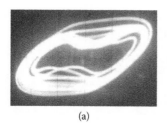
(a)

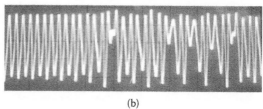
(b)

FIGURE 2.35
Intermittency near the boundary crisis region at $F = 400$ mV and $f = 1475$ Hz. (a) Trajectories in
$(v_1$-$v_2)$ phase space. (*Horizontal axis: 1 V/div and vertical axis: 0.5 V/div.*) (b) v_1 voltage with intermittent bursts strongly.

2.5 Concluding Remarks

In this chapter we presented briefly the main principles of a nonlinear
dynamical system. The descriptions of fixed points, bifurcation, and chaos
are explained. Further main routes to chaotic states of nonlinear systems
are obtained and derived from the mathematical point of view and then
experimental demonstration. In the last section Chua's nonlinear circuit is
illustrated experimentally using an inductor gyrator in association with
appropriate resistors and capacitors excited with the external sinusoidal
source so that the charging and discharging can happen simultaneously so
that chaotic behavior can be observed in both simulation and experiment.
This chaotic behavior and nonlinear dynamical evolution will be illustrated
in the remaining chapters of this book in several fiber optic lasers of different
feedback structures in the optical domain.

Bifurcation and chaotic dynamics of fiber laser systems will be illustrated
in Chapters 7 and 8 in which the envelopes of the lightwaves behave in nonlinear motions illustrated in the mathematical representations given in the
above sections.

The basic mathematical representations of other nonlinear optical systems
described by other chapters would be briefly described in their contents.

References

1. N. V. Dao, T. K. Chi, and N. Dung, *An Introduction to Nonlinear Dynamics and Chaos,*
   Vietnam National University Publishing House (in Vietnamese), Hanoi, 2005.
2. R. C. Hilborn, *Chaos and Nonlinear Dynamics: An Introduction for Scientists and
   Engineers,* Oxford University Press, New York, 1994.

3. E. N. Lorenz, "Deterministic Nonperiodic Flow," *Journal of the Atmospheric Sciences*, 20(2), 1963.
4. T. Kapitaniak, *Chaos for Engineers: Theory, Applications, and Control*, 2nd edition, Springer, Berlin., 1998
5. M. P. Kennedy, "ABC (Adventures in Bifurcations and Chaos): A Program for Studying Chaos," *J. Franklin Inst.*, 331B(6), 631–658, 1994.
6. K. Murali and M. Lakshmanan, "Effect of Sinusoidal Excitation on the Chua's Circuit," *IEEE Trans. Circuits and Systems–I: Fundam. Theory and Appl.*, 39(4), 264–270, 112, April, 1992.
7. B. Muthuswamy, T. Blain, and K. Sundqvist, "A Synthetic Inductor Implementation of Chua's Circuit," Technical Report No. UCB/EECS-2009-20, 2009. Access date: June 2011, http://www.eecs.berkeley.edu/Pubs/TechRpts/2009/EECS-2009-20.html
8. S. H. Strogatz, *Nonlinear Dynamics and Chaos: With Applications to Physics, Biology, Chemistry, and Engineering*, Westview Press, Boulder, CO, 2000.

3

Soliton Fiber Lasers

Le Nguyen Binh

Hua Wei Technologies, European Research Laboratories, Munich, Germany

Nguyen Duc Nhan

Institute of Technology for Posts and Telecommunications, Hanoi, Vietnam

CONTENTS

3.1 Introduction

This chapter describes the principles and formation mechanism of solitons in a lasing cavity formed in guided wave structures, especially the single-mode fibers. Mathematical equations essential for the generation and propagation in the cavity are given.

3.2 Nonlinear Schrödinger Equations

In order to appreciate the propagation of optical pulses in a nonlinear guided medium, Maxwell's equations are employed to derive the nonlinear wave evolution equation (see also Appendix A) that describes lightwave confinement and propagation in optical waveguides including optical fibers as follows [1]:

$$\nabla \times \nabla \times \vec{E} + \frac{1}{c^2}\frac{\partial^2 \vec{E}}{\partial t^2} = -\mu_0 \frac{\partial^2 \vec{P}}{\partial t^2} \tag{3.1}$$

Using the vector identity $\nabla \times \nabla \times \vec{E} = \nabla(\nabla \cdot \vec{E}) - \nabla^2 \vec{E}$ and $\nabla \cdot \vec{E} = 0$ for dielectric materials, the nonlinear wave propagation in nonlinear waveguide in the time-spatial domain in vector form can be expressed as

$$\nabla^2 \vec{E} - \frac{1}{c^2}\frac{\partial^2 \vec{E}}{\partial t^2} = \mu_0 \left(\frac{\partial^2 \vec{P}_L}{\partial t^2} + \frac{\partial^2 \vec{P}_{NL}}{\partial t^2} \right) \tag{3.2}$$

where $\vec{E}$ is the electric field vector of the optical wave, μ_0 is the vacuum permeability assuming a nonmagnetic wave-guiding medium, c is the speed of light in vacuum, $\vec{P}_L, \vec{P}_{NL}$ are, respectively, the linear and nonlinear parts of polarization vector $\vec{P}$, which are formed as

$$\vec{P}_L(\vec{r},t) = \varepsilon_0 \int_{-\infty}^{\infty} \chi^{(1)}(t-t') \cdot \vec{E}(\vec{r},t')dt' \tag{3.3}$$

$$\vec{P}_{NL}(\vec{r},t) = \varepsilon_0 \iint_{-\infty}^{\infty}\int \chi^{(3)}(t-t_1,t-t_2,t-t_3) \times \vec{E}(\vec{r},t_1)\vec{E}(\vec{r},t_2)\vec{E}(\vec{r},t_3) \tag{3.4}$$

where $\chi^{(1)}$ and $\chi^{(3)}$ are the first- and third-order susceptibility tensors. Thus, the linear and nonlinear coupling effects in optical waveguides can be described by Equation (3.2). In optical waveguides such as optical fibers, only the third-order nonlinearity is of special importance because it is responsible for all nonlinear effects. The second term on the right-hand side of Equation (3.2) is responsible for nonlinear processes including interaction between optical waves through the third-order susceptibility. If the nonlinear response is assumed to be instantaneous, the time dependence of $\chi^{(3)}$ is given by the product of three delta functions of the form $\delta(t - t_1)$ and then Equation (3.4) reduces to

$$\vec{P}_{NL}(\vec{r},t) = \varepsilon_0 \chi^{(3)} \vec{E}(\vec{r},t)\vec{E}(\vec{r},t)\vec{E}(\vec{r},t) \tag{3.5}$$

where $\chi^{(1)}$ and $\chi^{(3)}$ are the second- and fourth-rank tensors, respectively.

To simplify the analysis of Equation (3.2), the following assumptions will be made:

- Only one mode is present in the waveguide or a single-mode fiber is considered.
- The field is linearly polarized in the same direction and the polarization state remains unchanged during the propagation.
- The nonlinearity can be seen as a small perturbation because nonlinear change in the refractive index is $\Delta n/n < 10^{-6}$ in practice.
- The variation of the carrier wave is much faster than that of the envelope of the optical pulse. On the other hand, the bandwidth of the optical pulse $\Delta\omega$ is much smaller than the carrier frequency ω_0.

In this approximation of the slowly varying envelope, the electric field can be written in the form:

$$\vec{E}(\vec{r},t) = \tfrac{1}{2}\hat{x}[E(\vec{r},t)\exp(-j\omega_0 t) + c.c.] \tag{3.6}$$

where w_0 is the carrier frequency, $\hat{x}$ is the polarization unit vector, and $E(\vec{r},t)$ is a slowly varying function of time. $\vec{P}_L$, $\vec{P}_{NL}$ can also be expressed in a similar manner. Introducing (3.6) into (3.3) and (3.5) yields

$$P_L = \varepsilon_0 \chi_{xx}^{(1)} E \tag{3.7}$$

$$
\begin{aligned}
P_{NL} &= \frac{1}{8}\varepsilon_0 \chi_{xxxx}^{(3)} (Ee^{-j\omega_0 t} + E^* e^{j\omega_0 t})^3 \\
&= \frac{1}{8}\varepsilon_0 \chi_{xxxx}^{(3)} (E^3 e^{-j3\omega_0 t} + 3E^2 E^* e^{-j\omega_0 t} + 3EE^{2*} e^{j\omega_0 t} + E^{3*} e^{j3\omega_0 t}) \\
&= \frac{1}{8}\varepsilon_0 \chi_{xxxx}^{(3)} ((E^3 e^{-j3\omega_0 t} + c.c.) + 3|E|^2 (Ee^{-j\omega_0 t} + c.c.))
\end{aligned}
\tag{3.8}
$$

The first term in (3.8) describes the nonlinear part of the polarization at three times the original carrier frequency, which is responsible for the third harmonic generation (THG) and requires phase matching. The second term describes the nonlinear part of the polarization at the carrier frequency and is responsible for most of important nonlinear effects relating to the nonlinear refractive index.

The linear and nonlinear parts of polarization are related to the dielectric constant as

$$\varepsilon(\omega) = \varepsilon_L + \varepsilon_{NL} \tag{3.9}$$

where ε_L, ε_L are the linear and nonlinear contributions to the dielectric constant and are obtained from (3.7) and (3.8):

$$\varepsilon_L = 1 + \chi_{xx}^{(1)} \quad \text{and} \quad \varepsilon_{NL} = \frac{3}{4}\chi_{xxxx}^{(3)} |E|^2 \tag{3.10}$$

This dielectric constant can be used to define the refractive index $n(w)$ and the absorption coefficient $\alpha(w)$ of the nonlinear medium as follows:

$$\varepsilon(\omega) = \left[n(\omega) + \frac{j\alpha(\omega)c}{2\omega} \right]^2 \approx n^2(\omega) + j\frac{n(\omega)\alpha(\omega)c}{\omega} \tag{3.11}$$

Both $n(w)$ and $\alpha(w)$ relate to the linear and nonlinear parts of $\varepsilon(w)$ by introducing

$$n(\omega) = n_0 + n_2 |E|^2 \quad \text{and} \quad \alpha(\omega) = \alpha_0 + \alpha_2 |E|^2 \tag{3.12}$$

where the linear index n_0 and the absorption coefficient α_0 are related to the real and imaginary parts of $\chi_{xx}^{(1)}$, while the nonlinear index n_2 and the two-photon absorption coefficient α_2 are related to the real and imaginary parts of ε_{NL} by

$$n_2 = \frac{3}{8n}\text{Re}\left(\chi_{xxxx}^{(3)}\right), \quad \alpha_2 = \frac{3\omega_0}{4nc}\text{Im}\left(\chi_{xxxx}^{(3)}\right) \tag{3.13}$$

For some optical waveguides such as optical fibers, the coefficient α_2 is negligible and the nonlinear refractive index is responsible for the nonlinear response of the propagation medium.

3.2.1 Nonlinear Schrödinger Equation

The nonlinear Schrödinger equation (NSE) plays an important role in the description of nonlinear effects in optical pulse propagation. It can be derived from the wave Equation (3.2) after some algebra by using the method of separating variables (see Appendix A):

$$\frac{\partial A(z,t)}{\partial z} + \frac{\alpha}{2} A(z,t) - j \sum_{n=1}^{\infty} \frac{j^n \beta_n}{n!} \frac{\partial^n A(z,t)}{\partial t^n} = j\gamma \left(1 + \frac{j}{\omega_0} \frac{\partial}{\partial t} \right)$$

(3.14)

$$\times A(z,t) \int_{-\infty}^{\infty} R(t') |A(z,t-t')|^2 \, dt'$$

where $A(z,t)$ is the slowly varying complex envelope propagating along z in the propagation medium, the effect of propagation constant β around the optical carrier w_0 is the Taylor-series expanded, $R(t)$ is the nonlinear response function, and $\gamma = \omega_0 n_2 / cA_{eff}$ is the nonlinear coefficient. We note here that the optical frequency term has been removed and only the amplitude of the modulated lightwaves is involved whose complex term would contain the phase of the lightwave carrier. Thus, we will see later that the software package developed for the propagation of the lightwave envelope involves the complex amplitude. This gives advantages as the optical frequency is very high and it is impossible to sample the wave at this frequency in current computing systems.

In most cases in optical communications application, the optical pulses with the width larger than 100 fs are employed, Equation (3.14) can be further simplified as

$$\frac{\partial A}{\partial z} + \frac{\alpha}{2} A + \frac{j\beta_2}{2} \frac{\partial^2 A}{\partial \tau^2} - \frac{\beta_3}{6} \frac{\partial^3 A}{\partial \tau^3} = j\gamma \left[|A|^2 A + \frac{j}{\omega_0} \frac{\partial(|A|^2 A)}{\partial \tau} - T_R A \frac{\partial(|A|^2)}{\partial \tau} \right]$$

(3.15)

where a frame of reference moving with the pulse at the group velocity v_g is used by making the transformation $\tau = t - z/v_g \equiv t - \beta_1 z$, and the propagation constant is expanded up to the third-order term that includes the group velocity dispersion (β_2) and the third-order dispersion (β_3). In (3.15), the first moment of the nonlinear response function is defined as

$$T_R \equiv \int_0^{\infty} t R(t') \, dt'$$

(3.16)

which is responsible for the Raman scattering effect, and the second term in the right side of (3.15) is responsible for the self-steepening effect. However, if the width of optical pulses is of the order of picoseconds, the high-order effects such as self-steepening and Raman scattering can be ignored. Hence, Equation (3.15) becomes

$$\frac{\partial A}{\partial z} + \frac{\alpha}{2} A + \frac{j\beta_2}{2} \frac{\partial^2 A}{\partial \tau^2} - \frac{\beta_3}{6} \frac{\partial^3 A}{\partial \tau^3} = j\gamma |A|^2 A$$

(3.17)

This equation can describe the most important linear and nonlinear propagation effects of optical pulse in optical fibers.

If we further simplify Equation (3.17) by setting the attenuation factor and the third-order dispersion coefficient to zero, then the traditional NLS can be obtained as

$$\frac{\partial A}{\partial z} + \frac{j\beta_2}{2}\frac{\partial^2 A}{\partial \tau^2} = i\gamma |A|^2 A \tag{3.18}$$

Equation (3.18) is a well-known equation in nonlinear fiber optics which is employed to explain propagation of optical solitary waves in nonlinear dispersive medium.

3.2.2 Ginzburg–Landau Equation: A Modified Nonlinear Schrödinger Equation

Although the NSEs described above can be used to explain most nonlinear effects including higher-order effects, they only describe the pulse propagation in passive nonlinear media without gain. In a propagation medium with gain as fiber amplifiers, the gain effect is required to be included in the NLS equation. The general equation that governs the pulse propagation in active fibers is given by ignoring other effects for simplification as follows:

$$\frac{\partial A(z,\omega)}{\partial z} = \frac{1}{2}g(\omega)A(z,\omega) \tag{3.19}$$

where $g(w)$ is the gain coefficient of the active fiber. For an approximation of a homogeneously broadened system, the gain spectral shape takes a Lorentzian profile [2]:

$$g(\omega) = \frac{g_0}{1+(\omega-\omega_g)^2/\Delta\omega_g^2} \tag{3.20}$$

where g_0 is the maximum small signal gain, w_g is the atomic transition frequency, and $\Delta\omega_g$ is the gain bandwidth that relates to the dipole relaxation time. The gain spectrum can be approximated by the Taylor expansion in the neighborhood of ω_g given by

$$g(\omega) \approx g_0\left(1-(\omega-\omega_g)^2/\Delta\omega_g^2\right) \tag{3.21}$$

Thus, by substituting (3.21) into (3.19), and taking the inverse Fourier transform with an assumption of the carrier frequency ω_0 close to ω_g, the propagation equation with amplification is obtained as

$$\frac{\partial A(z,t)}{\partial z} = \frac{g_0}{2}\left[A(z,t)+\frac{1}{\Delta\omega_g^2}\frac{\partial^2 A(z,t)}{\partial t^2}\right] \tag{3.22}$$

However, in many cases of pulse propagation, especially in the mode-locked fiber laser systems, the gain saturation plays an important role in pulse amplification. Therefore, the saturation need to be included in (3.22) by replacing g_0 in g_{sat} [2]:

$$g_{sat}(z) = \frac{g_0(z)}{1 + P_{av}(z)/P_{sat}} \tag{3.23}$$

and (3.22) is modified by using (3.23) as

$$\frac{\partial A(z,t)}{\partial z} = \frac{g_{sat}}{2}\left[A(z,t) + \frac{1}{\Delta\omega_g^2}\frac{\partial^2 A(z,t)}{\partial t^2} \right]$$

$$\approx \frac{g_0}{2\left(1 + P_{av}(z)/P_{sat}\right)}\left[A(z,t) + \frac{1}{\Delta\omega_g^2}\frac{\partial^2 A(z,t)}{\partial t^2} \right] \tag{3.24}$$

where g_0 is assumed to be constant along the active fiber, P_{sat} is the saturated power of the gain medium, P_{avg} is the average power of the signal at position z in the active fiber as

$$P_{av}(z) = \frac{1}{T_m}\int_{-T_m/2}^{T_m/2} |A(z,t)|^2\, dt \tag{3.25}$$

For a full model of the pulse evolution in gain medium, other effects such as dispersion, nonlinear effects are also required to be included in (3.24). Hence, by adding the term of gain effect from (3.24) into (3.17), a modified NLS can be obtained:

$$\frac{\partial A}{\partial z} + \frac{j}{2}\left(\beta_2 + \frac{jg_{sat}}{\Delta\omega_g^2} \right)\frac{\partial^2 A}{\partial \tau^2} - \frac{\beta_3}{6}\frac{\partial^3 A}{\partial \tau^3} = i\gamma|A|^2 A + \frac{1}{2}(g_{sat} - \alpha)A \tag{3.26}$$

This equation is also called the Ginzburg–Landau (G–L) equation that can be derived from the wave equation [3]. In addition to the cubic G–L equation, an extended version is the quintic cubic G–L equation (QCGL), which has also attracted considerable attention [4]. The G–L equations play an important role in the description of nonlinear systems including nonlinear fiber optics [2,4,5] as well as fiber lasers [6].

3.2.3 Coupled Nonlinear Schrödinger Equations

For a birefringent fiber, there is more than one polarization state propagating in the fiber; therefore, (3.6) can be replaced by

$$\vec{E}(\vec{r},t) = \tfrac{1}{2}\left[\left(\hat{x}E_x + \hat{y}E_y\right)\exp(-j\omega_0 t) + c.c.\right] \tag{3.27}$$

where E_x, E_y are the orthogonal polarization components of the optical field. Then the polarization components of the nonlinear induced polarization are given by

$$P_{NL,x}(\vec{r},t) = -\varepsilon_0 2n_2 n(\omega_0)\left[\left(|E_x|^2 + \tfrac{2}{3}|E_y|^2\right)E_x + \tfrac{1}{3}\left(E_x^* E_y\right)E_y\right] \qquad (3.28)$$

$$P_{NL,y}(\vec{r},t) = -\varepsilon_0 2n_2 n(\omega_0)\left[\left(|E_y|^2 + \tfrac{2}{3}|E_x|^2\right)E_y + \tfrac{1}{3}\left(E_y^* E_x\right)E_x\right] \qquad (3.29)$$

By the same manner, two coupled equations for the slowly varying components of E_x and E_y can be derived as follows [7]:

$$\frac{\partial A_x}{\partial z} + \frac{\Delta\beta_1}{2}\frac{\partial A_x}{\partial \tau} + \frac{j}{2}\left(\beta_2 + \frac{jg_{sat}}{\Delta\omega_g^2}\right)\frac{\partial^2 A_x}{\partial \tau^2} - \frac{\beta_3}{6}\frac{\partial^3 A_x}{\partial \tau^3}$$
$$= \frac{1}{2}(g_{sat} - \alpha)A_x + j\gamma\left[\left(|A_x|^2 + \tfrac{2}{3}|A_y|^2\right)A_x + \tfrac{1}{3}A_x^* A_y^2 \exp(-2j\Delta\beta_0 z)\right] \qquad (3.30)$$

$$\frac{\partial A_y}{\partial z} - \frac{\Delta\beta_1}{2}\frac{\partial A_y}{\partial \tau} + \frac{j}{2}\left(\beta_2 + \frac{jg_{sat}}{\Delta\omega_g^2}\right)\frac{\partial^2 A_y}{\partial \tau^2} - \frac{\beta_3}{6}\frac{\partial^3 A_y}{\partial \tau^3}$$
$$= \frac{1}{2}(g_{sat} - \alpha)A_y + j\gamma\left[\left(|A_y|^2 + \tfrac{2}{3}|A_x|^2\right)A_y + \tfrac{1}{3}A_y^* A_x^2 \exp(+2j\Delta\beta_0 z)\right] \qquad (3.31)$$

where A_x, A_y are the slowly varying envelopes of orthogonal polarization components, $\Delta\beta_1 = \beta_{1,x} - \beta_{1,y} = \Delta n_g/c$ is the phase mismatching factor, and Δn_g is the group birefringence.

With highly birefringent fibers, the terms $\exp(-2j\Delta\beta_0 z)$ and $\exp(+2j\Delta\beta_0 z)$ can be neglected due to their rapid oscillations and equations (3.30) and (3.31) become

$$\frac{\partial A_x}{\partial z} + \frac{\Delta\beta_1}{2}\frac{\partial A_x}{\partial \tau} + \frac{j}{2}\left(\beta_2 + \frac{jg_{sat}}{\Delta\omega_g^2}\right)\frac{\partial^2 A_x}{\partial \tau^2} - \frac{\beta_3}{6}\frac{\partial^3 A_x}{\partial \tau^3} = j\gamma\left(|A_x|^2 + \tfrac{2}{3}|A_y|^2\right)A_x$$
$$+ \frac{1}{2}(g_{sat} - \alpha)A_x \qquad (3.32)$$

$$\frac{\partial A_y}{\partial z} - \frac{\Delta\beta_1}{2}\frac{\partial A_y}{\partial \tau} + \frac{j}{2}\left(\beta_2 + \frac{jg_{sat}}{\Delta\omega_g^2}\right)\frac{\partial^2 A_y}{\partial \tau^2} - \frac{\beta_3}{6}\frac{\partial^3 A_y}{\partial \tau^3} = j\gamma\left(|A_y|^2 + \tfrac{2}{3}|A_x|^2\right)A_y$$
$$+ \frac{1}{2}(g_{sat} - \alpha)A_y \qquad (3.33)$$

These equations are significant in the models relating to polarization states such as polarization mode dispersion in fiber transmission and nonlinear polarization rotation in passive-mode locking [7–9]. For an optical fiber with the length L, the phase variation of polarization components due to non-linearity can be derived by considering the nonlinear term only in (3.32) through (3.33) for simplicity:

$$\phi_{NL}^x = \gamma L \left(|A_x|^2 + \tfrac{2}{3} |A_y|^2 \right) \tag{3.34}$$

$$\phi_{NL}^y = \gamma L \left(|A_y|^2 + \tfrac{2}{3} |A_x|^2 \right) \tag{3.35}$$

and hence the angle of polarization rotation is given by

$$\varphi_{NPR} = \phi_{NL}^y - \phi_{NL}^x = \frac{\gamma L}{3} \left(|A_y|^2 - |A_x|^2 \right) \tag{3.36}$$

Note that this angle is zero when the light input is linearly polarized due to $|A_x|^2 = |A_y|^2$. On the other way, the polarization ellipse rotates with propagation in the fiber.

3.3 Optical Solitons

3.3.1 Temporal Solitons

The NLS equation (3.18) derived in Section 3.3.2 governs the propagation of optical pulse in nonlinear dispersive media such as optical waveguides or fibers. By using the transformation of variables as follows,

$$T = \tau / \tau_0, \quad \xi = z / L_D, u = \sqrt{\gamma L_D}\, A \tag{3.37}$$

where τ_0 is a temporal scaling parameter often taken to be the input pulse width and $L_D = \tau_0^2 / |\beta_2|$ is the dispersion length. Equation (3.18) can be normalized to the $(1 + 1)$-dimensional NLS equation as

$$j \frac{\partial u}{\partial Z} - \frac{s}{2} \frac{\partial^2 u}{\partial \tau^2} \pm |u|^2 u = 0 \tag{3.38}$$

where $s = sgn(b_2) = \pm 1$ stands for the sign of the group velocity delay (GVD) parameter that can be positive or negative, depending on the wavelength. The nonlinear term is positive (+1) for optical fibers but may become negative (–1) for waveguides made of semiconductor materials. Thus, there are two cases of basic propagation in optical fibers relating to the dispersion which is normally measured by the parameter

$$D = \frac{d}{d\lambda}\left(\frac{1}{v_g}\right) = -\frac{2\pi c}{\lambda^2}\beta_2 \tag{3.39}$$

where the dispersion parameter D is expressed in units of ps/(nm.km). For the standard single-mode fiber, D is positive or GVD is anomalous at wavelengths >1.3 mm, and D is negative or GVD is normal at wavelengths shorter than 1.3 mm.

Because of two different signs of the GVD parameter, optical fibers can support two different types of solitons that are solutions of Equation (3.38). In particular, Equation (3.38) has solutions in the form of dark temporal solitons in the case of normal GVD ($s = +1$) and bright temporal solitons in the case of anomalous GVD ($s = -1$). These solutions can be found by the inverse scattering method [10]. In case of anomalous GVD, Equation (3.38) takes the form

$$j\frac{\partial u}{\partial Z} + \frac{1}{2}\frac{\partial^2 u}{\partial \tau^2} + |u|^2 u = 0 \tag{3.40}$$

and the most interesting solution of this equation is the fundamental soliton with the general form given by

$$u(\xi,\tau) = \sec h(\tau)\exp(j\xi/2) \tag{3.41}$$

Thus, at the input of the fiber, the soliton is given as $u(0,\tau) = \sec h(\tau)$ and it can be converted into real units as follows:

$$A(0,T) = \sqrt{P_0}\,\sec h\left(\frac{T}{T_0}\right) = \left(\frac{|\beta_2|}{\gamma T_0^2}\right)^{\frac{1}{2}}\sec h\left(\frac{T}{T_0}\right) \tag{3.42}$$

Solution 3.42 indicates that if a hyperbolic-secant pulse with peak power P_0, which satisfies expression 3.42, it can propagate undistorted without change in temporal and spectral shapes in an ideal lossless fiber at arbitrary distances as shown in Figure 3.1. This important feature results from the balance between GVD and self-phase modulation (SPM) effects. When a peak power launched into the fiber is much higher, a higher-order soliton can be excited with the input form

$$u(0,\tau) = N\sec h(\tau) \tag{3.43}$$

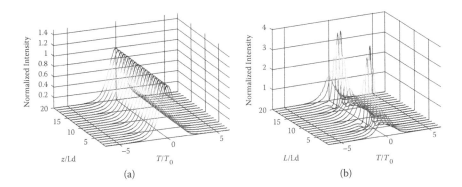

FIGURE 3.1
Evolution of (a) the first-order soliton and (b) the second-order soliton.

where N is an integer that represents the soliton order and is determined by

$$N^2 = \frac{L_D}{L_{NL}} = \frac{\gamma P_0 T_0^2}{|\beta_2|} \qquad (3.44)$$

Different from the fundamental soliton, the temporal and spectral shapes of higher-order solitons vary periodically during propagation with the period $\xi_0 = \pi/2$ or in real units $z_0 = \pi L_D/2$.

With $s = +1$ in the case of normal GVD, Equation (3.38) takes the form

$$j\frac{\partial u}{\partial Z} - \frac{1}{2}\frac{\partial^2 u}{\partial \tau^2} + |u|^2 \, u = 0 \qquad (3.45)$$

and solutions of (3.45) are dark solitons found by the inverse scattering method similar to the case of bright solitons. Solution of a fundamental dark soliton is given by [1,11]

$$u(\xi, \tau) = \cos\phi \tan h[\cos\phi(\tau - \xi\sin\phi) - j\sin\phi]\exp(j\xi) \qquad (3.46)$$

The features of the dark soliton are a high constant power level with an intensity dip and an abrupt phase change at the center depending on parameter ϕ. For $\phi = 0$, the dark soliton is a black soliton with the zero dip and a phase jump of π at the center. When $\phi \neq 0$, the dip intensity is nonzero and such solitons are called the gray solitons as shown in Figure 3.2.

In general, solitons in fiber have attracted a lot of interest in research from fundamentals to practical applications due to their unique features.

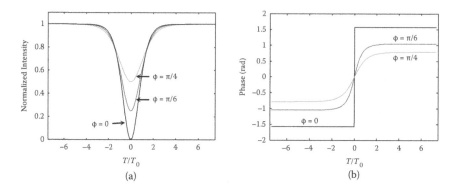

FIGURE 3.2
(a) Intensity and (b) phase profiles of dark solitons for various values of the internal phase *f*.

In communication systems, the understanding of solitons is of importance in both transmission and signal processing [2,12].

3.3.2 Dissipative Solitons

Optical solitons mentioned above only exist in an ideal propagation medium that has no perturbations such as loss and gain. In practical systems such as mode-locked fiber lasers, the optical signal is periodically amplified to compensate the loss that it experiences in the fiber cavity. Thus, there is a periodic variation of the pulse power, which can form an index grating and induce modulation stability. It was demonstrated that solitons are able to exist in these systems [2]. Therefore, propagation of the optical pulse in this case should be described by the complex cubic G–L equation (3.26) that includes the loss and gain effects. Similar to the NLS equation, it is useful to introduce the dimensionless variables and Equation (3.26) becomes

$$j\frac{\partial u}{\partial \xi} - \frac{1}{2}(s+jd)\frac{\partial^2 u}{\partial \tau^2} + |u|^2 \, u = \frac{j}{2}\mu u \qquad (3.47)$$

where

$$d = g_{sat}L_D T_2^2 / T_0^2, \quad \mu = (g_{sat} - \alpha)L_D \qquad (3.48)$$

Because Equation (3.47) is not integrable, so a solitary wave can be guessed to give [2]

$$u(\xi,\tau) = N_s\left[\sec h\!\left(\frac{\tau}{\tau_0}\right)\right]^{(1+jq)} \exp(j\kappa\xi) \qquad (3.49)$$

The parameters of the solution are determined by substituting it in Equation (3.47) and they are

$$N_s^2 = \frac{1}{2\tau_0^2}[s(q^2 - 2) + 3qd] \tag{3.50}$$

$$\tau_0^2 = -\frac{[d(1-q^2) + 2sq]}{\mu} \tag{3.51}$$

$$\kappa = -\frac{1}{2\tau_0^2}[s(1-q^2) - 2qd] \tag{3.52}$$

$$q = \frac{3s \pm \sqrt{9 + 8d^2}}{2d} \tag{3.53}$$

Thus, the width and the peak power of the soliton are determined by the system parameters such as loss, gain, and its finite bandwidth. Different from solitons supported by the NLS equation, influence of these parameters plays an important role in the existence of solitons in these fiber systems. Due to periodic perturbations during propagation, solitons dissipate their energy that plays an essential role for their formation and stabilization. Such a soliton is often called a dissipative soliton or an autosoliton due to its mechanism of self-organization [5]. In order to preserve the shape and energy, a balance between gain and loss mechanisms is required in addition to the balance between GVD and SPM. A frequency chirping can help to maintain the balance between gain and loss during propagation in a bandwidth-limited system, and this explains why dissipative solitons are normally chirped pulses.

3.4 Generation of Solitons in Nonlinear Optical Fiber Ring Resonators

3.4.1 Master Equations for Mode Locking

Mode locking is a technique to force axial modes of a laser in phase to generate short pulses. For mode-locked fiber ring lasers, an optical pulse in every round-trip would experience the same effects such as loss and gain as that in a transmission fiber span. Thus, the circulation of the pulse inside the ring resonators is approximately equivalent to a propagation of the pulse in a fiber transmission link with infinite length. Therefore, Equation (3.26) can be used to describe mode locking in fiber lasers. However, the variation of the pulse in the round-trip time scale is considered rather than that in the distance scale.

In addition, a modulation function M that is responsible for various mode-locking mechanisms is necessary to be added into (3.26). Then Equation (3.26) can be modified by introducing new variables $T = \xi/v_g$, where v_g is the group velocity of the optical pulse, to obtain

$$\frac{1}{v_g}\frac{\partial A(T,\tau)}{\partial T} + \frac{j}{2}\left(\beta_2 + \frac{jg_{sat}}{\Delta\omega_g^2}\right)\frac{\partial^2 A(T,\tau)}{\partial\tau^2} - \frac{\beta_3}{6}\frac{\partial^3 A(T,\tau)}{\partial\tau^3}$$

$$= j\gamma|A(T,\tau)|^2 A(T,\tau) + \frac{1}{2}\left(g_{sat} - \alpha\right)A(T,\tau) + M(A,T,\tau)$$

(3.54)

A further modification is implemented by multiplication of both sides in (3.54) with L_c, the ring cavity length, and setting $T_c = L_c/v_g$, which is the cavity period adding. Then (3.54) becomes

$$T_c\frac{\partial A(T,\tau)}{\partial T} + \frac{j}{2}\left(\beta_2 + \frac{jg_{sat}}{\Delta\omega_g^2}\right)L_c\frac{\partial^2 A(T,\tau)}{\partial\tau^2} - \frac{\beta_3 L_c}{6}\frac{\partial^3 A(T,\tau)}{\partial\tau^3}$$

$$= j\gamma L_c|A(T,\tau)|^2 A(T,\tau) + \frac{L_c}{2}(g_{sat} - \alpha)A(T,\tau) + M(A,T,\tau)$$

(3.55)

Equation (3.55) is the well-known master equation that was first derived by Haus to describe mode locking in the time domain [13]. Thus, there are two time scales in this equation: the time T is measured in terms of the cavity period or the round-trip time T_c, while the time τ is measured in terms of the pulse window. The term M, which depends on the mode-locking mechanism, passive or active, is a function of amplitude and time in terms of time scale τ in every round-trip. In most theoretical studies on mode locking, Equation (3.55) has been applied for the investigation of the pulse evolution in the cavity. We note that the parameters in the master equation are averaged over the cavity for analysis.

3.5 Passive Mode Locking

As mentioned in Chapter 1, there are three popular structures of the passive mode-locked fiber laser as shown in Figure 3.3. In the first structure, a saturable absorber acts as a passive mode locker to attenuate lower-intensity parts of a pulse, whereas higher-intensity parts of a pulse are minimally attenuated because they quickly saturate the absorber and pass through without loss (see Figure 3.3a). A saturable absorber is normally a semiconductor device

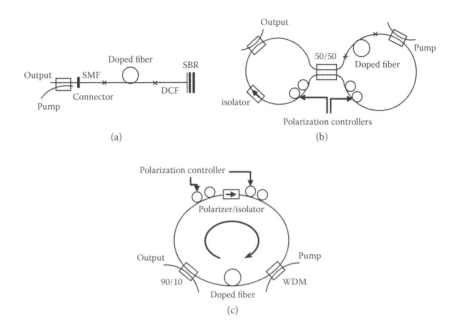

FIGURE 3.3
Typical configurations of passive mode-locked fiber laser: (a) a linear cavity configuration, (b) a configuration based on nonlinear fiber loop mirror, and (c) a ring configuration based on nonlinear polarization rotation.

that can be a bulk InGaAsP saturable absorber or a saturable Bragg reflector (SBR) based on InGaAs/InP multiple quantum wells. In some cases a mirror attached to the saturable absorber is also made using a periodic arrangement of thin layers that forms a grating and reflects light through Bragg diffraction. In practice, most SBRs are slow saturable absorbers because their response is much longer than the time scale of the pulse width. For the passive mode-locked fiber laser using SBR, the width of the mode-locked pulses is commonly of picoseconds time scale [14,15]; however, shorter pulses of less than 500 fs can be generated by careful dispersion management in the fiber cavity [16]. The second configuration is based on a nonlinear fiber loop mirror (NFLM), known as an all-optical switch. The NFLM is a Sagnac interferometer as described in Figure 3.4a whose intensity-dependent transmission can shorten optical pulses propagating inside the cavity. In passive mode-locked fiber lasers based on this configuration, the NFLM connects to a main fiber ring through a 3 dB coupler that splits the entering pulses into two equal counterpropagation parts as shown in Figure 3.3b. Because the optical amplifier is unequally located in the Sagnac ring, the counterpropagation pulses acquire different nonlinear phase shifts after a round-trip inside such a Sagnac loop. Moreover, the phase difference also depends on the temporal profile of the optical pulse; thus, the peak of the pulse is passed without loss while the pulse wings are reflected due to their lower intensity and smaller

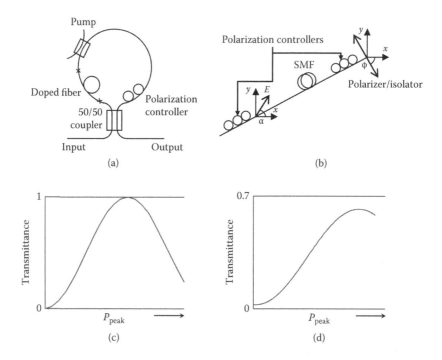

FIGURE 3.4
A description of operation and transmissivity of artificial fast saturable absorption based on (a,c) nonlinear fiber loop mirror (NFLM) and (b,d) nonlinear polarization rotation (NPR), respectively.

phase shift. It can be shown that the transmittance of the NFLM varies as a function of pulse power P via [17,18]:

$$T = 1 - 0.5\{1 + \cos[0.5\gamma(G-1)PL]\} \tag{3.56}$$

where G is the amplification factor and L is the length of the Sagnac loop. Thus, a complete transmission is implemented when the peak power P_p satisfies the condition:

$$P_p = \frac{2\pi}{\gamma L(G-1)} \tag{3.57}$$

In other words, the peak of the pulse experiences a higher net gain per round-trip than its wings to shorten the pulse. This mechanism is sometimes known as additive pulse mode locking (APM), and the NFLM can be considered as a fast or artificial saturable absorber as shown in Figure 3.4c. The fiber lasers using NFLM were first proposed for mode locking in 1991 with generated pulse-width of 0.4 ps [19] and a much shorter width of 290 fs was obtained from this laser by optimizing the dispersion and nonlinearity in the fiber cavity [20].

In the third configuration using nonlinear polarization evolution, the intensity-dependent change in the polarization state is explored for mode locking through a polarizing element. Figure 3.3c shows a setup of the passive mode-locked fiber laser based on this principle. The physical mechanism behind mode locking makes use of the nonlinear birefringence. A polarizer that can be also an isolator combined with two polarization controllers acts as a mode locker as described in Figure 3.4b. The polarizer makes the optical wave linearly polarized, and then the following polarization controller changes the polarization state of the wave to elliptical. The polarization state evolves nonlinearly during propagation of the pulse due to the nonlinear phase shift of two orthogonal polarization components in the birefringent fiber ring. The transmissivity of this structure is given by

$$T = \cos^2 \alpha \cos^2 \varphi + \sin^2 \alpha \sin^2 \varphi + \frac{1}{2} \sin 2\alpha \sin 2\varphi \cos(\Delta\phi_L + \Delta\phi_{NL}) \quad (3.58)$$

where α is the angle between the polarization directions of the input light and the fast axis of optical fiber, φ is the angle between the fast axis of optical fiber and the polarization direction of the polarizer, $\Delta\phi_L$ and $\Delta\phi_{NL}$ are linear and nonlinear phase differences between the two orthogonal polarization components, respectively. And they are given by

$$\Delta\phi_L = \phi_L^y - \phi_L^x = \frac{2\pi L}{\lambda}(n_y - n_x) \quad (3.59)$$

$$\Delta\phi_{NL} = \phi_{NL}^x - \phi_{NL}^y = -\frac{\gamma L P \cos 2\alpha}{3} \quad (3.60)$$

where n_x and n_y are the refractive indices of the fast and slow axes of the optical fiber, respectively, and L is the length of the fiber in the cavity. Because of the intensity dependence of the nonlinear phase shift in expression (3.60) as shown in Figure 3.4d, the state of polarization varies across the pulse profile. The polarization controller before the polarizer is adjusted to force the polarization to be linear in the peak of the pulse; hence, the high-intensity part passes the polarizer without loss while the lower-intensity wings are blocked. Thus, the pulse is shortened after every round-trip inside the fiber ring. This configuration can generate easily very narrow pulses of sub-100 fs by careful dispersion optimization [21–23]. The shortest pulse of 47 fs has been generated from the erbium-doped fiber laser by this technique [24].

In the above techniques, the ring configurations based on NFLM and NPR are normally applicable to soliton fiber lasers due to their fast response of the saturable absorption process as well as a possibility of self-initialization. A fast saturable absorption mode locking can be theoretically described by introducing the saturable loss modulation into the master equation (3.55). The modulation of a fast saturable absorber $M_{sa}(t)$ can be modeled by [25]

$$M_{sa}(t) = \frac{s_0}{1 + |A|^2/(A_{eff} I_{sat})} \tag{3.61}$$

where s_0 is the unsaturated loss, and I_{sat} is the saturation intensity of the absorber. In case of weak saturation ($|A|^2 \cong I_{sat}$), expression (3.61) can be Taylor expanded to give

$$M_{sa}(t) \approx s_0 - \frac{s_0}{A_{eff} I_{sat}} |A|^2 = s_0 - s_{SAM} |A|^2 \tag{3.62}$$

where s_{SAM} is called the self-amplitude modulation (SAM) coefficient. The master equation of passive mode locking thus can be derived by using (3.55) and (3.62):

$$T_c \frac{\partial A}{\partial T} + \frac{j}{2} \left(\beta_2 + \frac{j g_{sat}}{\Delta \omega_g^2} \right) L_c \frac{\partial^2 A}{\partial \tau^2} - \frac{\beta_3 L_c}{6} \frac{\partial^3 A}{\partial \tau^3} = j \gamma L_c |A|^2 A + \frac{L_c}{2} (g_{sat} - \alpha) A$$

$$- \frac{s_0}{2} A + s_{SAM} |A|^2 A \tag{3.63}$$

We can simplify Equation (3.63) by incorporating the unsaturated loss s_0 into the loss coefficient and ignoring the effects of dispersion and nonlinearity:

$$T_c \frac{\partial A}{\partial T} - \frac{g_{sat}}{2\Delta \omega_g^2} L_c \frac{\partial^2 A}{\partial \tau^2} = \frac{L_c}{2} (g_{sat} - \alpha') A + s_{SAM} |A|^2 A \tag{3.64}$$

This is the simplest case of passive mode locking where the pulse shaping is based on purely saturable absorption. The solution of (3.64) is a simple soliton pulse equation given by

$$A(\tau) = A_0 \sec h \left(\frac{\tau}{\tau_0} \right) \tag{3.65}$$

By substitution of (3.65) into (3.64), the pulse width and relations between parameters of the system can be obtained [25]:

$$\tau_0^2 = \frac{g_{sat} L_c}{s_{SAM} \Delta \omega_g^2 |A_0|^2} \tag{3.66}$$

and

$$g_{sat} - \alpha' = -\frac{g_{sat}}{\Delta\omega_g^2 \tau_0^2} \tag{3.67}$$

Expression (3.66) can explain why the pulse width in passive mode locking is much shorter than that in active mode locking due to the loss modulation and the curvature being proportional to $s_{SAM}|A_0|^2/\tau_0^2$, then the curvature of loss modulation increases faster when the pulse is shorter while it remains unchanged in active mode locking.

However, the effects of GVD and SPM are always significant to pulse shaping in practical passive mode-locked fiber lasers. Hence, the master equation needs to include these effects and is given by

$$T_c \frac{\partial A}{\partial T} + \frac{j}{2}\left(\beta_2 + \frac{jg_{sat}}{\Delta\omega_g^2}\right)L_c \frac{\partial^2 A}{\partial \tau^2} = j\gamma L_c |A|^2 A + \frac{L_c}{2}(g_{sat} - \alpha')A + s_{SAM}|A|^2 A \tag{3.68}$$

This equation has a simple steady-state solution as follows [26]:

$$A(T,\tau) = A_0 \left[\sec h\left(\tau/\tau_0\right)\right]^{1+jq} e^{j\kappa T} \tag{3.69}$$

By using (3.69) as an anzat and balancing terms, the following pulse parameters and relations can be obtained as

$$\tau_0^2 = \frac{L_c\left[g_{sat}(2-q^2)/\Delta\omega_g^2 + 3\beta_2 q\right]}{2s_{SAM}|A_0|^2} \tag{3.70}$$

$$q = -\frac{3g_{sat}}{2\beta_2\Delta\omega_g^2} \pm \sqrt{\left(\frac{g_{sat}}{2\beta_2\Delta\omega_g^2}\right)^2 + \frac{2(\beta_2 + \gamma\tau_0|A_0|^2)}{\beta_2}} \tag{3.71}$$

$$\kappa = \frac{L_c}{2\tau_0^2 T_c}\left(\beta_2(q^2 - 1) + \frac{g_{sat}q}{\Delta\omega_g^2}\right) \tag{3.72}$$

$$g_{sat} - \alpha' = \frac{1}{\tau_0^2}\left(\frac{g_{sat}(q^2 - 1)}{\Delta\omega_g^2} - 2\beta_2 q\right) \tag{3.73}$$

Equation (3.71) indicates that a combination of anomalous GVD ($\beta_2 < 0$) and the SPM effect can give a zero chirp solution and the shortest pulses can be obtained in this case. For a small SAM coefficient and weak filtering, a soliton can be formed via the balance of anomalous GVD and SPM. The SAM and filtering effects can be considered as weak perturbations in the fiber cavity. However, they play an important role in stabilization of the pulse against noise build-up in the intervals between the pulses [25].

When solitons are periodically perturbed by the gain, loss, filtering, and SAM effects inside the fiber ring cavity, they radiate or generate continuum or dispersive waves. If the continuum components shed by the soliton are phase matched from pulse to pulse, its energy can build up and sidebands appear in the spectrum where the frequency components with the relative phase of soliton and dispersive wave change by an integer multiple (n) of 2π per round-trip. These parasitic sidebands were first described and explained by Kelly [27]. This phenomenon is observed in most passive soliton fiber lasers. The positions of sidebands in the spectrum depend strongly on the dispersion of the fiber cavity via [28]

$$\Delta\lambda_n = \pm n\lambda_0 \sqrt{\frac{2n}{cDL_c} - 0.0787 \frac{\lambda_0^2}{(c\tau_{FWHM})^2}} \tag{3.74}$$

where n is the order of sideband, D is the fiber dispersion parameter in the cavity, t_{FWHM} is the full width at half-maximum of the pulse and $\lambda0$ is the center wavelength. Thus, from the positions of sidebands in the obtained optical spectrum the dispersion of the cavity can be estimated.

3.6 Active Mode Locking

3.6.1 Amplitude Modulation (AM) Mode Locking

In AM mode locking, the amplitude modulation provides a time-dependent loss. The pulse will form at time slots where the loss dips are below the gain level. The modulation of an amplitude modulator can be mathematically described by

$$M_A(T,\tau) = -m_{AM}[1 - \cos(\omega_m\tau)] \tag{3.75}$$

and the pulse evolution equation can be derived from general master equation (3.55) to describe AM mode locking as follows:

$$T_c \frac{\partial A}{\partial T} + \frac{j}{2}\left(\beta_2 + \frac{jg_{sat}}{\Delta\omega_g^2}\right)L_c \frac{\partial^2 A}{\partial\tau^2} - \frac{\beta_3 L_c}{6}\frac{\partial^3 A}{\partial\tau^3} = j\gamma L_c |A|^2 A$$

$$+ \frac{L_c}{2}(g_{sat} - \alpha)A - m_{AM}[1 - \cos(\omega_m\tau)]A \tag{3.76}$$

where m_{AM} is the modulation index, and $\omega_m = 2\pi f_m$ is the angular modulation frequency. For active AM mode locking, the modulation frequency is normally much higher and harmonics of the cavity fundamental frequency (f_c) $f_m = N f_c$, with N being the order of harmonic.

In case of purely AM mode locking, we ignore the effects of GVD and SPM. Additionally, we can Taylor expand the modulation function to second order in time due to the pulse being positioned only at the minimum of the modulation curve. Then Equation (3.76) becomes

$$T_c \frac{\partial A}{\partial T} - \frac{g_{sat} L_c}{\Delta \omega_g^2} \frac{\partial^2 A}{\partial \tau^2} = \frac{L_c}{2}(g_{sat} - \alpha)A - \frac{m_{AM}}{2}\omega_m^2 \tau^2 A \qquad (3.77)$$

The solution of this equation is a Gaussian pulse given by

$$A = A_0 \exp\left[-\frac{\tau^2}{2\tau_0^2}\right] \qquad (3.78)$$

where

$$\tau_0 = \sqrt[4]{\frac{g_{sat} L_c}{\Delta \omega_g^2 m_{AM} \omega_m^2}} \qquad (3.79)$$

This is the result predicted by Kuizenga–Siegman [29], which shows the pulse width is proportional to the inverse of the gain bandwidth and the modulation frequency. The eigenvalue of the equation can give the important condition of the mode-locked laser through the expression

$$g - l = \sqrt{\frac{m_{AM}\omega_m^2 g}{2\Delta\omega_g^2}} \qquad (3.80)$$

where $g = \frac{g_{sat} L_c}{2}$ and $l = \frac{\alpha L_c}{2}$; G and l parameters are considered as the gain and the loss within one round-trip of the fiber cavity. Thus, the expression (3.80) also indicates that the gain must be fixed at a certain value higher than the loss to compensate for the excess loss caused by the modulator and the filtering from the limited gain bandwidth. This condition requires that the gain be sufficient for compensating the loss of the ring cavity. Optical fiber amplifiers are thus preferred to operate in the saturation region in the cases when the ring is either under modulation or no modulation states, so as to achieve stability of the total energy distributed in the ring.

With presence of the GVD effect in the fiber lasers, the evolution of the pulse inside the fiber ring cavity satisfies the following equation:

$$T_r \frac{\partial A}{\partial T} + \frac{j}{2}\left(\beta_2 + \frac{jg_{sat}}{\Delta\omega_g^2}\right)L_c \frac{\partial^2 A}{\partial \tau^2} = \frac{L_c}{2}(g_{sat} - \alpha)A - \frac{m_{AM}}{2}(\omega_m \tau)^2 A \qquad (3.81)$$

This is a Hermite's differential equation, and a stable solution of this equation takes the form [25]

$$A = A_0 \exp\left[-\frac{\tau^2(1+jq)}{2\tau_0^2}\right]\exp(j\kappa T) \qquad (3.82)$$

This is a chirped Gaussian pulse with the pulse parameters obtained by balancing terms in (3.81):

$$\tau_0 = \sqrt[4]{\frac{L_c[g_{sat}(1-q^2) + 2q\beta_2\Delta\omega_g^2]}{m_{AM}\omega_m^2\Delta\omega_g^2}} \qquad (3.83)$$

$$q = -\frac{g_{sat}}{\Delta\omega_g^2\beta_2} \pm \sqrt{\left(\frac{g_{sat}}{\Delta\omega_g^2\beta_2}\right)^2 + 1} \qquad (3.84)$$

$$\kappa = -\frac{L_c}{\tau_0^2 T_c}\left(\frac{g_{sat}q}{\Delta\omega_g^2} - \beta_2\right) \qquad (3.85)$$

The result shows that if $\beta_2 = 0$ (ignoring GVD effect), then the chirp factor $q = 0$ and $k = 0$, and subsequently, the pulse width in (3.83) returns to (3.79). Thus, the presence of GVD can cause the generated pulse chirped.

When the fiber cavity is pumped with sufficiently high power, the SPM effect is not negligible and included in the master equation as fully described in (3.76). With the addition of sufficient negative GVD and SPM, the solitary pulse formation can be obtained, and the solution of (3.76) is assumed to be a chirped secant hyperbole pulse [30]:

$$A = A_0\left[\sec h\left(\frac{\tau}{\tau_0}\right)\right]^{(1+jq)} \qquad (3.86)$$

In a nonlinear regime, the pulse is shortened by a combination of SPM and negative GVD similar to passive mode locking. However, in order to shorten the pulse by soliton compression, the following two conditions must be satisfied [31]:

- The synchronization between the modulator and the pulse train must be maintained or the solitons must be exactly retimed on each round-trip. This is a common condition for stable operation of mode-locked fiber lasers against the thermal drift of the cavity length. This condition also ensures the phase matching requirement that is the total phase of the lightwaves circulating in the ring cavity must be a multiple number of 2π.

- The excess loss of the continuum determined by the eigenvalue of (3.76) must be higher than that of the soliton. The resulting condition is [32]

$$\text{Re}\sqrt{\frac{m_{AM}\omega_m^2}{2L_c}\left[\frac{g_{sat}}{\Delta\omega_g^2}-j\beta_2\right]} > \frac{\pi^2}{24}m_{AM}\omega_m^2\tau_0^2 + \frac{g_{sat}}{3\Delta\omega_g^2\tau_0^2} \quad (3.87)$$

In (3.87), on the right side is the loss experienced by the soliton, and on the left side is the loss of the continuum. In addition, the modulation must not drive the soliton unstable. The condition for suppression of energy fluctuations of the soliton can be obtained from soliton perturbation theory [31–33]:

$$\frac{g_{sat}}{3\Delta\omega_g^2\tau_0^2} > \frac{\pi^2}{24}m_{AM}\omega_m^2\tau_0^2 \quad (3.88)$$

From the above conditions, the mode-locked pulse can be compressed with the width much shorter than that predicted by Kuizenga–Siegman. The factor of pulse width shortening R can be found as follows [34]:

$$R \leq 1.37\left(\frac{\beta_2 L}{g_{sat}/(\Delta\omega_g)^2}\right)^{1/4} \quad (3.89)$$

The condition (3.89) determines the lower limit of the pulse width with the help of SPM and negative GVD. The possible pulse width reduction is proportional to the fourth root of dispersion that indicates the need for an excessive amount of dispersion to maintain a stable soliton while suppressing the continuum.

3.6.2 Frequency Modulation (FM) Mode Locking

In contrast to AM mode locking, FM mode locking is based on a periodic chirping caused by phase modulation. When the modulation frequency is exactly harmonics of the fundamental frequency, the phase matching condition is satisfied for resonance of optical waves in the cavity. In the frequency domain, the sidebands generated by phase modulation are matched to the axial modes of the cavity. While in the time domain, the pulses are built up at the extremes of the modulation cycles. At these temporal positions, the optical waves are not chirped and thus the optical waves are constructively summed when they are in phase, while they are destructively interfered at other temporal positions due to repeated linear chirping in the cavity.

A phase modulation of the optical field can be given by the function

$$M_F(T,\tau) = jm_{FM}\cos(\omega_m\tau) \tag{3.90}$$

where m_{FM} is the phase modulation index. When using the expression of phase modulation for a mode locker, an FM mode locking can be described by the master equation as follows:

$$T_c\frac{\partial A}{\partial T} + \frac{j}{2}\left(\beta_2 + \frac{jg_{sat}}{\Delta\omega_g^2}\right)L_c\frac{\partial^2 A}{\partial\tau^2} - \frac{\beta_3 L_c}{6}\frac{\partial^3 A}{\partial\tau^3} = j\gamma L_c\,|A|^2\,A + \frac{L_c}{2}(g_{sat}-\alpha)A$$

$$+ jm_{FM}\cos(\omega_m\tau)A \tag{3.91}$$

Because the pulse is formed in only the narrow part of the modulation period, the modulation function can be approximated to the second order of the Taylor expansion. Then Equation (3.91) can be simplified by ignoring the nonlinear effect:

$$T_c\frac{\partial A}{\partial T} + \frac{j}{2}\left(\beta_2 + \frac{jg_{sat}}{\Delta\omega_g^2}\right)L_c\frac{\partial^2 A}{\partial\tau^2} = \frac{L_c}{2}(g_{sat}-\alpha)A + jm_{FM}\left(1 - \frac{\omega_m^2\tau^2}{2}\right)A \tag{3.92}$$

Therefore, Equation (3.92) describes the FM mode locking in a linear regime. Similar to AM mode locking, the solution of this equation is a chirped Gaussian pulse with the form

$$A = A_0\exp\left[-\frac{\tau^2(1+jq)}{2\tau_0^2}\right]\exp(j\kappa T) \tag{3.93}$$

By the same techniques as in previous sections, the pulse parameters can be obtained:

$$\tau_0 = \sqrt[4]{\frac{L_c\left[\beta_2\Delta\omega_g^2(1-q^2)+2qg_{sat}\right]}{m_{FM}\omega_m^2\Delta\omega_g^2}} \qquad (3.94)$$

$$q = \frac{\beta_2\Delta\omega_g^2}{g_{sat}} \pm \sqrt{\left(\frac{\beta_2\Delta\omega_g^2}{g_{sat}}\right)^2+1} \qquad (3.95)$$

$$\kappa = \frac{m_{FM}}{T_c} - \frac{L_c}{2\tau_0^2 T_c}\left(\frac{g_{sat}q}{\Delta\omega_g^2}-\beta_2\right) \qquad (3.96)$$

$$g_{sat} - \alpha = \frac{1}{\tau_0^2}\left(\beta_2 q + \frac{g_{sat}}{\Delta\omega_g^2}\right) \qquad (3.97)$$

In the simplest case—that is, the pure FM mode locking with $\beta_2 = 0$, the generated pulse is always chirped with $q = \pm 1$ due to phase modulation. On the other way, the pulses generated from a FM mode-locked laser can be located at either extreme up-chirp ($q > 0$) or down-chirp ($q < 0$) of the modulation cycle in the absence of dispersion and nonlinearity that can create a switching between these two states in a random manner [29]. However, in the presence of the dispersion effect, this switching can be suppressed as possibly indicated in (3.97), which is the expression of the excess loss of the cavity. If $\beta_2 < 0$ (anomalous dispersion), the excess loss at up-chirp half cycle is lower than that at down-chirp cycle ($q > 0$) and vice versa. Thus, the pulses located at positive half-cycles are preferred in the anomalous dispersive fiber cavity while those located at negative half-cycles are preferred in the normal dispersive fiber cavity. On the other hand, the pulse in the up-chirp cycle is compressed by the dispersion while the down-chirp pulse is broadened in the anomalous GVD cavity. This shortened up-chirp pulse experiences less chirp after passing the phase modulator, which reduces the loss due to the filtering and gains bandwidth limitation effects. The up-chirp pulse is finally dominant in the anomalous GVD cavity. This stability of the mode-locked pulse in the FM fiber ring laser has been theoretically demonstrated [35].

In a nonlinear regime, the SPM is also significant in pulse shaping. Similar to AM mode locking, the soliton is formed inside the fiber cavity with the balance between GVD and SPM effects. To generate stable solitons, the required gain for noise must be higher than that for the soliton, and the condition for stability can be obtained by the perturbation soliton theory [36]:

$$\text{Re}\sqrt{\frac{jm_{FM}\omega_m^2}{2L_c}\left[\frac{g_{sat}}{\Delta\omega_g^2}+j\beta_2\right]} > \frac{2g_{sat}}{3\Delta\omega_g^2\tau_0^2} \qquad (3.98)$$

In Equation (3.98) the left side is the loss of the amplification stimulated emission (ASE) noise and the right side is the loss of the soliton. Tamura and Nakazawa [35,37] also indicated that the pulse energy equalization occurs in the support of SPM and the filtering effect when the dispersion of the cavity is anomalous. This stability of the soliton in the presence of the third-order dispersion in an FM mode-locked fiber laser has been numerically investigated in [38].

Besides the synchronous mode locking in an FM mode-locked fiber laser, another mechanism for mode locking based on asynchronous phase modulation has been proposed [39]. In this scheme, asynchronous modulation is obtained by detuning of modulation frequency with a proper amount. However, in order to generate a stable soliton train, the detuning is required to remain within a limit that satisfies the following condition [36]:

$$|\Delta f_{\lim}| \ll \frac{1}{T_c} \text{Re} \sqrt{\frac{jm_{FM}\omega_m^2}{2L_c}\left[\frac{g_{sat}}{\Delta\omega_g^2} + j\beta_2\right]} \tag{3.99}$$

With a small detuning within this limit, the soliton can overcome the frequency shift to remain the mode-locked state. When the detuning exceeds the above limit, the noise can build up and destroy the solitons. The fiber laser will operate in an FM oscillation state if the modulation frequency is moderately detuned. In this regime, the output has a constant intensity in the time domain but with periodical chirp, and its optical spectrum is largely broadened [40]. The transition from FM oscillation state to phase mode locking has complex behavior at smaller detuning where the relaxation oscillations can occur with different properties [41]. In this state, the noise can build up faster than the soliton to become the new pulse and replace the old one. The cause of relaxation oscillation is the change of the cavity loss when the modulation frequency is detuned. The pulse passes through the modulator with a small time shift from the extremes of the modulation cycles that increase the loss of the pulse while the ASE noise gets more gain at the extremes of modulation cycles where they have the lowest loss in the cavity. When the detuning becomes larger, a new pulse can build up from the ASE noise while the old pulse decays and disappears finally. This process is periodically repeated and this repetition can satisfy the phase-matching condition for resonance in the fiber ring cavity, which leads to the relaxation oscillation. In this state, the output is periodically strong spikes but with lower average power because more power is stored in the cavity from resonance. The relaxation oscillation can occur several times due to the central mode hopping in the detuning process. The supermode noise can be dramatically enhanced between these transitions by getting the excess energy from the cavity through matching between the relaxation oscillation frequency and the frequency of beatings between the modulation sidebands and lasing modes [42]. All interesting phenomena exist only in the FM mode-locked lasers, and those have been theoretically and experimentally investigated [40–42].

3.6.3 Rational Harmonic Mode Locking

In active mode locking, there is one way to increase the repetition rate— use the rational harmonic mode locking with detuning the modulation frequency to a rational number of the cavity fundamental frequency [43]:

$$f_m = Nf_c \pm \frac{f_c}{M} \qquad (3.100)$$

where N, M are integers; N is the harmonic order, and M can be considered as the multiplication factor.

To understand the rate multiplication in rational harmonic mode locking, a simple description in the frequency domain is shown in Figure 3.5a. In the frequency domain, the harmonics of modulation frequency would only be matched to the different multiples of the cavity modes when the modulation frequency is detuned with amount of f_c/M. Then the repetition rate of the output pulse can be multiplied by a factor of M. The mechanism can be understood in the time domain as described in Figure 3.5b. When the fiber laser is detuned by a ratio f_c/M, the difference between the cavity round-trip time and N times the modulation frequency is equal to the time delay experienced by a pulse after one round-trip. In another way, the phase delay of a pulse between consecutive round-trips is proportional to $2piN/M$ and the pulse returns to its original positions after M round-trips. As a result, there are M sets of pulses in one round-trip window resulting in a multiplied repetition rate.

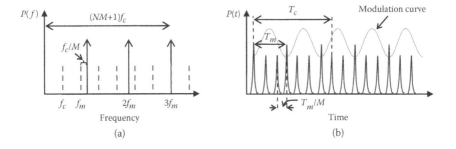

FIGURE 3.5
A description of rational harmonic mode locking (a) in frequency domain, (b) in time domain with $N = 2$ and $M = 3$ as an example.

Because of the pulse distribution over a nonuniform modulation window, the output pulses suffer from large amplitude fluctuations that limit the application of rational harmonic AM mode locking in practical systems. Several methods such as nonlinear optics methods [44,45] and modulator transmittance adjustment [46–48] have been proposed for pulse amplitude equalization.

For the FM mode locking, the situation is different from the AM mode locking when it has been shown that the rational harmonic mode locking is due to the contributions of harmonics of the modulation frequency in the amplified electrical driving signal [49]. The higher-order harmonics are generated from the power amplifier operating in a saturated scheme. Therefore, the phase of the optical field at the output of the phase modulator can be modulated via

$$E_{out} = E_{in} e^{j\varphi(t)}$$
(3.101)

where E_{in}, E_{out} are the optical field at the input and output of the phase modulator, respectively, and $j(t)$ varies corresponding to the driving signal. In case of the amplified signal, this variation can be represented by a summation of a series of cosine functions as follows:

$$\varphi(t) = \sum_{k=1}^{\infty} m_k \cos(k 2\pi f_m t + \theta_k)$$
(3.102)

where f_m is the modulation frequency, m_k and θ_k are the modulation index and the phase delay bias for each frequency component, respectively. From analysis in [49], the field of the optical signal experiences average phase modulation after M round-trips by

$$\overline{\varphi}_M(t) = \sum_{k=1}^{\infty} M m_{kM} \cos(kM 2\pi f_m t + \theta_{kM})$$
(3.103)

Thus, when the modulation frequency f_m is detuned as specified in Equation (3.100), the modulation effect of lower-order harmonics of the f_m are cancelled, and only the frequency component of M times the modulation frequency is enhanced and becomes dominant after M round-trips. In another way, the effective phase modulation curve with the number of cycles multiplied by M times in the same transmission window is proven in Figure 3.6. Because of mode locking based on frequency chirping, the generated pulse train in FM rational mode locking does not suffer from an unequal amplitude problem.

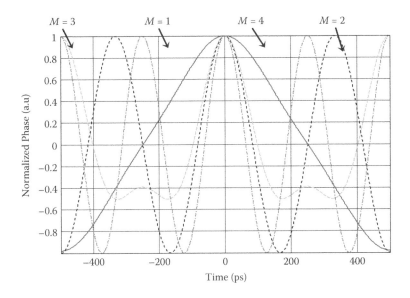

FIGURE 3.6
Average phase modulation profile at different detuning f_d/M with $M = 1$, 4 to achieve rational harmonic mode locking in the frequency modulation mode-locked fiber laser. The driving signal of the phase modulator is modeled with magnitudes of higher harmonic components as follows: $m_2 = 0.008\ m_1$, $m_3 = 0.06\ m_1$, $m_4 = 0.001\ m_1$ and $m_5 = 0.0008\ m_1$.

3.7 Actively Frequency Modulation Mode-Locked Fiber Rings: Experiment

3.7.1 Experimental Setup

By using an electro-optic phase modulator as a mode locker, the active mode-locked fiber ring laser can generate a high-speed pulse train with low jitter. Moreover, it is simple in synchronization between the fiber cavity and other electronic devices. In this section, an active FM mode-locked fiber ring laser will be constructed for demonstration of soliton generation at a high repetition rate in which the erbium gain medium is used for amplification inside the cavity.

Figure 3.7 shows the experimental setup of the FM mode-locked fiber ring laser. In this setup, an erbium-doped fiber amplifier (EDFA) pumped at 980 nm is used in the fiber ring to moderate the optical power in the loop for mode locking. This amplifier operates in the saturation region, and the output power can be adjusted by varying the pump power or the current of the pump laser diode. A 50 m Corning SMF-28 fiber is inserted after the EDFA with the aim of enhancing the nonlinear phase shift through the SPM effect

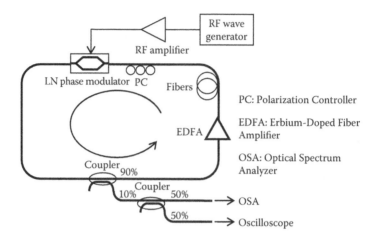

FIGURE 3.7
Experimental setup of the active frequency modulation mode-locked fiber laser.

as well as ensuring that the average dispersion in the fiber ring is anomalous, which is important in designing a stable soliton fiber laser. By measuring the input and output power of the EDFA, the total loss of the cavity can be estimated. This loss, consisting of the insertion loss of the phase modulator and connections, is approximately 10.5 dB to 11 dB in our setup. We note that the loss can vary due to the change in the polarization state.

An electro-optic integrated phase modulator PM-315P of Crystal Technology, Inc. of Sunnyvale, California, United States, assumes the role as a mode locker and controls the states of locking in the fiber ring. At the input of the phase modulator, a polarization controller (PC) consisting of two quarter-wave plates and one half-wave plate is used to control the polarization of light, which is required to minimize the loss cavity and influences the stable formation of solitons. The phase modulator with half-wave voltage of 9 V is driven by a sinusoidal signal generated from a signal generator HP-8647A in the region of 1 GHz frequency. The radio-frequency (RF) sinusoidal wave is amplified by a broadband RF power amplifier DC7000H with 18 dB gain, which can provide a maximum saturated power of approximately 19 dBm at the output. Thus, the phase modulation index of ~1 radian can be achieved for mode locking. The fundamental frequency of the fiber laser cavity is determined by tuning the modulation frequency to lock the fiber laser at different harmonics. In our setup the fundamental frequency of the fiber cavity is 2.2862 MHz, which is equivalent to the 90 m total length of the fiber ring.

The outputs of the mode-locked laser extracted from the coupler 90:10 are monitored by an optical spectrum analyzer HP-70952B and an oscilloscope Agilent DCA-J 86100C with an optical bandwidth of 65 GHz. Because of the fiber laser operating at only 1 GHz, bandwidth of the oscilloscope is wide enough for pulse width measurement larger than 10 ps. In case of the pulse with the

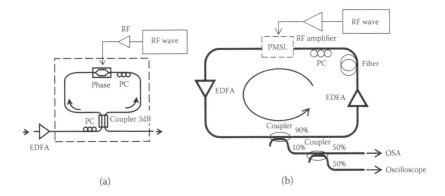

FIGURE 3.8
Experimental setup of the active mode-locked fiber laser using phase-modulated Sagnac loop (PMSL). (a) A whole fiber ring laser. (b) Detailed diagram of the PMSL.

width less than 10 ps, the rise time of the oscilloscope, which is 7.4 ps, should be considered in the estimation of the pulse width as $\tau_p = \sqrt{\tau_{meas}^2 - \tau_{equi}^2}$, where τ_p, τ_{meas} and τ_{equi} are the estimated and measured pulse widths and the rise time of the oscilloscope, respectively. An RF spectrum analyzer FS315 is also used to determine the stability of the generated pulse train.

Besides the conventional ring structure as described above, another setup of the active mode-locked fiber ring laser using electro-optic (EO) phase modulator, which is based on the Sagnac loop interferometer polarization maintaining Sagnac loop (PMSL) and also was implemented in our experiment. When the phase modulator is placed at the middle of a fiber Sagnac loop as shown in Figure 3.8, phase modulation is converted into amplitude modulation by interference between clockwise light and counterclockwise light, which have a phase difference between them. This effect comes from the fact that the phase modulator is optimized for only one transmission direction. The transmission of the PMSL output is given by [50]

$$T(t) = \sin^2\left[\frac{\Delta\varphi(t) + \phi}{2}\right] \tag{3.104}$$

where $D\varphi(t)$ is the differential phase caused by the driving signal between two passage directions, and ϕ is the bias differential phase.

Thus, in this configuration of the active mode-locked fiber laser, the PMSL can be considered as an amplitude modulator without the bias drift problem and possibly polarization dependence [50,51]. Expression (3.104) shows that the intensity modulation of the PMSL can operate at double modulation frequency. However, the mode-locked pulses at peaks of the transmission acquire residual chirp from phase modulation [52]. This chirp will affect the pulse characteristics as well as stability in the same way as in the active FM mode-locked fiber ring laser. Similar to the rational mode locking, the repetition rate multiplication can be archived by detuning with a different rational number of harmonics. The detuning allows the pulses to experience the opposite chirp in consecutive round-trips and consequently, a uniform intensity transmission window can be obtained. In fact, the intensity modulation curve of the PMSL depends strongly on the phase modulation profile. If the phase modulation curve is distorted, the higher-order harmonics in the resulted intensity modulation signal are also strongly enhanced to facilitate the rational harmonic mode locking.

3.7.2 Results and Discussion

3.7.2.1 Soliton Generation

By setting the saturated power of the EDFA of 5 dBm, a stable pulse train is generated by tuning the modulation frequency at the 438th-order harmonics of the fundamental frequency. Figures 3.9a through 3.9c show the time trace and spectrum of the generated mode-locked pulse. The temporal and spectral widths of the pulse are 12.5 ps and 0.23 nm, respectively. Thus,

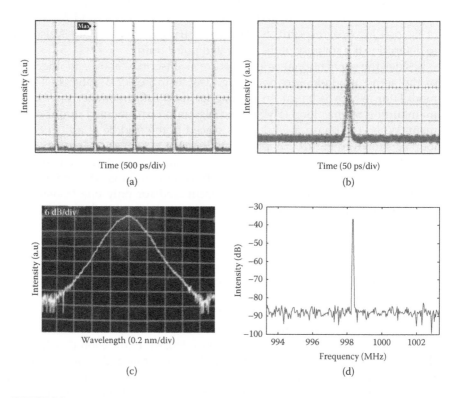

FIGURE 3.9
Time traces of (a) the pulse train, and (b) the single pulse, (c) corresponding optical spectrum, (d) radio-frequency spectrum of the mode-locked pulse generated from the frequency modulation mode-locked fiber ring.

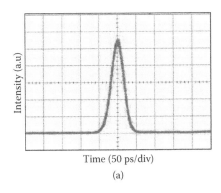
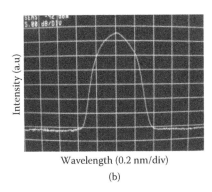

Time (50 ps/div) (a) Wavelength (0.2 nm/div) (b)

FIGURE 3.10
(a) Time trace and (b) spectrum of the mode-locked pulse generated from the FM mode-locked fiber ring laser using the phase modulator Mach-40-27.

this result indicates that the pulse is transform limited with the product of time-bandwidth of 0.36. Because no optical band-pass filter is used in the setup, the emission wavelength of the fiber laser is around 1560 nm where the gain of the EDFA is maximized and flat after optimizing the polarization state of the cavity. Stability of the mode-locked pulse train is demonstrated by the RF spectrum analysis as shown in Figure 3.9d. The sideband suppression ratio (SSR) of higher 50 dB was achieved without any feedback circuit for stabilization.

With the aim of increasing in the phase modulation index, we replaced the phase modulator PM-315P by the phase modulator Covega's Mach-40-27, which has a half-wave voltage of only 4 V at 1 GHz modulation frequency. It is surprising that a larger width of generated pulses at the higher modulation index of 2.3 rads was obtained in this setup with the same intracavity optical power as in the previous setup. Figure 3.10 shows the time trace and the corresponding optical spectrum of the mode-locked pulse generated from this setup. The pulse broadening in this case indicates a limitation of pulse spectrum or gain bandwidth in the cavity that can relate to the property of the phase modulators and would be examined to explain more detail in Chapter 4.

In the configuration of the mode-locked fiber laser using PMSL, the generated pulse train has a high stability but wide pulse width. Figure 3.11 shows the time trace of the pulse train at 1 GHz generated from this configuration. Because of the insertion of the 3 dB coupler in the PMSL, the total loss of the ring cavity is about 16 dB, which is much higher than that of the conventional FM mode-locked fiber ring laser. High cavity loss reduces the efficiency of the soliton compression effect in the fiber laser. Additionally, the PMSL is equivalent to an amplitude modulator, so that the modulation index depends on the intensity modulation curve converted from the phase modulation. With a small phase modulation index, the modulation index of the PMSL is

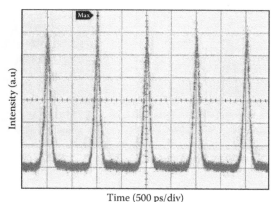

Time (500 ps/div)

FIGURE 3.11
Time trace of the pulse train generated from the mode-locked fiber laser using phase-modulated Sagnac loop.

also small. Therefore, the width of the generated pulse is about 100 ps, and the pulse shape is a Gaussian pulse rather than a soliton.

3.7.2.2 Detuning Effect and Relaxation Oscillation

When the modulation frequency f_m is detuned, the fiber laser can experience various regimes. Especially, the transition state shows complex behaviors such as relaxation oscillation and excess noise enhancement. In our setup, the fiber laser experiences three main regimes that consist of the mode-locked regime, FM oscillation, and transition regimes depending on detuning amount $|\Delta f_m| \cdot f$Because the range of $|\Delta f_m|$ in each regime depends on the cavity length or the cavity dispersion, we inserted 100 m SMF-28 fiber into the fiber ring (case A) beside 50 m SMF-28 fiber in the original setup (case B). By detuning the FM, the important regimes in both cases are identified as follows:

- When $|\Delta f_m| > 2$kHz for the case A and $|\Delta f_m| > 4$kHz for case B, the fiber ring laser operates in an FM oscillation regime that can be identified by its optical spectrum. At large $|\Delta f_m|$, the optical spectrum is similar to a CW signal due to the limitation of resolution in optical spectrum analyzer (OSA), while the RF spectrum cannot identify the first harmonic component of f_m as shown in Figure 3.14a. When the effective modulation index is sufficient by decreasing $|\Delta f_m|$, the optical spectrum is broadened with a double-peak shape due to the energy going to the optical frequencies far from the center carrier mode as shown in Figure 3.12a. The spectrum keeps broadening when the $|\Delta f_m|$ decreases close to 2 kHz and 4 kHz for the cases of $\beta_2 > 0$ and $\beta_2 < 0$, respectively. Figure 3.14b also shows the typical RF spectrum in this regime, which indicates a strong supermode noise and a broad

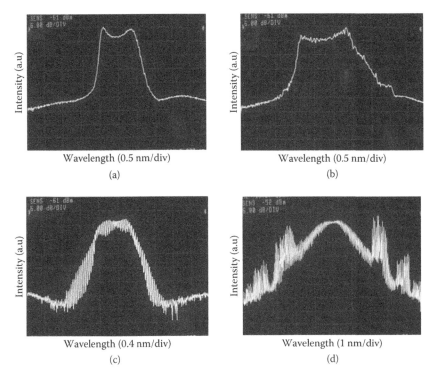

Wavelength (0.5 nm/div)

(a)

Wavelength (0.5 nm/div)

(b)

Wavelength (0.4 nm/div)

(c)

Wavelength (1 nm/div)

(d)

FIGURE 3.12

Optical spectra at different frequency detuning regimes: (a) frequency modulation oscillation, (b) entering the transition regime, (c) enhanced relaxation oscillation in the transition regime, and (d) relaxation oscillation at higher optical power level.

line-width of each side mode due to the beating between the cavity modes and the modulation frequency. Moreover, the strength of the first harmonic is increased according to the reduction of $|\Delta f_m|$.

- When $0.3\,\text{kHz} < |\Delta f_m| < 2\,\text{kHz}$ for case A and $0.8\,\text{kHz} < |\Delta f_m| < 4\,\text{kHz}$ for case B, the fiber laser enters a transition regime where many complex behaviors such as the relaxation oscillations as well as the enhanced supermode noise status can occur [42]. The double-peak spectrum broadening becomes maximum before the energy at the main carrier grows up according to the reduction of detuning as shown in Figure 3.12b. Between the FM oscillation and mode-locking regimes, the relaxation oscillations (RO) as well as the enhanced supermode noise status have also been observed. In the first stage, the time trace shows a high constant intensity that varies continuously and rapidly in the time domain. When the detuning deceases further, the envelope of high intensity is more deeply modulated as shown in Figure 3.13a. Because the supermode noise is still the dominant noise source, the RF spectrum exhibits behavior similar to that shown in Figure 3.14b. In a narrower resolution bandwidth, the beat

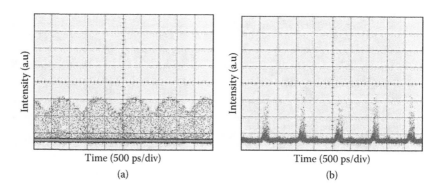

FIGURE 3.13
Time traces in the transition regime: (a) at initial stage and (b) at latter stage when the relaxation oscillation is enhanced.

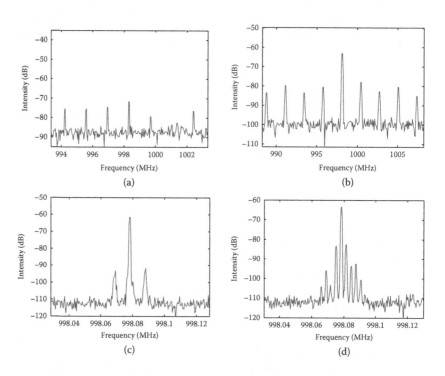

FIGURE 3.14
Radio-frequency spectra at different regimes: (a) continuous-wave-like regime, (b) frequency modulation oscillation regime and transition regime in span of 20 MHz, (c) initial stage of transition regime in span of 100 kHz, and (d) enhanced relaxation oscillation in transition regime in span of 100 kHz.

noise between the modulation sidebands and the cavity modes of about 10 kHz can be observed as in Figure 3.14c. In the latter stage of the transition regime, when the detuning is decreased to around 500 Hz, the RO becomes stronger as an enhanced excess noise. In this state, the building up of new pulses and the decay of old pulses can occur at the same time that a rapid variation of both amplitude and time position is exhibited. Therefore, in this state the time trace of the signal is observed as noisy pulses as shown in Figure 3.13b and the corresponding optical spectrum show ripples in the envelope as shown in Figure 3.12c. If the optical power in the cavity is further increased, the optical spectrum can exhibit sidebands as seen in Figure 3.12d, which indicates the existence of ultrashort pulses with high peak power in this stage. Figure 3.14d shows the RF spectrum with strong sidebands of 2 kHz formed by beating between the modulation frequency and the RO frequency.

- When $|\Delta f_m| < 0.3$ kHz for the case A and $|\Delta f_m| < 0.8$ *kHz* for the case B, the mode-locked state can be achieved. When the detuning is decreased to a small amount that is within a specific limitation as specified in (3.99) or no detuning, the mode locking can be achieved to generate the stable pulse train as shown in Figure 3.9. By providing a sufficient gain in the cavity the RO noise suppression greater than 40 dB and the supermode noise suppression greater than 45 dB can be achieved in our setup without using any stabilization technique.

In the transition regime, it is really interesting that the existence of ultra-short pulses with very high peak power is observed in this regime when the detuning is about 500 Hz incorporating with the adjustment of the polarization controller. Figure 3.15 shows the spectrum and the time trace of this

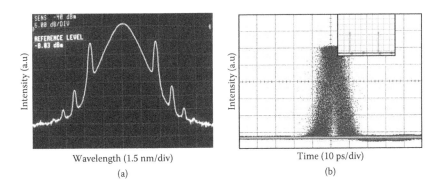

Wavelength (1.5 nm/div)

(a)

Time (10 ps/div)

(b)

FIGURE 3.15
Spectrum and time trace of the hybrid mode-locking state in the case A with an insertion of 100-m SMF-28 fiber into the ring cavity: (a) observed spectrum and (b) recorded domain pulse and jittering.

state in case of A. In this figure the time trace shows the generated pulses with very narrow width and fixed high peak power but strong timing jitter, while the optical spectrum shows a broad spectral width and Kelly sidebands generated by the resonance of dispersive waves and the generated pulses. These results indicate clearly the existence of solitons in this state. Based on the spectral width of 1.58 nm, the pulse width is approximately 1.8 ps, which is impossible to be obviously resolved by the oscilloscope. From the sideband locations, the total cavity dispersion estimated by (3.74) is of about −0.0172 ps^2/m. High stability of the optical spectrum also demonstrates that solitons are stably formed inside the cavity. We believe that the passive mode locking based on NPR plays a key role in this state. This process can be explained as follows: when the cavity is slightly detuned, the relaxation oscillation occurs in which pulses in the form of spikes acquire such high peak power that the NPR becomes significant in the weak birefringence cavity. In a favorable condition by changing the settings of the polarization controller, the passive mode locking based on NPR can be achieved to shape the pulse circulating in the cavity. Thus, the fiber laser in this state operates similar to a hybrid passive-active mode-locked laser [53–55]. In order to verify this passive mechanism the RF modulation signal was turned off; however, the optical spectrum with sidebands remained at least 5 minutes before it disappeared. Owing to the detuning, solitons experience a frequency shift that results in temporal variation of the pulses or timing jitter. Moreover, this state operates in the additive-pulse mode locking (APM) regime, and it is easily prone to dropout as demonstrated in Figure 3.15b by the baseline at the bottom of the pulse trace. This state is also observed in case B, although it is more difficult for adjustment due to an insufficient NPR effect in a shorter fiber cavity. By carefully adjusting the polarization controller, solitons generated by this mechanism in case B can be obtained as shown in Figure 3.16.

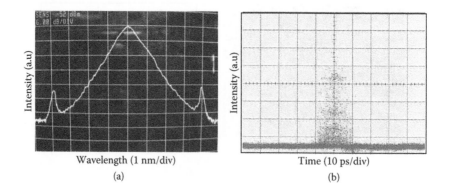

FIGURE 3.16

Spectrum and time trace of the hybrid mode-locking state in the case B with an insertion of 50-m SMF-28 fiber into the ring cavity: (a) observed spectrum and (b) time domain pulse.

With the first sideband spacing of 3.95 nm, the estimated average GVD of the cavity is about −0.0144 ps²/m. The estimation of the average GVD is valid due to a reduction of the dispersion in case B of −1.1 ps², which is exactly equivalent to the 50 m SMF-28 fiber.

3.7.2.3 Rational Harmonic Mode Locking

As described before, a rational harmonic mode locking can be implemented in the active FM mode-locked fiber laser by the higher-order harmonics of the amplified driving signal. By using an RF spectrum analyzer, the magnitude of the harmonics at the output of the power amplifier is measured as a function of the RF input power. Figure 3.17 shows the measured results that indicate a strong increase of the second harmonics while the magnitude of the first-order harmonic remains unchanged at saturated value of 19 dBm at the RF input power higher than 2 dBm. Higher-order harmonics such as the third- and fourth-order harmonics are slightly enhanced at the input power higher than 5 dBm. The magnitude of high-order harmonics determines the modulation index of the rational harmonic mode locking at corresponding orders.

When the modulation frequency is detuned by the amount of $\pm f_c/2$ and $\pm f_c/3$ from the 438th harmonics of the cavity frequency, pulse trains at the repetition rate of double and triple modulation frequency are generated as shown in Figure 3.18 at the RF input power of 7 dBm. The pulse train at the second-order rational harmonic mode locking shows a better performance than that at the third-order rational harmonic mode locking due to the higher modulation index of the second harmonic component. Other rational harmonic mode

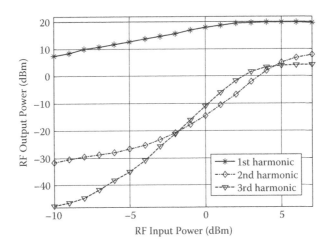

FIGURE 3.17
Output power of the first, second, and third harmonics as a function of the radio-frequency input power.

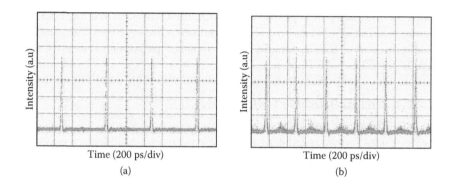

Time (200 ps/div)   Time (200 ps/div)

(a)   (b)

FIGURE 3.18
Time traces of the pulse trains in (a) the second- and (b) the third-order rational harmonic mode locking.

locking such as the fourth and the fifth orders also can be obtained by an appropriate detuning but give very poor performance due to weak strength of the corresponding harmonic components in the driving signal.

Similarly, the rational harmonic mode locking can be achieved in the mode-locked fiber laser using PMSL. Although the phase modulator operates in the small signal modulation region, the higher-order harmonic components in amplitude modulation of the PMSL can be easily enhanced by the distortion of phase modulation through phase-amplitude conversion. By detuning the modulation frequency of $\pm f_c/M$ with M from 2 to 6, the repetition rate of the pulse train is multiplied by a factor M as shown in Figure 3.19. However, the amplitude of pulses is unequal because of the nonuniformity of the intensity modulation profile.

Thus, the pulse trains at the output of the mode-locked fiber ring laser using PMSL show the problem of nonuniform amplitude, which is a disadvantage of the rational harmonic mode locking using amplitude modulation. The uniform pulse trains are always generated by phase modulation in rational harmonic mode locking if the strength of higher-order harmonic components is sufficiently high in the driving signal.

3.8 Simulation of Active Frequency Modulation Mode-Locked Fiber Laser

3.8.1 Numerical Simulation Model

Although an FM mode locking can be theoretically described by the master equation with averaged parameters, it is difficult in solving this equation to find an analytical solution with involvement of all important effects.

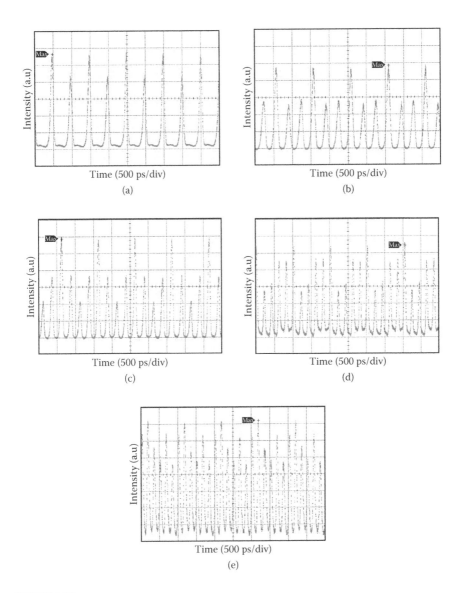

FIGURE 3.19

Time traces of (a) the second-order, (b) the third-order, (c) the fourth-order, (d) the fifth-order, and (e) the sixth-order rational harmonic mode locking in the fiber ring laser using phase-modulated Sagnac loop.

Therefore, in order to understand the physical processes occurring inside the fiber ring cavity, a numerical model is developed in this chapter. The model of a mode-locked fiber laser with a ring configuration consists of basic components similar to that in the experimental setup. Figure 3.20 shows the block diagram of the numerical model for an active FM mode-locked

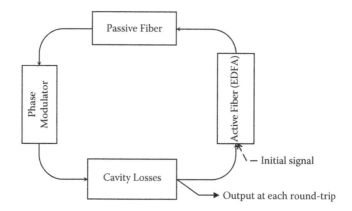

FIGURE 3.20
Numeric model for an active frequency modulation mode-locked fiber ring laser.

fiber ring laser. In this model, a slowly varying envelope of the optical pulse passes through each component once in each round-trip. This process will be repeated until a desired solution is obtained. Thus, the envelope function of the pulse at the nth round-trip can be given by

$$A^n(t) = \hat{L}\hat{M}\hat{F}\hat{F}_g \cdot A^{n-1}(t) \tag{3.105}$$

where A^n, A^{n-1} are the complex amplitudes of the pulse at the nth and the $(n-1)$th round-trips, respectively; $\hat{L}, \hat{M}, \hat{F}$ and $\hat{F}_g$ are the operators representing the loss of the cavity, the modulation mechanism, and the passive and active fibers, respectively. Based on well-understood physical mechanisms, the operators or the models of the components inside the ring cavity can be exactly as described and the effect of each component is individually considered in the model.

First, the operators $\hat{F}$ and $\hat{F}_g$ that describe propagation of the optical field in optical fibers can be modeled by the generalized NLS equation (3.26), which can be rewritten for convenience as follows:

$$\frac{\partial A}{\partial z} + \frac{j}{2}\left(\beta_2 + \frac{jg_{sat}}{\Delta\omega_g^2}\right)\frac{\partial^2 A}{\partial\tau^2} - \frac{\beta_3}{6}\frac{\partial^3 A}{\partial\tau^3} = i\gamma|A|^2 A + \frac{1}{2}(g_{sat} - \alpha)A \tag{3.106}$$

In case of passive fiber, the gain factor is set to zero, but the gain factor with saturation is nonzero in active fiber for amplification in the EDFA and modeled by using Equations (3.23) and (3.25). The ASE noise generated from the EDFA is also included at the end of the active fiber. The ASE noise is modeled as an additive complex Gaussian-distributed noise with a variance given by

$$\sigma_{ASE}^2 = h\nu n_{sp}(G-1)B_{ASE} \tag{3.107}$$

where h is Planck constant, is the optical carrier frequency, G is the gain co-efficient and B_{ASE} is the optical noise bandwidth, and n_{sp} is the spontaneous factor that relates to the noise figure NF of the EDFA as follows:

$$n_{sp} = \frac{NF.G-1}{2(G-1)} \tag{3.108}$$

Equation (3.106) for both active and passive fibers can be solved by using the well-known split-step Fourier method in which the fiber is split into small sections of length and the linear and nonlinear effects are alternatively evaluated between two Fourier-transform domains, respectively [1].

Second, the operator $\hat{M}$ for an EO phase modulator is given by

$$\hat{M} = \exp[jm\cos(\omega_m(\tau + \Delta\tau_s) + \phi_0)] \tag{3.109}$$

where m is the phase modulation index, f_0 is the initial phase, and $\omega_m = 2\pi f_m$ is the angular modulation frequency, assumed to be a harmonic of the fundamental frequency of the fiber ring, Dt_s is the time shift caused by detuning and given by

$$\Delta\tau_s = \frac{T_m - T_h}{T_c} T \tag{3.110}$$

where $T_m = 1/f_m$ is the modulation period, $T_h = T_c/N$ with T_c is the cavity period and N is the harmonic order, and T is the time scale in terms of the round-trip scale.

Besides the attenuation of optical fibers, there are some losses inside the cavity such as coupling loss and insertion losses of the modulator and connectors. All these losses need to be included in the simulation and combined into the total cavity loss factor. The influence of this loss is given as

$$\hat{L} = 10^{l_{dB}/20} \tag{3.111}$$

where l_{dB} is the total loss of the cavity in dB. Thus, this effect in turn determines the required gain coefficient of the EDFA to ensure that the gain is sufficient to compensate the total loss in a single round-trip for stable lasing operation.

Equations (3.106) through (3.111) provide a full set of equations for numerical simulation of an active FM mode-locked fiber ring laser. Due to the recursive nature of pulse propagation in a ring cavity, operation of the operators is repeatedly applied to the complex envelope of the pulse to find a stable solution. The complex amplitude of the output is used as the input of the next round-trip and stored for display and analysis. In simulation of pulse formation, a complex Gaussian-distributed noise of –10 dBm is used as a seeding signal. Depending on the strength of the effects in the model, a stable pulse

TABLE 3.1

Parameter Values Used in Simulations of the Frequency Modulation (FM)
Mode-Locked Fiber Laser

$\beta_2^{SMF} = -21.7 \text{ ps}^2/\text{km}$ for $\bar{\beta}_2 < 0$,	$\beta_2^{EDF} = 19 \text{ ps}^2/\text{km}$,	$\Delta\omega_g = 16 \text{ nm}$,
$\beta_2^{SMF} = +21.7 \text{ ps}^2/\text{km}$ for $\bar{\beta}_2 > 0$	$\gamma^{EDF} = 0.0023 \text{ W}^{-1}/\text{m}$,	$f_m \approx 1\text{GHz}$
$\gamma^{SMF} = 0.0014 \text{ W}^{-1}/\text{m}$	$\alpha^{EDF} = 0.5 \text{ dB/km}$	$m = 0.05\pi \div 1\pi \text{ rad}$
$\alpha^{SMF} = 0.2 \text{ dB/km}$	$P_{sat} = 5 \div 8 \text{ dBm}$	$\lambda = 1559 \text{ nm}$
$L_{SMF} = 80 \text{ m}$	$g_0 = 0.315 \text{ m}^{-1}$	
$L_{EDF} = 10 \text{ m}$	$NF = 5 \text{ dB}$	

Notes: SMF, standard single-mode fiber; EDF, erbium-doped fiber; NF, noise figure of
the erbium-doped fiber amplifier (EDFA).

can be found in 500 round-trips or even up to 10,000 round-trips. The number of samples in the simulation window as well as the step-size in the spatial domain is properly chosen to minimize numerical errors in calculation.

3.8.2 Results and Discussion

3.8.2.1 Mode-Locked Pulse Formation

By using the numerical model described above, the pulse formation in the
FM mode-locked fiber ring laser can be investigated. Table 3.1 summarizes
all parameters used in the simulations. Figure 3.21a shows an evolution of the
mode-locked pulse built up from the noise at $m \sim 1$ radian in the ring cavity
with an anomalous $\bar{\beta}_2$ of $-0.0171 \text{ ps}^2/\text{m}$. Figure 3.21b plots the peak power as
a function of round-trip number. The steady state is only reached after 5000
round-trips, and a damped oscillation occurs in the initial stage of the pulse
formation process. Figure 3.22 shows the time trace and the spectrum of mode-
locked pulse at steady state. We note that the pulse with the width of 11.6 ps is

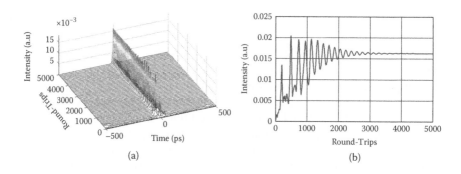

FIGURE 3.21
(a) Numerical simulated evolution of the mode-locked pulse formation from the noise, (b) variation of the peak power 5000 round-trips in the anomalous average dispersion cavity.

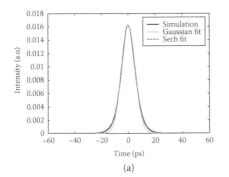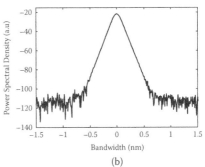

FIGURE 3.22
(a) Numerical simulated time trace and (b) the corresponding spectrum of mode-locked pulse at steady state in the anomalous average dispersion cavity.

well fitted to a secant hyperbolic pulse rather than a Gaussian pulse. However, its spectrum exhibits no sideband due to weak dispersive waves in the cavity.

The effect of the phase modulation index m on the mode-locked pulse is also numerically investigated by varying the index in a range from 0.175 rad to π rad, and the results are shown in Figure 3.23. The increase in modulation index shortens the pulse width. At the index lower than 0.5 rad, the rate of pulse shortening is higher due to the soliton compression effect that results from the dominant SPM effect in pulse shaping. However, at the modulation index higher than 0.5 rad, where the active phase modulation becomes stronger in pulse shaping, the reduction of the pulse width is slow and almost linear. The chirp of pulse is increased with the increase in the phase modulation index, and the pulse diverges from the secant hyperbolic profile and

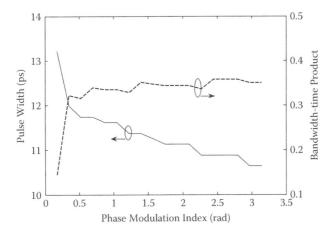

FIGURE 3.23
Variation of mode-locked pulse parameters as a function of the phase modulation index in the anomalous average dispersion cavity.

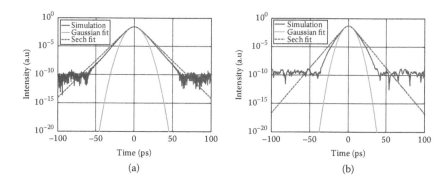

FIGURE 3.24
Numerical simulated waveforms in steady state of mode-locked pulses at two different modulation indices (a) $m = 0.87$ radian, and (b) $m = \pi$ radian.

closer to a Gaussian pulse at a higher index. Figure 3.24 shows the waveforms of the generated pulse at two different modulation indices with the Gaussian fit and secant hyperbolic fit curves.

Instead of an anomalous dispersion cavity, the sign of dispersion in the fibers is reversed to provide a normal dispersion cavity with $\bar{\beta}_2$ of +0.0213 ps²/m. With the same conditions of mode locking, the mode-locked pulse in the normal dispersion cavity is wider than that in the anomalous dispersion cavity. The parameters of the mode-locked pulse as a function of the modulation index in the normal dispersion cavity are depicted in Figure 3.25.

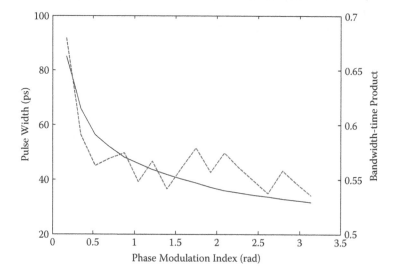

FIGURE 3.25
Variation of mode-locked pulse parameters as a function of the phase modulation index in the normal average dispersion cavity.

The pulse is also shortened with the increase in the phase modulation index. Evolution of the mode-locked pulse in the normal average dispersion cavity at $m \sim 1$ radian is shown in Figure 3.26a. Figure 3.27 shows the pulse profile in the time domain and its spectrum is steady state, which indicates a parabolic pulse rather than a soliton or Gaussian pulse. Furthermore, in the initial stage of pulse formation, there is an existence of dark soliton embedded into the background pulse growing up as shown in Figure 3.26b. It is understood that in this stage the accumulated phase modulation is relatively weak, so that the SPM effect is dominant due to the high gain from the EDFA. The dark soliton formation occurs from the balance of normal GVD and SPM effects, yet the dark soliton is unstable. It experiences a periodic variation of time position (dotted line in Figure 3.26b) and decays due to the chirping caused by active phase modulation. Figure 3.28 shows the waveform with a dip at near center of the background pulse and its phase profile with a phase change of about $\pi/2$ at the dip at the 2900th round-trip. Because the intensity of the dip that also varies along the evolution is nonzero, the formed dark soliton is referred to as the gray soliton rather than the black soliton. The existence of a dark soliton remains only until the accumulated phase shift is sufficient to lock a pulse at the minimum of modulation cycle.

In the FM mode-locked fiber laser, the pulse formation at up-chirping or down-chirping cycles depends on the dispersion of the cavity, which is anomalous or normal. On the other way, the pulse switching between positive and negative modulation cycles is suppressed due to the presence of the GVD effect. Figure 3.29 shows evolutions of the mode-locked pulses built up from noise in normal and anomalous dispersive fiber rings in one modulation period. The modulation curve is phase shifted by $\pi/2$ to display both up-chirp and down-chirp cycles in the same simulation window. The dashed lines in the graphs indicate the phase modulation curve applied in every round-trip. Thus, the pulse is built up only at the extreme of the up-chirp cycle in the anomalous dispersion cavity or only at the extreme of the down-chirp cycle in the normal dispersion cavity.

3.8.2.2 Detuning Operation

In the detuning operation, the modulation frequency is moved away from the harmonic of the cavity frequency. On the other hand, the pulse passes through the modulator at different positions each round-trip that are not only at the extreme of the modulation cycle and experiences a frequency shifting. During propagation through the dispersive fiber, the frequency shift is converted into a temporal position variation in the simulation window. However, a large detuning can destroy a stable mode-locking state due to a fast varying of the modulation cycle between successive round-trips, and the fiber laser falls into the FM oscillation regime that generates a highly chirped signal. Figure 3.30 shows the numerical simulated results in time and frequency domains when the modulation frequency is detuned by 6 kHz. The evolution

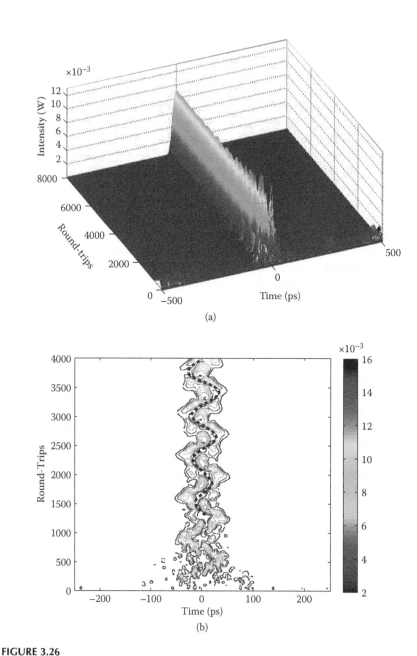

(a)

(b)

FIGURE 3.26
(a) Numerical simulated evolution of mode-locked pulse formation from noise in the normal average dispersion cavity over 8000 round-trips. (b) The evolution in the first 4000 round-trips showing the dark solution formation (black dotted line) embedded in the background pulse.

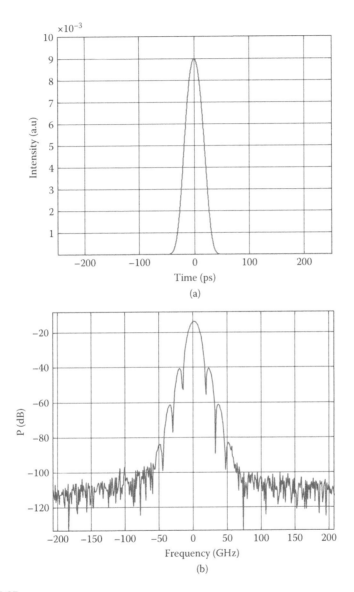

FIGURE 3.27
(a) Simulated waveform and (b) spectrum of mode-locked pulse at steady state in the normal average dispersion cavity.

in Figure 3.30a indicates patterns like noise changing from one round-trip to the next. An example of unstable waveform in the time domain at the 5000th round-trip is shown in Figure 3.30b and its spectrum in Figure 3.30c is broadened with two peaks as observed in the experiment. By taking the average of the waveforms over the last 500 round-trips, a waveform with the envelope modulated at f_m can be observed in Figure 3.30d.

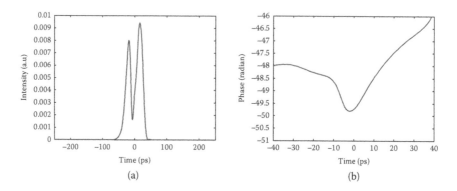

FIGURE 3.28
(a) Simulated waveform and (b) its phase profile at the 2900th round-trip showing a gray solution embedded in the building-up pulse.

When the detuning is slightly moderate, the phase variation of the modulation cycle between successive round-trips is sufficiently slow to enable the pulse to build up in the cavity with adequately high gain. However, the built-up pulses experience the frequency shift induced by detuning that leads to the variation of temporal position and higher loss to be decayed. Relaxation oscillation behavior occurs in this state as shown in Figure 3.31a at the detuning of 1 kHz. Repetition of the process consisting of the pulse decay and the pulse building up exhibits turbulence-like behavior as seen in the contour plot view in Figure 3.31b. Figure 3.31c shows a typical time trace of this state at the 5000th round-trip in which there are three pulses existing simultaneously in the same modulation cycle. In this figure, the lowest pulse close to

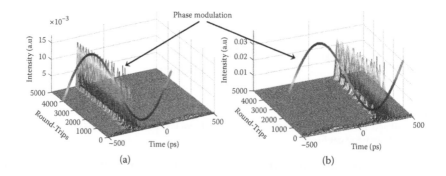

FIGURE 3.29
Numerical simulated evolution of mode-locked pulse built up from noise in (a) anomalous dispersion cavity and (b) normal dispersion cavity at the same phase of modulation curve.

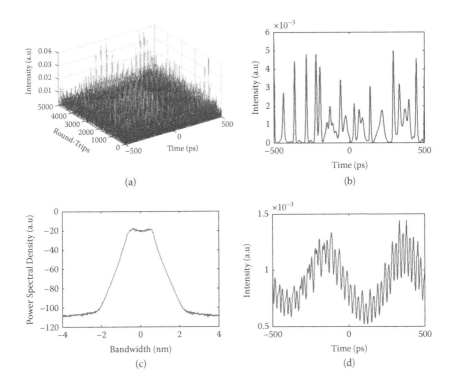

FIGURE 3.30
(a) Numerical simulated evolution of the signal circulating in the cavity, (b) the time trace of the output over 5000 round-trips, (c) the spectrum, and (d) the time trace averaged over the last 500 round-trips when the modulation frequency is detuned by 6 kHz.

the extreme of the modulation cycle is the new pulse building up from noise, the middle pulse with highest peak power and narrow width is experiencing the time shift, while the last pulse with lower peak, which is far from the extreme is decaying due to higher loss. The built-up pulses can survive in around 500 to 1000 round-trips. Because very narrow pulses of 2.5 ps are generated in this case, the corresponding spectrum exhibits sidebands as shown in Figure 3.31d.

When the detuning is small enough, the cavity remains in the mode-locked state as in asynchronous mode locking. However, the mode-locked pulse can experience a variation of temporal position induced by the frequency shift and the dispersion of the cavity. Figure 3.32 shows the behavior of the mode-locked pulse when the modulation frequency is detuned by 0.25 kHz. The contour plot in Figure 3.32b indicates that the pulse still can overcome the detuning problem to stabilize in the cavity.

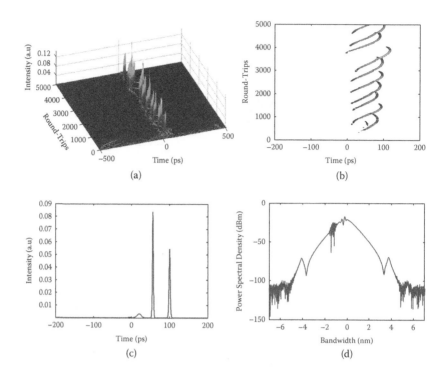

FIGURE 3.31
(a) Numerical simulated evolution of the signal circulating in the cavity, (b) contour plot view of the evolution, (c) the time trace of the output over 5000 round-trips, (d) the spectrum averaged over last 500 round-trips when the modulation frequency is detuned by 1 kHz with a higher gain factor $g_0 = 0.315$ m^{-1} and $P_{sat} = 8$ dBm.

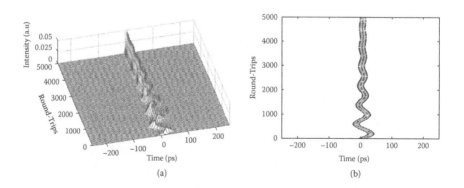

FIGURE 3.32
(a) Numerical simulated evolution of the mode-locked pulse circulating in the cavity. (b) Contour plot view of the evolution when the modulation frequency is detuned by 250 Hz.

3.9 Concluding Remarks

In this chapter, the fundamentals of optical pulse propagation and mode-locking mechanisms have been reviewed. In order to obtain a stable pulse train from active mode locking, some conditions have been summarized and explained. In an active mode-locked fiber laser with sufficiently high gain, the SPM effect becomes significant for pulse compression inside the anomalous dispersion average cavity to generate solitons.

Active mode-locked fiber ring lasers using the EO phase modulator have been experimentally demonstrated for the generation of a sequence of short pulses. The obtained results also indicate that the FM mode-locked fiber laser offers better performance in term of pulse width and stability than the mode-locked fiber laser using PMSL acting as an amplitude modulator. The rational harmonic mode locking has been achieved by the enhancement of higher-order harmonics in the electrical driving signal applied to the optical modulator. Moreover, the detuning effect in the fiber laser has been characterized through the time traces, optical spectrum, and RF spectrum analysis. All obtained results indicated complex behaviors in the transition regime in which the hybrid passive-active mode locking can be achieved to generate solutions of narrow pulse width.

Furthermore, a numerical model of the FM mode-locked fiber ring laser was also developed to investigate the physical processes occurring inside the fiber cavity. The formation of mode-locked pulses in various conditions of the cavity has been numerically studied to demonstrate that indeed, short pulses generated from the FM mode-locked fiber laser system are chirped solitons. The model has also been used to reproduce all possible operations of the FM mode-locked fiber laser including the detuning effect in the experiment to demonstrate the validity of the model, which would be referred to in the next chapters.

References

1. G. P. Agrawal, *Nonlinear Fiber Optics*, 3rd. ed. Academic Press, San Diego, 2001.
2. G. P. Agrawal, *Applications of Nonlinear Fiber Optics*, Academic Press, San Diego, CA, 2001.
3. G. P. Agrawal, Optical Pulse Propagation in Doped Fiber Amplifiers, *Phys. Rev. A*, 44, 7493, 1991.
4. S. C. Mancas and S. R. Choudhury, The Complex Cubic-Quintic Ginzburg-Landau Equation: Hopf Bifurcations Yielding Traveling Waves, *Math. Comput. Simul.*, 74, 281–291, 2007.

5. N. Akhmediev and A. Ankiewicz, Eds., *Dissipative Solitons* (Lecture Notes in Physics 661). Springer, Berlin, 2005.
6. H. A. Haus et al., Structures for Additive Pulse Mode Locking, *J. Opt. Soc. Am. B*, 8, 2068–2076, 1991.
7. C. Menyuk, Nonlinear Pulse Propagation in Birefringent Optical Fibers, *IEEE J. Quantum Electron.*, 23, 174–176, 1987.
8. D. Y. Tang et al., Subsideband Generation and Modulational Instability Lasing in a Fiber Soliton Laser, *J. Opt. Soc. Am. B*, 18, 1443–1450, 2001.
9. K. M. Spaulding et al., Nonlinear Dynamics of Mode-Locking Optical Fiber Ring Lasers, *J. Opt. Soc. Am. B*, 19, 1045–1054, 2002.
10. V. E. Zakharov and A. B. Shabat, Exact Theory of Two-Dimensional Self-Focusing and One-Dimensional Self-Modulation of Waves in Nonlinear Media, *Soviet Physics-JETP*, 34, 669, 1972.
11. Y. S. Kivshar and B. Luther-Davies, Dark Optical Solitons: Physics and Applications, *Phys, Rep.*, 298, 81–197, 1998.
12. L. N. Binh, *Photonic Signal Processing: Techniques and Applications*, CRC Press, Boca Raton, FL, 2008.
13. H. Haus, A Theory of Forced Mode Locking, *IEEE J. Quantum Electron.*, 11, 323–330, 1975.
14. M. Zirngibl et al., 1.2 ps Pulses from Passively Mode-Locked Laser Diode Pumped Er-Doped Fibre Ring Laser, *Electron. Lett.*, 27, 1734–1735, 1991.
15. D. Abraham et al., Transient Dynamics in a Self-Starting Passively Mode-Locked Fiber-Based Soliton Laser, *Appl. Phys. Lett.*, 63, 2857–2859, 1993.
16. B. C. Collings et al., Stable Multigigahertz Pulse-Train Formation in a Short-Cavity Passively Harmonic Mode-Locked Erbium/Ytterbium Fiber Laser, *Opt. Lett.*, 23, 123–125, 1998.
17. N. J. Doran and D. Wood, Nonlinear-Optical Loop Mirror, *Opt. Lett.*, 13, 56–58, 1988.
18. M. E. Fermann et al., Nonlinear Amplifying Loop Mirror, *Opt. Lett.*, 15, 75754, 1990.
19. I. N. Duling, III, Subpicosecond All-Fibre Erbium Laser, *Electron. Lett.*, 27, 544–545, 1991.
20. M. Nakazawa et al., Low Threshold, 290 fs Erbium-Doped Fiber Laser with a Nonlinear Amplifying Loop Mirror Pumped by InGaAsP Laser Diodes, *Appl. Phys. Lett.*, 59, 2073–2075, 1991.
21. K. Tamura et al., 77-fs Pulse Generation from a Stretched-Pulse Mode-Locked All-Fiber Ring Laser, *Opt. Lett.*, 18, 1080–1082, 1993.
22. K. Tamura et al., Technique for Obtaining High-Energy Ultrashort Pulses from an Additive-Pulse Mode-Locked Erbium-Doped Fiber Ring laser, *Opt. Lett.*, 19, 46–48, 1994.
23. G. Lenz et al., All-Solid-State Femtosecond Source at 1.55 μm, *Opt. Lett.*, 20, 1289–1291, 1995.
24. D. Y. Tang and L. M. Zhao, Generation of 47-fs Pulses Directly from an Erbium-Doped Fiber Laser, *Opt. Lett.*, 32, 41–43, 2007.
25. H. A. Haus, Mode-Locking of Lasers, *IEEE J. Selected Topics in Quantum Electron.*, 6, 1173–1185, 2000.
26. O. E. Martinez et al., Theory of Passively Mode-Locked Lasers Including Self-Phase Modulation and Group-Velocity Dispersion, *Opt. Lett.*, 9, 156–158, 1984.

27. S. M. J. Kelly, Characteristic Sideband Instability of Periodically Amplified Average Soliton, *Electron. Lett.,* 28, 806–807, 1992.

28. M. L. Dennis and I. N. Duling, III, Experimental Study of Sideband Generation in Femtosecond Fiber Lasers, *IEEE J. Quantum Electron.,* 30, 1469–1477, 1994.

29. D. Kuizenga and A. Siegman, FM and AM Mode Locking of the Homogeneous Laser—Part I: Theory, *IEEE J. Quantum Electron.,* 6, 694–708, 1970.

30. H. Sotobayashi and K. Kikuchi, Design Theory of Ultra-Short Pulse Generation from Actively Mode-Locked Fiber Lasers, *IEICE Trans. Electron.,* E81-C, 201–207, 1998.

31. D. J. Jones et al., Subpicosecond Solitons in an Actively Mode-Locked Fiber Laser, *Opt. Lett.,* 21, 1818–1820, 1996.

32. H. A. Haus and A. Mecozzi, Long-Term Storage of a Bit Stream of Solitons, *Opt. Lett.,* 17, 1500–1502, 1992.

33. H. A. Haus and A. Mecozzi, Noise of Mode-Locked Lasers, *IEEE J. Quantum Electron.,* 29, 983–996, 1993.

34. F. X. Kärtner et al., Solitary-Pulse Stabilization and Shortening in Actively Mode-Locked Lasers, *J. Opt. Soc. Am. B,* 12, 486–496, 1995.

35. K. Tamura and M. Nakazawa, Pulse Energy Equalization in Harmonically FM Mode-Locked Lasers with Slow Gain, *Opt. Lett.,* 21, 1930–1932, 1996.

36. H. A. Haus et al., Theory of Soliton Stability in Asynchronous Modelocking, *J. Lightwave Technol.,* 14, 62627, 1996.

37. M. Nakazawa et al., Supermode Noise Suppression in a Harmonically Modelocked Fibre Laser by Selfphase Modulation and spe, *Electron. Lett.,* 32, 461, 1996.

38. N. G. Usechak et al., FM Mode-Locked Fiber Lasers Operating in the Autosoliton Regime, *IEEE J. Quantum Electron.,* 41, 753–761, 2005.

39. C. R. Doerr et al., Asynchronous Soliton Mode Locking, *Opt. Lett.,* 19, 1958–1960, 1994.

40. S. Longhi and P. Laporta, Floquet Theory of Intracavity Laser Frequency Modulation, *Phys. Rev. A,* 60, 4016, 1999.

41. S. Yang et al., Experimental Study on Relaxation Oscillation in a Detuned FM Harmonic Mode-Locked Er-Doped Fiber Laser, *Opt. Commun.,* 245, 371–376, 2005.

42. S. Yang and X. Bao, Experimental Observation of Excess Noise in a Detuned Phase-Modulation Harmonic Mode-Locking Laser, *Phys. Rev. A,* 74, 033805, 2006.

43. Z. Ahmed and N. Onodera, High Repetition Rate Optical Pulse Generation by Frequency Multiplication in Actively Modelocked fib, *Electron. Lett.,* 32, 455, 1996.

44. M.-Y. Jeon et al., Pulse-Amplitude-Equalized Output from a Rational Harmonic Mode-Locked Fiber Laser, *Opt. Lett.,* 23, 855–857, 1998.

45. C. G. Lee et al., Pulse-Amplitude Equalization in a Rational Harmonic Mode-Locked Semiconductor Ring Laser Using Optical Feedback, *Opt. Commun.,* 209, 417–425, 2002.

46. Y. Kim et al., Pulse-Amplitude Equalization in a Rational Harmonic Mode-Locked Semiconductor Fiber Ring Laser Using a Dual-Drive Mach-Zehnder Modulator, *Opt. Express,* 12, 907–915, 2004.

47. G. Zhu and N. K. Dutta, Eighth-Order Rational Harmonic Mode-Locked Fiber Laser with Amplitude-Equalized Output Operating at 80 Gbits/s, *Opt. Lett.,* 30 (17), 2212–2214, 2005.

48. F. Xinhuan et al., Pulse-Amplitude Equalization in a Rational Harmonic Mode-Locked Fiber Laser Using Nonlinear Modulation, *Photonics Technol. Lett., IEEE,* 16, 1813–1815, 2004.
49. Y. Shiquan and B. Xiaoyi, Rational Harmonic Mode-Locking in a Phase-Modulated Fiber Laser, *Photonics Technol. Lett., IEEE,* 18, 1331334, 2006.
50. M. L. Dennis et al., Inherently Bias Drift Free Amplitude Modulator, *Electron. Lett.,* 32, 547–548, 1996.
51. M. L. Dennis and I. N. Duling, III, Polarisation-Independent Intensity Modulator Based on Lithium Niobate, *Electron. Lett.,* 36, 1857–1858, 2000.
52. Y. Shiquan and B. Xiaoyi, Repetition-Rate-Multiplication in Actively Mode-Locking Fiber Laser by Using Phase Modulated Fiber Loop Mirror, *IEEE J. Quantum Electron.,* 41, 1285–1292, 2005.
53. H. A. Haus et al., Additive-Pulse Modelocking in Fiber Lasers, *IEEE J. Quantum Electron.,* 30, 200–208, 1994.
54. T. F. Carruthers et al., Active-Passive Modelocking in a Single-Polarisation Erbium Fibre Laser, *Electron. Lett.,* 30, 1051–1053, 1994.
55. C. R. Doerr et al., Additive-Pulse Limiting, *Opt. Lett.,* 19, 31–33, 1994.

4

Multibound Solitons

Le Nguyen Binh

Hua Wei Technologies, European Research Center, Munich, Germany

Nguyen Duc Nhan

Institute of Technology for Posts and Telecommunications, Hanoi, Vietnam

CONTENTS

4.1 Introduction

Optical solitons have attracted considerable attention in research and practical applications on not only their generation techniques but also dynamics. Understanding the dynamics of solotonic waves is of much significance for soliton applications in communications and signal processing. Therefore, extensive studies on soliton interactions must be conducted. For practical dissipative systems such as mode-locked fiber lasers, the interaction between solitons can be theoretically described by the complex Ginzburg–Landau equation (CGLE) instead of the nonlinear Schrödinger equation, which is only valid in conservative systems. Hence, interaction states of solitons based on CGLE have attracted considerable attention in theoretical approaches. Malomed analyzes the interaction of slightly overlapping CGLE solitons and first predicted about the formation of effectively stable dual- and multipulse bound states of solutions [1,2]. Then Akhmediev et al. numerically investigated the interaction of the CGLE solitons by using a two-dimensional (2-D) phase space approach [3]. They also found the stable solution of dual and multipulse bound solitons with a $\pi/2$ phase difference between them. These theoretical results opened a new research frontier on the mutual relationship between the states and the dynamics of short pulses in the mode-locked fiber lasers.

It had taken one decade since Malomed's prediction to experimentally demonstrate the existence of the bound solitons in the passive mode-locked fiber laser using nonlinear polarization rotation (NPR) technique [4]. Then, the bound solitons were also reported in the passive mode-locked fiber laser using a nonlinear fiber loop mirror (NFLM) [5]. Most numerical and experimental investigations on various operational modes and dynamics of bound solitons have been implemented in the passive mode-locked fiber lasers [4–18]. However, the question is whether or not it is possible to generate bound solitons in active mode-locked fiber lasers. Bound soliton pairs, but not multibound solitons, were experimentally observed for the first time in an active frequency modulation (FM) mode-locked fiber laser [19]. Then, multibound solitons with an order higher than two were experimentally demonstrated [20]. In this chapter, we describe the generation and the formation mechanism of multibound solitons from active FM mode-locked fiber lasers. The characteristics of multibound solitons are described and analyzed.

4.2 Bound Solitons in Passive Mode Locking

4.2.1 Multipulsing Operation

Bound soliton states can be considered as one of the multiple soliton opera-tion modes of the passive mode-locked fiber laser. On the other hand, the formation of bound solitons relates to multipulse operation and effective interactions between solitons inside the fiber cavity. In a strongly nonlinear regime of operation, passive mode-locked fiber lasers exhibit a multipuls-ing behavior. The existence of multiple solitons in the cavity also refers to the soliton energy quantization effect that limits the peak power of solitons under some specific conditions. Efforts in finding mechanisms of multipulse formation that is important for optimization and design of solution fiber lasers have been achieved [21–23].

There are three main mechanisms of multipulse formation that limit the peak power of the pulse circulating in the cavity. The first mechanism relates to the pulse splitting due to the limited gain bandwidth [22]. Under high nonlinearity due to strong pumping, the mode-locked pulse is compressed by the self-phase modulation (SPM) effect so that its spectrum is broadened so wide that some frequency components can exceed limited bandwidth of the gain medium. Consequently, the pulse experiences higher extra loss of the cavity and splits into two pulses with wider pulse width or smaller spec-tral width to avoid this loss.

According to the above mechanism, a pulse would experience a compres-sion process until its width is so narrow that it splits into two pulses. In con-ditions of the fiber laser, however, solutions are considerably chirped by SPM and gain bandwidth limitation effect. A generation of subpulses from an initial pulse can occur during circulating in the high-power cavity because of strong chirping after a number of round-trips [21]. These subpulses can keep growing to have the same width and amplitude. According to dissipa-tive solution theory, this second mechanism of the multipulse generation can also be explained as follows: Due to the dissipative nature of the fiber lasers, the dispersive waves are generated by shedding the energy of the solution during circulating in the cavity. In a strongly pumping gain medium, a part of dispersive waves acquires sufficient gain to grow and evolve into a new dissipative solution with the width and amplitude specified by the fiber cav-ity parameters [24].

Another mechanism of multipulsing is based on the cavity effect that relates to the transmittance of the cavity formed by NPR or NFLM. As explained in Chapter 3, the transmittance of the fiber cavity in both tech-niques using NPR and NFLM is a sinusoidal function of the nonlinear phase delay (modulation). Thus, there are two distinct regimes of operation in a period of the transmittance curve (see Figure 3.1). In the first half, the transmittance increases with increasing intensity, while the transmittance decreases with increasing intensity in the second half. The maximum of the transmittance curve is also the transition point between two regimes. For mode locking, the cavity operates in the first regime due to the characteristic of saturable absorption. When the nonlinear phase delay is sufficient in solution operation of the fiber laser, the cavity can be dynamically switched from the saturable absorption mode to the saturable amplification mode that limits the peak of solutions generated inside the cavity. For NPR technique, the switching point between two modes can be adjusted by changing the setting of polarization controllers to change the linear phase delay. As a result of the cavity feedback, the peak power of solution is limited, which is responsible for the multipulse operation in the cavity [25].

4.2.2 Bound States in a Passive Mode-Locked Fiber Ring

Although the multipulsing operation was experimentally observed sometime ago [26], only until 2001 were the bound states of solitons confirmed for the first time in an experiment on soliton formation in a passive mode-locked fiber laser using the nonlinear polarization rotation technique [4]. This observation activated stimulating research interest in new states of solitons in the mode-locked fibers. Grelu et al. also reported the experimental observation of two, three, and multibound solitons with the separation between adjacent pulses varying under different conditions [6,9]. By using a strictly dispersion-managed fiber cavity, multipulse solitons can be formed with a fixed pulse separation [15]. In the bound states, the multipulse bound solitons function as a unit, and they also behave in similar dynamics to a single soliton in the cavity such as forming states of bound multipulse solitons or period-doubling, period-quadrupling, and chaotic states [16]. The investigation of bound solitons was extended in fiber lasers operating in large normal cavity dispersion regimes where the pulse shape is parabolic rather than hyperbolic secant [27]. Bound soliton pairs were also observed in nonlinear optical loop mirror (NOLM) figure-eight fiber lasers under the condition of nonbalancing in the total dispersion [8]. Research efforts on bound solitons have been in progress, and the latest report has shown the observation of the bound state of 350 pulses or a "soliton crystal" in an NPR mode-locked fiber laser, a record in the number of pulses in a bound state [28].

The formation of bound states following the pulse-splitting process in passive mode locking can be affected by various interaction mechanisms among solitons inside the cavity. Depending on the setup of the fiber laser, one or more than one interaction mechanisms become stronger than the others and determine the formation and the dynamics of bound solitons. In passive mode-locked fiber lasers, there are a variety of possible interaction mechanisms that relate to different modes of bound solitons as follows:

- Gain depletion and recovery: The interaction between pulses with the transient depletion and recovery dynamics of the gain medium. The pulses effectively repel each other due to a group velocity drift caused by the time-dependent gain depletion acting in conjunction with the gain recovery. This mechanism is significant in the stabilization of pulse spacing in harmonic mode locking [29].

- Acoustic effect and electrostriction: Due to the intense electric field, the optical pulses can interact with the materials forming the fiber-guided medium to generate acoustic waves in the propagation of the consecutive pulses. This electrostriction induces a small frequency shift between the lightwaves under the pulses leading to effective pulse-to-pulse attraction [30].

- CW component and soliton interaction: In some given settings of the passive mode-locked fiber laser using NPR, a CW component can coexist with the solitons in the cavity. It has been experimentally shown that the CW component causes the central frequency shift in each soliton [12]. When the CW component becomes unstable, the solitons acquire different frequency shifts that result in their various relative velocities in the cavity. This interaction is also responsible for motion mode and harmonic mode locking in passive mode-locked lasers [31].

- Soliton–soliton interaction: This direct interaction between solitons is from the nature of fundamental solitons that can attract or repel each other depending on their relative phases. A repulsive force appears between *quadrature* solitons, that is when their phase difference is in a multiple number of pi/2, while an attraction occurs when they are in phase. In general, this interaction is considerably effective only when the solitons are sufficiently close together [32].

- Soliton-dispersive wave interaction: During the circulating in the cavity, solitons radiate dispersive waves or continuum due to periodically varying perturbations of the cavity such as losses, bandwidth limited gain [32]. The dispersive waves create a local interaction that can generate an attractive or repulsive force between solitons [33].

When solitons interact together through these mechanisms, different bound states can be formed inside the fiber cavity. Depending on the binding strength between solitons, bound solitons can be classified into two basic types: loosely and tightly bound solitons. For the former type, solitons are often bound together into a bunch with the temporal separation between the pulses much larger than their pulse width. Therefore, the loosely bound solitons have a weak binding energy, and they would be easily destroyed by environmental perturbations. Due to wide separation between solitons, the long-range interactions such as the two first mechanisms play a dominant role in formation of loosely bound solitons. Self-stabilization of this state can be achieved by a balance of the interplay between these mechanisms [34].

For tightly bound solitons, two last mechanisms play a major role in the formation of the bound states. Because solitons in this type are closely separated, local interactions are attributed to the binding energy between individual solitons that determines the characteristics and the dynamics of bound solitons. It has been shown that the direct soliton interaction mainly contributes to the formation of tightly bound solitons with discrete fixed separations [15]. We note that pulses formed in the fiber lasers are often chirped solitons, the direct interaction among them is not like that of nonlinear Schrödinger equation (NLSE) solitons. Therefore, the overlap of oscillating tails of solitons results in an effective binding between them to obtain a stable bound state. However, it was believed that the dispersive wave-soliton interaction also affects the ability to generate various dynamics of bound solitons [35]. Due to strong binding energy, the tightly bound soliton behaves as a unit and exhibits all features of a single soliton pulse such as collision, harmonic mode locking, and evenly bound state.

Summarily, all various interactions among solitons always coexist in a passive mode-locked fiber laser. However, only one or two of them play a key role in the formation and characteristic of bound solitons. Furthermore, the existence of other interactions also contributes to the dynamics of their bound states. In other words, the bound soliton falls into only one of multipulse operations under some specific conditions of the cavity settings.

4.3 Bound Solitons in Active Mode Locking and Binding Conditions

In active mode locking, although the multipulsing operation is more difficult to achieve than that in passive mode locking due to higher loss of the cavity, it still occurs under strong pumping mode of the gain medium. Mechanisms of pulse splitting in active mode locking are also similar to those for passive mode locking. However, the width of the pulse in active mode locking is much wider than that in the passive mode locking, and the contribution of the limited bandwidth of the gain medium of the peak power limiting effect is negligible, unless when either a band-pass filter is inserted into the ring or an intrinsically filtering effect from the ring cavity sufficiently shortens the gain-bandwidth product. Therefore, splitting of a single pulse into multipulses can only occur in an active fiber cavity when the power in the fiber ring increases above a certain mode-locking threshold. At a higher-power, higher-order solitons can be excited. In addition, the accumulated nonlinear phase shift in the loop must be sufficiently high so that a single pulse can split into many pulses [36]. The number of split pulses depends on the optical power preserved in the ring, so there is a specific range of power for each splitting level. The fluctuation of pulses may occur at a region of power where there is a transition from the lower splitting level to the higher. Moreover, the chirping caused by the phase modulator in the loop also makes the process of pulse conversion from a chirped single pulse into multipulses taking place more easily [37,38]. In hybrid mode-locked fiber lasers using NPR for pulse shaping, the peak power limiting effect is caused by the additive pulse limiting (APL) effect [39]. By a proper setting of the cavity parameters, the cavity can be switched from an additive pulse mode (APM) locking regime to an APL regime that can clamp the peak of solitons at a certain level.

Because the pulse splitting only occurs at some specific positions around the minima of the modulation curve in the cavity, the split pulses are closely separated. The tightly bound states are thus assuming the only mode existed in the active mode-locked fiber lasers. Hence, there are effective interactions among these pulses to stabilize the multibound states in the cavity. In other words, the multibound solitons can be formed and stabilized by a balanced interplay between these interactions. The direct soliton interaction, which is influenced by the relative phase difference of the optical carrier under the envelope of adjacent pulses, is believed to contribute significantly in the formation of bound solitons. However, for an actively FM mode-locked fiber laser, another interaction among these pulses needs to be considered due to the linear chirping effect caused by active phase modulation. Particularly, the multibound soliton sequence is only stably formed at the up-chirping half-cycle for the case of an anomalous path-averaged dispersion fiber ring. On the other hand, the bound soliton pulses should be symmetrically distributed around the extreme of the positive phase modulation half-cycle where the pulses acquire up-chirping as described in Figure 4.1. Thus, the group velocity of the lightwave, induced by anomalous dispersion, contained within the first pulse of the soliton bunch is decreased after passing through the phase modulator, while the last pulse of the bunch receives the opposite effect—enhancement of group velocity. The variation of the group velocity between pulses creates an attractive force that pulls them to the extreme of a modulation half-cycle similar to the jittering control in soliton transmission systems [40]. Thus, another interaction between the solitons is required to balance the effective attraction by linear chirping, and it is the direct soliton interaction. A repulsive force sufficient

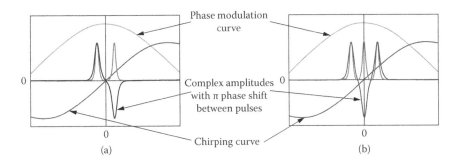

FIGURE 4.1
Effective interactions in multibound solitons: (a) dual-soliton bound state and (b) triple-soliton bound state.

enough to balance with the effective attractive force appears only in case of the π phase difference between adjacent pulses [41–43]. For other phase differences, there is an energy exchange or an attraction between pulses that are possibly not conforming to the stable existence of multibound pulse sequence in an frequency modulation mode-locked fiber laser (FM-MLFL). Thus, we can generate the multisoliton bound states using an appropriate phase modulation profile that can even be the high-order harmonics of the modulation frequency. When stable pulse sequence is formed, the pulses have traveled at least 500 rounds of the fiber ring. Thus, significant delays occur between the lightwave carriers under the bound soliton pulses. If longer distance is traveled by the pulse sequence, there would be breaking up of the sequence, and any nonlinear phase shifts can push the sequence into chaotic states. Currently it is difficult to compensate for the different delays between the lightwaves contained within the pulses. Therefore, in a specific mode-locked fiber laser setup, beside the optical power level and net dispersion of the fiber cavity, the modulator-induced chirp or the phase modulation index determine not only the pulse width but also the temporal separation of bound-soliton pulses at which the interactive effects cancel each other.

The presence of a phase modulator in the cavity is to balance the effective interactions among bound-soliton pulses, which is similar to the use of this device in a long haul soliton transmission system to reduce the timing jitter [44,45]. For this reason a simple perturbation technique can be applied to determine the role of phase modulation on the formation mechanism of multibound solitons. The optical field of a multisoliton bound state can be described as

$$u_{bs} = \sum_{i=1}^{N} u_i(z,t) \tag{4.1}$$

and

$$u_i = A_i \sec h\{A_i[(t - T_i)/T_0]\} \exp(j\theta_i - j\omega_i t) \tag{4.2}$$

where N is the number of solitons in the bound state; T_0 is the pulse width of soliton; and A_i, T_i, θ_i, ω_i represent the amplitude, position, phase, and frequency of soliton, respectively. In the simplest case of the multisoliton bound state, N is equal to 2 or we consider the dual-soliton bound state with the identical amplitude of pulse and the phase difference of π value ($\Delta\theta = \theta_{i+1} - \theta_i = \pi$). The ordinary differential equations for the frequency difference and the pulse separation can be derived by using the perturbation method [44,45]:

$$\frac{d\omega}{dz} = -\frac{4\beta_2}{T_0^3} \exp\left[-\frac{\Delta T}{T_0}\right] - 2\alpha_m \Delta T \tag{4.3}$$

$$\frac{d\Delta T}{dz} \beta_2 \omega \tag{4.4}$$

where β_2 is the averaged group-velocity dispersion of the fiber loop, ΔT is temporal separation between two adjacent solitons ($T_{i+1} - T_i = \Delta T$) and $\alpha_m = m\omega_m^2/(2L_{cav})$, L_{cav} is the total length of the ring, m is the phase modulation index, and ω_m is the angular modulation frequency. Equation (4.4) shows the evolution of frequency difference and position of bound solitons in the fiber ring in which the first term on the right-hand side represents the accumulated frequency difference of two adjacent pulses during a round-trip of the fiber ring, and the second represents the relative frequency difference of these pulses when passing through the phase modulator. At a steady state, the pulse separation is constant and the induced frequency differences cancel each other. On the other hand, if setting Equation (4.3) to zero, we have

$$-\frac{4\beta_2}{T_0^3}\exp\left[-\frac{\Delta T}{T_0}\right]-2\alpha_m\Delta T = 0 \quad \text{or} \quad \Delta T\exp\left[\frac{\Delta T}{T_0}\right]=-\frac{4\beta_2}{T_0^3}\frac{L_{cav}}{m\omega_m^2} \qquad (4.5)$$

We can see from (4.5) the effect of phase modulation to the pulse separation, and β_2 and α_m must have opposite signs, which means that in an anomalous dispersion fiber ring with negative value of β_2, the pulses should be up-chirped. With a specific setup of the active FM fiber laser, when the magnitude of chirping increases, the bound pulse separation can decrease subsequently. The pulse width is also reduced according to the increase in the phase modulation index and modulation frequency, so that the ratio $\Delta T/T_0$ cannot change substantially. Thus, the binding of solitons in the FM mode-locked fiber laser is assisted by the phase modulation. Bound solitons in the ring experience periodically the frequency shift, and hence their velocity in response to changes in their temporal positions by the interactive forces in the equilibrium state.

In principle, multibound solitons can also be generated in any active mode-locked fiber laser including amplitude mode locking with an appropriate frequency chirping in the cavity. Controllability of linear chirping in the amplitude modulator is required in order to maintain a balance in the steady state. Moreover, a frequency shifting facilitates a pulse splitting due to broadening the pulse spectrum in a limited gain bandwidth [38].

4.4 Generation of Multibound Solitons in Frequency Modulation Mode-Locked Fiber Laser

4.4.1 Experimental Setup

An active FM-MLFRL has been set up to generate multibound solitons. Initially, we used the experimental setup of the FM mode-locked fiber laser as shown in Figure 4.2a. Because the erbium-doped fiber amplifier (EDFA) in

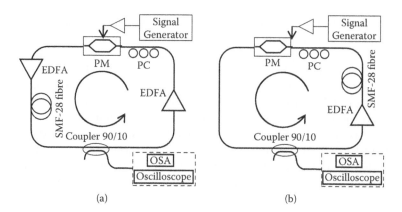

FIGURE 4.2
The experimental setups for multibound soliton generation (a) using two EDFAs with low gain and saturated power and (b) using one EDFA with high gain and saturated power.

this setup has low gain and saturated power that was insufficient for multi-pulse operation, two EDFAs with 12 dBm maximum output were used for the experiment of multibound soliton generation. However, this setup was limited in the number of bound solitons due to the limitation of the saturated power of the EDFA. Moreover, the amplification stimulated emission(ASE) noise was enhanced by the presence of two EDFAs in the ring that degraded the performance of the bound solitons. Therefore, we replaced this setup with the setup similar to that in the generation of a single-soliton train as described in Chapter 2. We show again the experimental setup of the active FM mode-locked fiber laser for generation of multibound solitons in Figure 4.2b. In this setup, only one EDFA with higher gain and maximum saturated power of 17 dBm is used. The modulation frequency of about 1 GHz is selected to generate the bound solitons. The output is characterized by the instruments in Chapter 2 to estimate the performance of the pulse train.

4.4.2 Results and Discussion

In our initial setup, the generation of multibound solitons in the active MLFL was achieved [46], yet the number of bound solitons was limited up to the quadruple state due to the limitation of the saturated power of the EDFA. We then extended the experimental investigation with the bound states of up to the sixth order by adjusting the polarization states in association with the total circulating optical power in the cavity. The multisoliton bound states are depicted in Figure 4.3. When the average optical power is suffi-ciently increased to a certain level, the dual bound states are correspond-ingly switched to higher-order states. A slight adjustment of the polarization controller is necessary to stabilize the bound states. Figures 4.3a through 4.3e show the traces and spectra of the lowest- to highest-order bound states, the

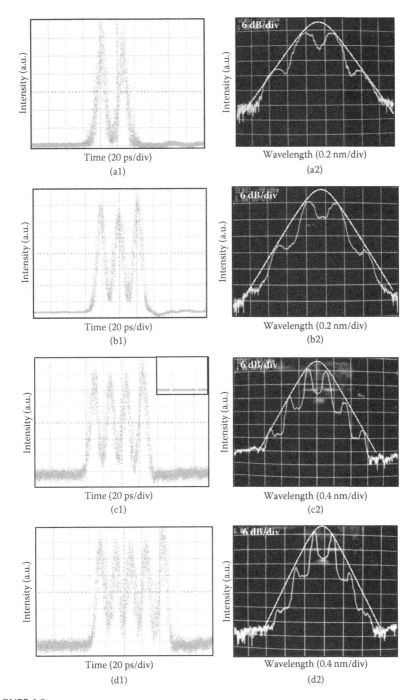

FIGURE 4.3

Time-domain oscilloscope traces (1) and optical spectra (2) of (a) dual-soliton, (b) triple-soliton, (c) quadruple-soliton, (d) quintuple-soliton, and (e) sextuple-soliton bound states.

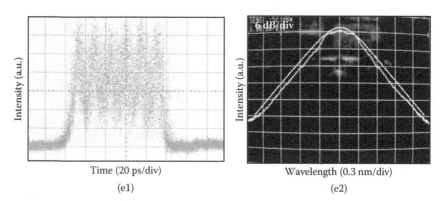

Time (20 ps/div)   Wavelength (0.3 nm/div)
(e1)   (e2)

FIGURE 4.3
(Continued)

sextuple-soliton bound state, as observed in our experiment. The significant advantage of the active fiber laser is the ease of generation of multisoliton bound sequence at moderately high modulation frequency as shown in the inset of Figure 4.3c1).

In the tightly bound states, solitons are closely separated, and then the overlap between solitons causes their spectra modulated with the shape and the symmetry depending on their relative phase difference. Hence the existence of the bound states is also confirmed through the shapes of modulated optical spectra as compared to the conventional spectrum of the single soliton state. When the multibound states appear after the optical power in the cavity is increased to an appropriate level, there is a sudden change of the optical spectrum. The symmetry of spectra in Figures 4.3a2 through 4.3e2 indicates a relative phase difference of π between two adjacent bound solitons. The dashed-dot lines show the envelope of modulated spectra that correspond to the spectrum of a single soliton pulse. The suppression of the carrier at the center of the pass-band spectrum further confirms the π phase difference between adjacent pulses, especially the pair of dual-bound solitons. The distance between the two spectral main lobes is exactly correlated to the temporal separation between two adjacent pulses in the time domain. The specific shape of the spectrum depends on the number of solitons in the bound states, which can be odd or even. In the case of even-soliton bound states, such as dual-soliton and quadruple-soliton bound states, there is always a dip at the center of the spectrum, while there is a small hump in the case of the odd-soliton bound state, such as the triple-soliton bound state and quintuple-soliton bound state. The small hump at the center of the spectrum is formed by the far interaction between the next neighboring pulses that are in-phase in bound states. This is similar to the case for the quadrature phase shift keying modulation format, which is well known in the field of digital communications [47]. When the phase difference moves away from the π value, the modulation and symmetry of spectrum vary accordingly. For the sextuple-soliton bound state, its spectrum is weakly modulated due to the variation of phase

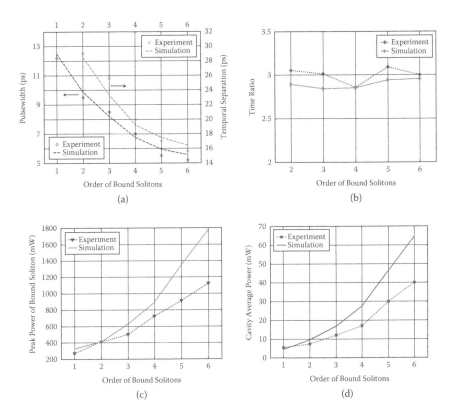

FIGURE 4.4
The variation of parameters with the number of solitons in bound states: (a) pulsewidth and temporal separation, (b) ratio between pulsewidth and separation, (c) peak power, and (d) corresponding average power of the cavity.

difference induced by enhanced ASE noise. The change in relative phase locking influences the interaction of adjacent pulses or the performance of bound soliton output as in Figure 4.3e.

All important parameters of the multibound solitons experimentally measured are summarized in Figure 4.4 that plots the average values of the pulse width ($\bar{\tau}$) and the time separation between pulses ($\Delta\tau$), the maximum peak power of a soliton pulse (P_0), and average optical power ($\bar{P}$) inside the fiber ring against the number of bound solitons. Figure 4.4a shows the variation of the pulsewidth decreasing with respect to the increase in the number of pulses in bound states. This is due to the pulse compression effect, the self-phase modulation, enhanced by the higher optical power level in the cavity at higher-order bound states. The temporal separation between pulses in the multibound state also accordingly decreases to remain a ratio between separation and pulsewidth that is nearly unchanged, which is approximately three as shown in Figure 4.4b. The onset power level of the energy stored in

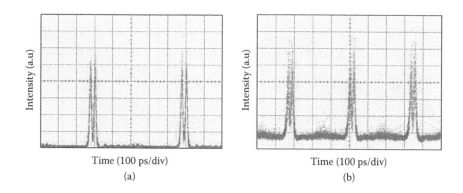

Time (100 ps/div)   Time (100 ps/div)
(a)   (b)

FIGURE 4.5
Time traces of dual-bound solitons in rational harmonic mode-locking schemes: (a) the second-order rational harmonic and (b) the third-order rational harmonic.

the ring is 7.5, 12, 17, 30, and 40 mW for the generation of dual, triple, quadruple, quintuple, and sextuple states, respectively.

Similar to the single-soliton state, the multibound solitons can be generated in rational harmonic mode locking. Figure 4.5 shows the time traces of dual- and triple-bound solitons at the second and third rational harmonic mode locking after increasing properly the intracavity saturated power as well as the radio-frequency (RF) input power to enhance the second harmonic of modulation frequency. With an appropriate phase modulation profile formed by the detuning amount of $\pm f_c/2$, the multibound solitons in rational harmonic mode locking are generated in the same interaction mechanism with the same characteristics as those in the conventional harmonic mode locking. However, the temporal separation between solitons in this case is smaller than that in conventional harmonic mode locking due to the higher chirp rate of higher-order harmonics of modulation frequency. Thus, the multibound solitons can exist in various regimes of operation similar to the conventional single-pulse mode.

When the higher-order bound solitons operate at a higher power level, they are more sensitive to the change in the polarization state and an even multiwavelength operation can occur. This effect is disadvantageous to the multibound soliton operation due to the reduction of energy of the operating wavelength from other excited wavelengths. On the other hand, higher-order bound solitons are more sensitive to fluctuations of the environmental condition. In experimental conditions, these polarization effects are usually controlled by a polarization controller, especially at the input of an integrated optical modulator so only one polarized state can be preferred through the modulator and hence the forcing of the matching condition of this polarized state. Moreover, the increase in saturated power in the fiber cavity allows generation of higher-order bound soliton states, yet the ASE noise is also enhanced under a strong pumping scheme in the EDFAs. If the phase noise induced by ASE noise is sufficient, it affects not only the phase matching

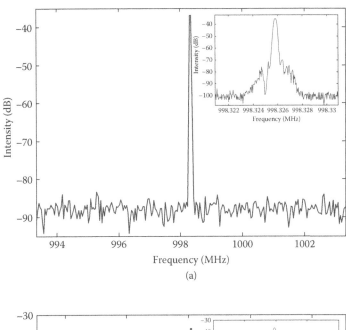

(a)

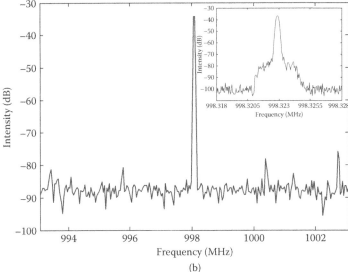

(b)

FIGURE 4.6
The radio-frequency (RF) spectra of (a) the dual-bound soliton and (b) the quadruple-bound soliton in the span of 10 MHz. (Inset: The RF spectra with high resolution in span of 10 kHz.)

conditions of the fiber ring but also the phase locking between adjacent pulses as discussed above. Figure 4.6 shows the RF spectra of the dual- and quadruple-bound soliton trains to estimate the stability of the multibound soliton train. From the results of RF spectrum analysis, the sideband to carrier ratio (SCR) is higher 45 dB for the dual-bound soliton, but it is reduced to 40 dB for the quadruple-bound soliton. Obviously, there are more fluctuations in higher-order bound states or they are more sensitive to the environmental conditions. However, there are small fluctuations in amplitude and temporal position of solitons in bound states as indicated by a broadening of the spectral line in insets of Figure 4.6. In the experiment, it has been found that the tuning of the polarization controller becomes harder to obtain a stable bound state when the number of solitons in the bound state is larger.

Although the electro-optic (EO) phase modulator plays a key role as a mode locker in our active mode-locked fiber laser, we should note that the polarization effect shows an important influence on multibound solitons. At high-power levels of the cavity, the polarization-dependent mechanisms such as polarization-dependent loss (PDL) and polarization-dependent gain (PDG) are enhanced due to the phase modulator also being a polarizing element [48–51]. Hence, the variation of the polarization state changes the total loss or the gain of the cavity that affects the shaping, the formation, as well as the parameters of multibound solitons. We have found that a stable multibound state is only obtained by proper polarization settings. If changing the setting of the polarization controller, beside the stable multibound states observed above, a noise-like pulse regime has been obtained. Figure 4.7 shows the time traces of the noise-like "square" pulses at the power levels of the quadruple- and quintuple-bound solitons, respectively. Although the pulses cannot be resolved in the traces, the widths of these pulses are exactly the same width of the bunch of corresponding bound solitons. This regime seems to be a multipulsing operation but is unstable. Optical spectra of these states are slightly modulated and varied similar to Figure 4.3e2. The pulses in the bunch, which might be in phase, move and even collide together or oscillate very fast around the extreme of the modulation cycle. As a result the pulses cannot be clearly isolated in the bunch. Therefore, the traces of these states on oscilloscope are seen like noisy waveforms.

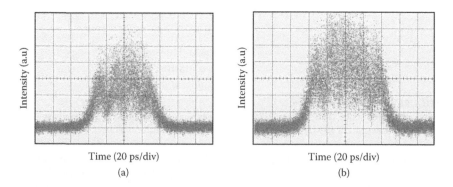

FIGURE 4.7
Time traces of the noise-like pulse at the optical power levels of (a) quadruple-bound and (b) quintuple-bound solitons, respectively.

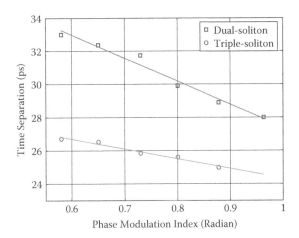

FIGURE 4.8
Variation of time-separation of dual and triple bound soliton states with respect to the phase modulation index.

Different from passive mode-locked fiber lasers, the active phase modulation contributes significantly to the formation and the stability of multibound soliton states. Figure 4.8 shows experimental results of the influence of chirping, which is directly proportional to the phase modulation index, on temporal separation between adjacent pulses at different bound states. When the chirp rate increases, the relative variation of group velocity between adjacent pulses is enhanced. Thus, the increase in frequency chirping reduces the temporal separation necessary to keep the frequency shift between these pulses unchanged in the cavity with a specific average dispersion. The decrease in the time separation in experimental results shows a nearly linear function of the phase modulation index that should be an exponential function as theoretically analyzed in [46]. This can be due to the small range of the phase modulation index. In addition, because the pulse width of the triple-bound soliton is narrower than that of the dual-bound soliton, the pulses acquire a smaller chirp that results in the lower rate of decrease in the time separation in the triple-bound soliton as seen in Figure 4.8.

4.4.3 Simulation

4.4.3.1 Formation of Multisoliton Bound States

To understand the operation of multibound solitons in the active FM mode-locked fiber laser, we numerically investigated the process of multibound soliton generation in the FM mode-locked fiber ring laser by using the numerical simulation model described in Chapter 2 for the active FM mode-locked fiber laser. First, we simulated the formation process of bound states in the

TABLE 4.1

Simulation Parameters of the Multibound Soliton Formation

$\beta_2^{SMF} = -21$ ps^2/km	$\beta_2^{ErF} = 6.43$ ps^2/km	$\Delta\omega_g = 16$ nm
$\gamma_{SMF} = 0.0019$ W^{-1}/m	$\gamma_{EDF} = 0.003$ W^{-1}/m	$f_m \approx 1$ GHz
$\alpha_{SMF} = 0.2$ dB/km	$\alpha_{EDF} = 0.5$ dB/km	$m = \pi/3$
$L_{SMF} = 80$ m	$P_{sat} = 8 - 14$ dBm	$\lambda = 1559$ nm
$L_{EDF} = 10$ m	$g_0 = 0.35 - 0.45$ m^{-1}, $NF = 5$ dB	$l_{cav} = 11$ dB

FM mode-locked fiber laser whose main parameters are shown in Table 4.1. The lengths of the active fiber and passive fiber are chosen to get the cavity's average dispersion $\bar{\beta}_2 = -10.7$ ps^2/km. Figure 4.9 shows a simulated dual-soliton bound state building up from initial Gaussian-distributed noise as an input seed over the first 2000 round-trips with the P_{sat} value of 8 dBm and g_0 of 0.36 m^{-1}. The built-up pulse experiences transitions with large fluctuations of intensity, position, and pulse width during the first 1000 round-trips before formation of the bound soliton state. Figures 4.9b and 14.9c show the time waveform and spectrum of the output signal at the 2000th round-trip.

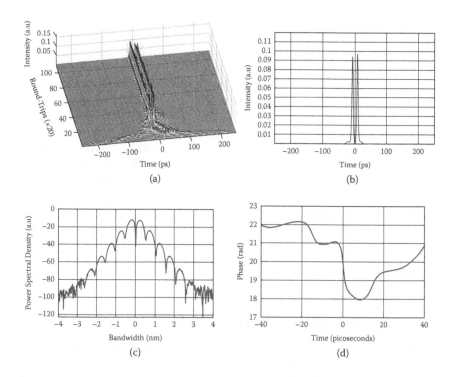

FIGURE 4.9
(a) Numerically simulated evolution of the dual-soliton bound state formation from noise, (b) the waveform, (c) the spectrum, and (d) the phase at the 2000th round-trip.

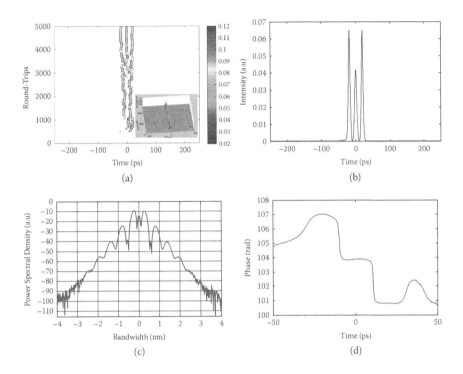

FIGURE 4.10
(a) Numerically simulated evolution of the triple-soliton bound state formation from noise, (b) the waveform, (c) the spectrum, and (d) the phase at the 5000th round-trip.

The bound states with a higher number of pulses can be formed at a higher gain of the cavity; hence, when the P_{sat} and gain g_0 are increased to 11 dBm and 0.38 m^{-1}, respectively, which is enhancing the average optical power in the ring, the triple-soliton bound steady state is formed from the noise seeded via simulation as shown in Figure 4.10. In the case of higher optical power, the fluctuation of the signal at initial transitions is stronger, and it needs more round-trips to reach a more stable triple-bound state. The waveform and spectrum of the output signal from the FM mode-locked fiber laser at the 5000th round-trip are shown in Figures 4.10b and 4.10c, respectively.

Although the amplitude of pulses is not equal indicating the bound state can require a larger number of round-trips before the effects in the ring balance, the phase difference of pulses accumulated during circulating in the fiber loop is approximately of π value, which is indicated by strongly modulated spectra. In particular from the simulation result, the phase difference between adjacent pulses is 0.98π in case of the dual-pulse bound state and 0.89π in case of the triple-pulse bound state. These simulation results reproduce the experimental results (shown in Figure 4.9d and Figure 4.10d) discussed above to confirm the existence of multisoliton bound states in an FM mode-locked fiber laser.

It is realized that the cavity requires a higher gain to form the multibound states than that in stable or steady states. By adjusting the parameters of gain medium after multipulse state formed in the cavity, stable multibound soliton states can be generated after at least 10,000 round-trips. Figure 4.10a shows the contour of the dual-bound soliton formation in 10,000 round-trips of the ring cavity. In this simulation, the gain factor is reduced from 0.37 m^{-1} to 0.35 m^{-1} at the 2000th round-trip (the dash line). In the first 2000 round-trips, the dual-bound soliton is formed, yet unstable with strong fluctuation of the peak power. Then the fluctuation is reduced by the reduction of gain factor, and the generated dual-bound soliton converges to a stable state after 10,000 round-trips as shown in Figure 4.10b. Figures 4.10c and 4.10d show the waveform and the phase of the output at the 10,000th round-trip. Similarly, the parameters of gain medium are also adjusted to a obtain stable higher-order multibound solitons. Figures 4.11 and 4.12 show the simulated results of the triple-bound soliton and the quadruple-bound soliton formation processes, respectively. For the triple-bound soliton, the gain factors are the same values as those in the case of the dual-bound

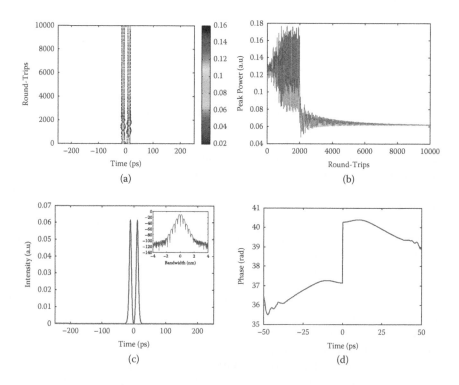

FIGURE 4.11
(a) Contour plot of simulated evolution of the dual-soliton bound state formation from noise, and (b) variation of the peak power with the gain switching at the 2000th round-trip, (c) the waveform (Inset: the corresponding spectrum), and (d) the phase at the 10,000th round-trip.

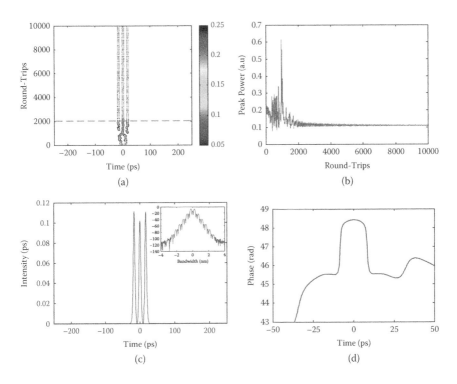

FIGURE 4.12

(a) Contour plot of simulated evolution of the triple-soliton bound state formation from noise, and (b) variation of the peak power with the gain switching at the 2000th round-trip, (c) the waveform (Inset: the corresponding spectrum), and (d) the phase at the 10,000th round-trip.

soliton, but the saturated power P_{sat} is increased to 11 dBm instead of 8 dBm. Variation of peak power and evolution of the triple-bound soliton formation with the gain switching at the 2000th round-trip are shown in Figures 4.12a and 4.12b. For the quadruple-bound soliton, both the initial gain and adjusted gain factors are increased to higher values that are 0.4 m^{-1} and 0.377 m^{-1}, respectively, at the P_{sat} of 11 dBm. Figure 4.12b shows a damping of the peak power variation after the gain adjustment to reach to a stable triple-bound soliton state; however, the peak power of the quadruple-bound soliton still oscillates evenly after 10,000 round-trips as seen in Figure 4.13b. Thus, it indicates more sensitivity of the higher-order multibound soliton to the operating condition. The sensitivity also exhibits through the uniform of pulse intensity in the bound state and the relative phase difference that is decreased from π in the dual-bound soliton to 0.87π in the quadruple-bound soliton as shown in Figures 4.11c and 4.11d, Figures 4.12c and 4.12d, and Figures 4.13c and 4.13d.

We note that a stable multibound soliton is difficult to be formed in the cavity without the adjustment of the gain parameters. With the high gain, the

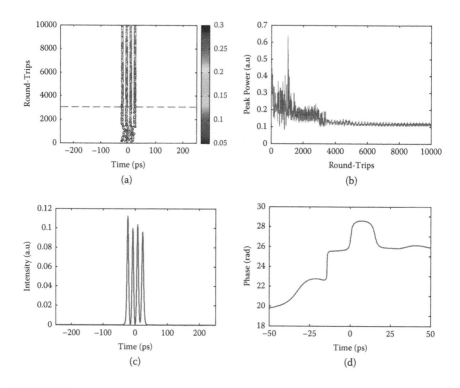

FIGURE 4.13

(a) Contour plot of simulated evolution of the quadruple-soliton bound state formation from noise. (b) Variation of the peak power with the gain switching at the 3000th round-trip. (c) The waveform (Inset: the corresponding spectrum). (d) The phase at the 10,000th round-trip.

multibound soliton can be generated but unstable or becomes quasi-stable with a periodic variation of soliton parameters in the bound state. With the low gain, the nonlinear phase shift is not sufficient to generate a desirable higher-order multipulse state, while it is too strong for the stable lower-order multibound state. Different from passive mode locking, the pulse splitting in our system is caused by the excitation of higher-order soliton and nonlinear chirping rather than an energy quantization mechanism.

4.4.3.2 Evolution of the Bound Soliton States in a Frequency Modulation Fiber Loop

Obviously, a multibound soliton can be stably generated in the phase-modulated fiber cavity. On the other hand, the bound soliton with the relative difference of π given by (4.1) and (4.2) is considered as a stable solution of the active FM mode-locked fiber laser. By using the multisoliton waveform in (4.1) and (4.2) as input, we have simulated the stability of multibound solitons in the active FM mode-locked fiber laser. Figure 4.14 shows the

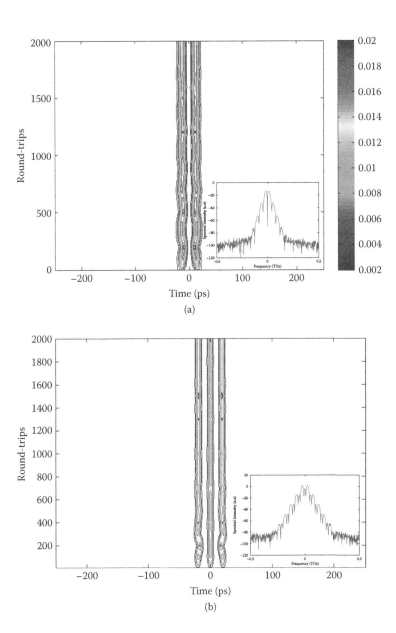

FIGURE 4.14

Evolution of (a) dual-bound soliton and (b) triple-bound soliton with a relative phase difference of π in the frequency modulation fiber ring cavity.

evolutions in 2000 round-trips of the dual- and triple-bound solitons in the cavity. Because the bound solitons are really chirped pulses, while the input in simulation is unchirped, there is a damping oscillation of bound solitons in the initial stage that is considered as a transition of solitons to adjust their own parameters to match the parameters of the cavity. However, the multibound solitons easily reach stable states after only a few hundreds of round-trips.

Simulation results on parameters of multibound solitons after 5000 round-trips are also shown in Figure 4.4 for comparison by using the experimental parameters of multibound solitons as the initial parameters. Generally, the simulated results agree with the experimental results. However, as observed in Figures 4.4c and 4.4d, there are discrepancies of the peak and average power levels between the simulated and experimental pulses, especially at higher-order soliton bound states. The level of discrepancy varies from 0.1 dB to 3 dB for the peak power of 1000 to 1800 mW, respectively. While the experimental results show a nearly linear dependence of the peak power on the order of the bound state, the simulation results show an exponential variation. Hence, there is also a difference of average power of the cavity between them. However, both sets of results indicate an exponential dependence of the cavity average power on the order of the bound soliton state. There are some reasons for taking into account the discrepancy as follows: First, the parameters of the fiber cavity used in the simulation are not totally matched to those in our experiment; second, when the real pulse width at a higher-order bound soliton state is narrower, the accuracy in the pulse width measured on the oscilloscope is reduced, although the influence of rise-time of the oscilloscope was considered in estimation. Furthermore, the variation of the polarization state becomes stronger at a higher power level of the cavity, which also increases the error in the power measurement. Hence, the error between simulation and experimental results increases at a higher order of the soliton bound state. In addition, the higher-order multibound soliton is more sensitive to the cavity parameter settings. Figure 4.15a shows an unstable state evolving in the FM fiber ring cavity. Multiple pulses are generated in the cavity, yet it is difficult to acquire the phase difference of π and uniform between pulses. The balance in the effective interaction between pulses is difficult to be achieved; the pulses can therefore collide and vary rapidly in both amplitude and time position. However, this rapid variation only occurs in a limited time window around the extreme of the modulation cycle as indicated by the dashed lines in Figure 4.15b. By taking an average over the last 2000 round-trips, the waveform and its spectrum, which can be represented for a dynamical state, are shown in Figures 4.15c and 4.15d, respectively. The waveform, which is like a noisy pulse, and the spectrum, which is slightly modulated, are similar to what is observed in the experiment (see Figure 4.7).

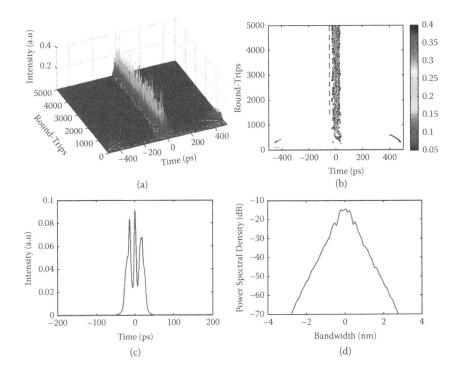

(a)

(b)

(c)

(d)

FIGURE 4.15
(a) Simulated evolution of multibound soliton in unstable condition over 5000 round-trips. (b) Contour plot view of the evolution. (c) The waveform, and (d) corresponding spectra averaged over last 2000 round-trips.

4.5 Relative Phase Difference of Multibound Solitons

4.5.1 Interferometer Measurement and Experimental Setup

The relative phase difference plays a key factor in stability as well as the determination of various modes of the bound soliton states. For passive mode locking, although the relative phase between the bound solitons of π has been confirmed in some experimental works [8,11], other relative phases have also been demonstrated [6,8,10]. The relative phase difference is of importance in the determination of dynamics of the bound solitons in the fiber laser system. In particular, the bound solitons with a $\pi/2$ phase difference can collide either elastically or inelastically with a single soliton depending on its initial phase [9]. On the other hand, the bound solitons can change their relative phase when the setting of the cavity changes.

Different from passive mode locking, only the bound solitons with a π phase difference are stably generated in the active mode-locked fiber laser

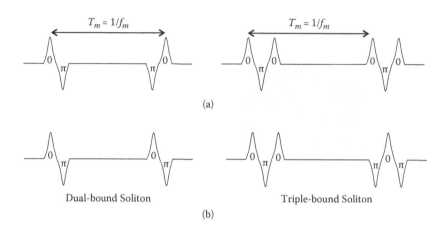

FIGURE 4.16
Descriptions of two possibilities of phase difference between neighboring pulses. (a) Solitons between neighboring bunches are in phase. (b) Solitons between neighboring bunches are out of phase.

system, which has been confirmed by their symmetrically modulated spectra. In each bunch of bound solitons, adjacent pulses are out of phase to form a stable bound state through the balanced interplay of effective interactions. However, if the multibound soliton is considered as a unit like the single pulse state, there might be two possibilities of the multibound soliton trains: one in which solitons between neighboring bunches are in phase and another in which solitons between neighboring bunches are out of phase as described in Figure 4.16. On the other hand, there may be a phase inversion between bunches of bound solitons. To check the dynamics of the relative phase difference in a multibound soliton train, which is impossible to be identified through the optical spectrum measurement, an interferometer measurement has been proposed and implemented. Figure 4.17 shows the schematic of the measurement based on an asymmetrical Mach–Zehnder

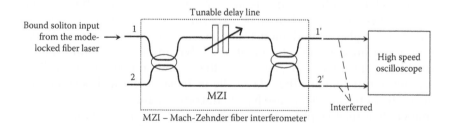

FIGURE 4.17
Experimental setup and principle of the interferometer measurement.

fiber interferometer (MZI). The intensities of the interference patterns at two output ports of the asymmetrical MZI can be given by

$$I^{1'} = \left|E_{out}^{1'}\right|^2 = \left|\tfrac{1}{2} j^2 E_{in}(t) + \tfrac{1}{2} E_{in}(t - T_d)\right|^2 \tag{4.6}$$

$$I^{2'} = \left|E_{out}^{2'}\right|^2 = \left|\tfrac{1}{2} j E_{in}(t) + \tfrac{1}{2} j E_{in}(t - T_d)\right|^2 \tag{4.7}$$

where E_{in} is the input field of the MZI, which is the field of the multibound soliton train, and T_d is the variable time delay between two arms. Depending on adjustment of the time delay, the pulses of two multibound solitons on two arms would be interfered constructively or destructively over the overlapped positions at the output coupler.

Thus, if there is no dynamic phase inversion between bunches of bound solitons or the phase difference of $\pm\pi$ between multibound solitons remains unchanged, the interference patterns of two outputs would be fixed and contrary to each other when the multibound solitons between two arms are overlapped over one or two pulses. In case of the triple-bound solitons as an example, the calculated patterns of constructive and destructive interferences with two overlapped pulses at the outputs are shown in Figure 4.18. On the contrary, the interference patterns are dynamically varied between two output ports depending on the phase states of multibound solitons between two arms.

In the experimental setup as shown in Figure 4.17, the asymmetric fiber interferometer is built by two 3 dB couplers connected together by optical fibers to form two arms of MZI. A tunable delay line of 80 ps delay time is inserted into an arm of MZI to sufficiently provide an overlapping of multibound solitons (MBS) between two arms at the output coupler. The amplitude of overlapped pulses at the output ports (1′ and 2′) of MZI depends

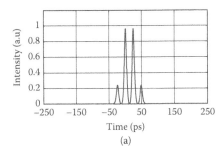
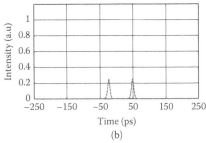

FIGURE 4.18
Calculated patterns of the triple-bound solitons overlapped over two pulses at two outputs of Mach–Zehnder fiber interferometer. (a) Constructive interference. (b) Destructive interference.

on the phase difference of solitons in bound states. This determines either constructive or destructive interferences in the time domain at overlapped positions. The interference patterns at two output ports are simultaneously monitored by two ports on the high-speed oscilloscope.

4.5.2 Results and Discussion

The initial phase delay between two arms of the MZI is adjusted by the fiber length difference between two arms. To put it another way, the time difference between multibound solitons in two arms is initially about 75 ps as shown in Figure 4.19a. By adjusting the tunable delay line, a specific overlapping between two triple-bound solitons can be achieved. Figures 4.19b and 4.19c show, as an example, the interference patterns of triple-bound soliton state over two overlapped pulses at ports 1′ and 2′, respectively. The results also indicate that the practical value of phase difference is not equal to π due to the peak power of overlapped pulses at port 1′ and port 2′, which is only three times higher and lower than that of input pulses, respectively. However, the interference patterns are not steady but alternatively changed between two ports. The alternating change of the patterns between two ports indicates that a phase inversion periodically occurs. For a single-pulse scheme in active mode locking, only the pulse train with

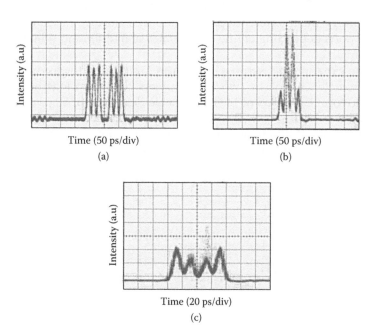

FIGURE 4.19
The time traces of triple-bound solitons (a) before overlapped, (b) overlapped at port 1′, and (c) overlapped at port 2′ of the Mach–Zehnder fiber interferometer.

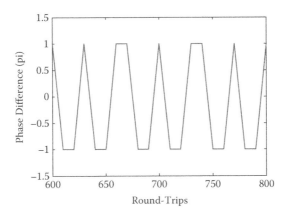

FIGURE 4.20
Periodic variation of the phase difference of a triple-bound soliton over 200 round-trips that is simulated in Figure 4.14b.

neighboring pulses out of phase is stable in the cavity [52]. Therefore, in multibound states, it is understandable when solitons are out of phase not only in the bunch but also between the bunches. Because there is a phase shift in each round-trip that is accumulated during circulation in the cavity, the phase inversion occurs after the accumulated phase shift is an integer multiple of 2π. This is also confirmed by simulation results as shown in Figure 4.20 to indicate a periodic variation of the phase difference between $-\pi$ and π.

4.6 Remarks

In the above sections, we reviewed the important mechanisms as well as interactions in bound soliton formation of mode-locked fiber lasers. Formation of multibound solitons in active mode locking has also been explained to show the role of phase modulation in the balanced interplay between interactions of multibound solitons.

The generation of stable multisoliton bound states in an FM mode-locked fiber laser has been experimentally and numerically demonstrated. We demonstrated that it is possible to generate the bound states from dual to sextuple states provided that there is sufficient optical energy circulating in the fiber ring. Multibound soliton states in a phase-modulated fiber ring are rigorously explained based on the phase matching and chirping effects of the lightwave and the velocity variation of the optical pulses in a dispersive fiber ring. Experimental and numerical results have confirmed the stable existence of multibound solitons with a phase difference of $\pm\pi$ between neighboring solitons. However, it is more sensitive to the cavity settings as well as the external perturbations at the higher-order multibound soliton states.

4.7 Effect of Phase Modulation on Multibinding of Soliton Pulses

4.7.1 Electro-Optic Phase Modulators

The integrated-optic modulators including the electro-optic (EO) phase modulators have become essential components in optical transmission systems and photonic signal processing because of their advantages, such as small size and compatibility with a single-mode optical fiber [53]. For an active mode-locked fiber laser, an EO phase modulator acts as a mode-locker, hence its characteristics significantly influence the performance as well as the operation modes of the fiber cavity. In our experimental platform we used two models of integrated EO phase modulators: one is the model PM-315P of Crystal Technology of Sunnyvale, California, United States and another is the model Mach-427 of Covega of Jessup, Maryland, United States. Depending on the geometry of the electrodes, an integrated $LiNbO_3$ modulator can belong to one of two types: lumped-type modulator or traveling-wave-type modulator.

4.7.1.1 Lumped-Type Modulator

A $LiNbO_3$ phase modulator consists of a titanium-diffused mono-mode waveguide and a pair of oriented electrodes. Titanium waveguides can support both transverse electric (TE) and transverse magnetic (TM) optical polarizations. Due to the crystal symmetry in $LiNbO_3$, there are two useful crystal orientations, Z-cut and X-cut, which take advantage of the strongest EO coefficient. For an X-cut device, the waveguide is symmetrically located between two electrodes as shown in Figure 4.21. In a Z-cut device with the optical axis perpendicular to the surface, one of the electrodes is

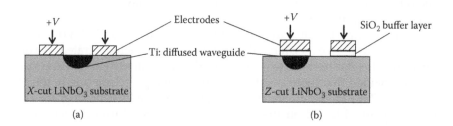

FIGURE 4.21
Geometry of $LiNbO_3$ phase modulators: (a) X-cut and (b) Z-cut orientation substrates.

directly placed on the waveguide, and an optical isolation layer is inserted between the waveguide and the electrode to avoid increased optical losses. Application of the electrical driving voltage to the electrodes causes a small change in the refractive index of the waveguide, and the phase of the optical signal passing through the modulator can be consequently changed as

$$\Delta\varphi = \pi \frac{V}{V_\pi} = \frac{\pi L}{\lambda} r_{eff} n_{eff}^3 \frac{V}{d} \Gamma \qquad (4.8)$$

where V is the applied voltage, L is the length of the electrode, λ is the wavelength, r_{eff} is the appropriate electro-optic coefficient, n_{eff} is the unperturbed refractive index, d is the separation between electrodes, and Γ is the overlap factor between the electric field and the optical mode field, which is an important parameter to optimize the modulator design.

Depending on the type of electrode, the EO phase modulators can be classified into two types: the lumped-type modulator and the traveling-wave modulator. A lumped-type EO phase modulator is illustrated in Figure 4.22.

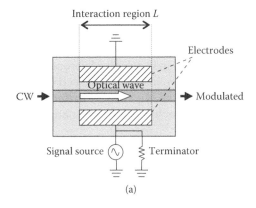

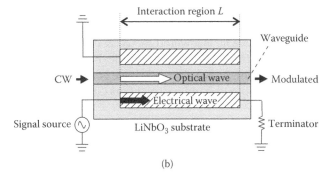

FIGURE 4.22
Two types of electro-optic phase modulator: (a) lumped-type, (b) traveling-wave type.

In this configuration, the RF driving voltage is directly fed to the electrodes whose lengths are small compared to the drive-signal wavelength, and the modulation bandwidth is limited by the *RC* time constant of the electrode capacitance (*C*) and the parallel matching resistance (*R*) as follows:

$$f_{3dB} = \frac{1}{\pi RC} \tag{4.9}$$

The matching resistance is normally set to 50 Ω to allow broadband matching to a 50 Ω driving source. For low operating voltage, the modulator of longer interactive lengths is required, but the capacitance of the electrode increases, which limits the maximum frequency of operation. It is difficult to fabricate a lumped-type EO phase modulator with a broad bandwidth and low operating voltage. Lumped-type EO phase modulators can operate at a frequency of a few GHz with the expense of low V_π.

4.7.1.2 Traveling-Wave Modulator

In order to increase the modulation bandwidth, traveling-wave electrodes are preferably used in the modulator design. Traveling-wave electrodes are designed as transmission lines, fed at one end, and terminated with a resistive load at the other end as shown in Figure 4.22b. The optical signal in the waveguide and the electrical signal in the electrode propagate in the same direction. Because the effective refractive index of the electrical wave n_m is about twice that of the optical wave n_o, there is a velocity mismatch between the electrical wave and the optical wave that limits the modulation bandwidth as follows:

$$f_{3dB} = \frac{1.4c}{[L(n_o - n_m)]} \tag{4.10}$$

where *c* is the velocity of light in a vacuum, and *L* is the length of the electrode.

Much effort has been put into designing traveling-wave modulators to match the velocities of the electrical and optical waves, and thus to improve the frequency response characteristics. To obtain the velocity matching, the effective index of the electric wave n_m is lowered by using effectively a dielectric buffer layer between the electrode and the waveguide. Based on this concept, many structures of the traveling-wave electrode have been proposed to produce the EO modulators with a broad bandwidth of tens GHz and low V_π [54–56].

4.7.2 Characterization Measurements

4.7.2.1 Half-Wave Voltage

In some options for half-wave voltage (V_π) measurement, the direct optical spectrum analysis offers an accurate and simple solution in case the modulation sidebands can be resolved by the optical spectrum analyzer (OSA).

When a CW light passes through a phase modulator, its spectrum is broadened by generation of modulation sidebands. The strength of sidebands is proportional to the modulation index and the sideband positions. An optical field in the frequency domain of a phase-modulated CW signal can be mathematically expressed by Fourier expansion as

$$E_o(\omega) = E_i \exp(j\omega_0 t) \sum_{k=-\infty}^{\infty} (j)^k J_k(m) \exp(jk\omega_m t) \qquad (4.11)$$

where ω_0, ω_m are the carrier frequency and RF modulation frequency, respectively; $\omega = \omega_0 + k\omega_m$; k is the sidemode index that is an integer; J_k is the kth-order Bessel function of first kind; and E_o, E_i are the output and input fields of the phase modulator, respectively. Thus, the intensity of each sidemode measured on an OSA is

$$I_o(\omega_0 + k\omega_m) = I_i J_k^2(m) \qquad (4.12)$$

where m is the modulation index and related to the V_π by (4.8) and $I = |E|^2$. Expression (4.12) shows a relationship between the modulation index and the intensity at a specific sideband that varies as a Bessel function as shown in Figure 4.23. Based on optical spectrum analysis, there are some approaches for the V_π measurement such as the carrier nulling method [57] and the relative sideband/carrier intensity ratio method [58]. In the power limitation of RF amplifiers, the relative first sideband/carrier intensity ratio is the most suitable method for small signal modulation that is used for our measurement.

The measurement setup and the components are depicted in Figure 4.24. The phase modulators consisting of the model PM-315P and the model

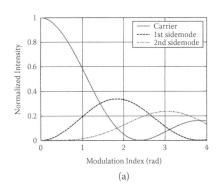
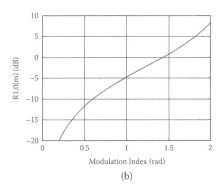

(a)
(b)

FIGURE 4.23
Variation of (a) normalized optical intensity for carrier and sidemodes, and (b) ratio $R_{1,0}$ as a function of phase modulation index m.

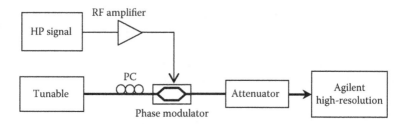

FIGURE 4.24
The OSA-based V_π measurement setup.

Mach427 Covega were characterized by OSA at a modulation frequency of 1 GHz. For a conventional OSA, the resolution is limited to resolve the modulation sidemodes at such a low frequency. Fortunately, we borrowed a high-resolution spectrum analyzer Agilent 83453B for this measurement. By the OSA with resolution of <0.008 pm (~1 MHz), the intensity of sidemodes with spacing of 1 GHz can be clearly displayed as shown in Figure 4.25. At a specific RF driving power the modulation index m is calculated from the relative intensity ratio between the carrier and the first sidemode $R_{1,0}$:

$$R_{1,0}(m) = \frac{I_o(\omega_0 \pm \omega_m)}{I_o(\omega_0)} = \frac{J_1^2(m)}{J_0^2(m)} \tag{4.13}$$

which is measured by the OSA. By varying the RF driving voltages, the V_π of two phase modulators can then be determined from the slope of the linear fit line as shown in Figure 4.26. The estimated V_π of PM-315 and Mach427 are 3.93 V and 8.87 V, respectively. These results agree with the specifications provided by producers.

4.7.2.2 Dynamic Response

The dynamic response of the EO phase modulator is an important characteristic that indicates the variation of modulation efficiency over a range of frequencies. Because optical phase modulators only modulate the phase of the optical carrier, it is impossible to measure directly the modulator response. A conversion of phase modulation into amplitude modulation needs to be done for this measurement. Several techniques were proposed to implement this operation consisting of using a Mach–Zehnder interferometer [57], using a Fabry–Perot interferometer as an optical discriminator [59], and using a Sagnac loop configuration [60]. However, two former techniques are limited in use due to their complexity and reliability, the last technique provides an efficient and simple way of measuring the dynamic response for all types of

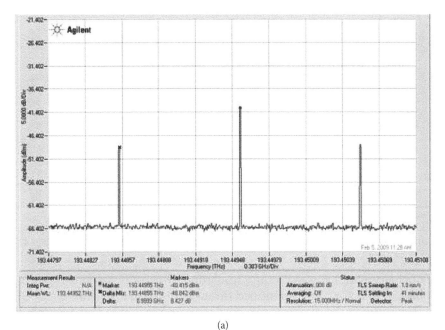

(a)

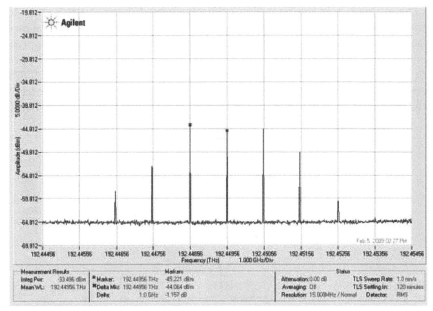

(b)

FIGURE 4.25
Optical spectra of the signal modulated by phase modulators: (a) PM-315P, (b) Mach-427 at the same radio-frequency driving level of 19 dBm.

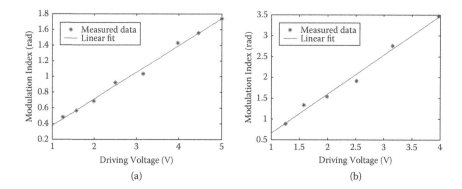

FIGURE 4.26
Phase modulation index calculated from measured relative intensity ratio $R_{1,0}$ at 1 GHz as a function of radio-frequency driving voltage for two models: (a) PM-315P and (b) Mach-427.

EO phase modulators with high resolution. Especially, various types of the phase modulator can be identified through this measurement.

The setup of the dynamic response measurement based on a Sagnac loop is shown in Figure 4.27. This configuration is similar to the configuration of phase modulation Sagnac loop (PMSL) described in Chapter 2. The only main difference is the position of the phase modulator in the Sagnac loop that is off-centered. Two counterpropagating optical waves are phase modulated and then coherently summed at the output 3 dB coupler. This process converts phase modulation into intensity modulation, so that the response of the phase modulator can be measured by a network analyzer HP8753D and an S-parameter test set HP85046A. The polarization is optimized by the polarization controller to maximize the optical power passing through

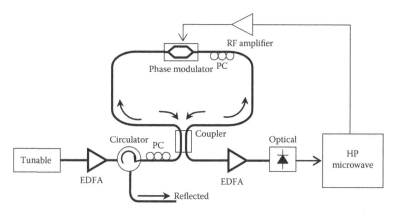

FIGURE 4.27
The measurement setup of the dynamic response.

the lithium niobate optical waveguide. Thus, the transfer function for this structure, which can be detected by the network analyzer, is given by [60]

$$H(f) = \frac{1}{2} l_{PM}^2 K_0(f) \Re[(1 + \eta^2(f)) - 2\eta(f)\cos\phi(f)]R_0 \qquad (4.14)$$

where l_{PM} is the insertion loss of phase modulator, $K_0(f)$ is the modulation response parameter, R_0 is the load resistance, $\Re$ is the detector responsivity, $\phi(f)$ is the phase difference, and $\eta(f)$ is the ratio of backward-to-forward phase modulation index that relates to the signal transit time t_L [61] as follows:

$$\eta(f) = \frac{\sin(2\pi f \tau_L)}{2\pi f \tau_L} \qquad (4.15)$$

This structure is also a notch filter applied in photonic signal processing [62], so that the dynamic response of the phase modulator should be the envelope of the periodic notch filter response. The free spectral range (FSR) of the notch response relates to the phase difference by

$$\phi(f) = 2\pi f \tau \quad \text{and} \quad \tau = \frac{1}{FSR_{notch}} = \frac{\Delta L}{c/n} \qquad (4.16)$$

where τ, ΔL are the time delay and the length difference between two sides of the Sagnac loop, respectively. In order to measure accurately the dynamic response of the modulator, FSR of the notch response must be as small as possible to give a high resolution, which is the reason why the phase modulator must be located off-center of the loop. In our setup, the length difference ΔL is about 3.5 m that gives a FSR of 58 MHz.

With a frequency range up to 3 GHz, the network analyzer may not cover the whole bandwidth of the phase modulators. However, this measurement allows an identification of different types of the phase modulator. Figure 4.28 shows the dynamics responses for two phase modulators and indicates the difference between them. For a lumped-type modulator such as PM-315P, the ratio η keeps unchanged over the bandwidth of the phase modulator, so the notch response shows deep notches and a flat passband within the bandwidth as shown in Figure 4.28a. The envelope of this response gives exactly the dynamics response of the phase modulator with the measured 3 dB bandwidth of 2.7 GHz. For a traveling-wave modulator, the velocity mismatch effect causes the notch depth to disappear at certain frequencies f_k, which are related to the transit time τ_L as follows [61]:

$$f_k = \frac{k}{2\tau_L} \quad \text{with } k = 1, 2, \dots \qquad (4.17)$$

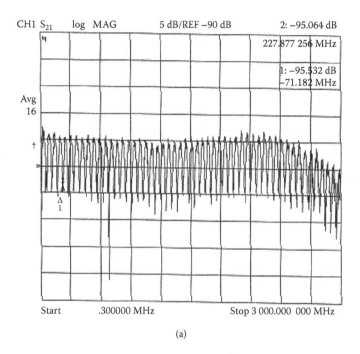

(a)

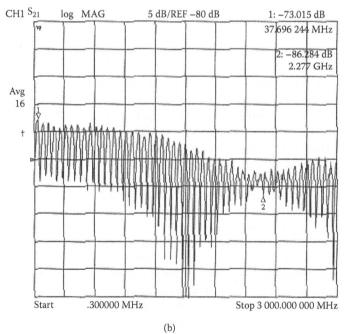

(b)

FIGURE 4.28
Measured frequency notch response of the phase modulators (a) lumped-type PM-315P, (b) traveling-wave-type Mach-427 within measured frequency range up to 3 GHz.

The measured notch response of the phase modulator Mach-427 with the null frequency f_1 of 2.27GHz is shown in Figure 4.28 that indicates a response of the traveling-wave modulator. And the net dynamic response of the modulator can be also obtained by a correction of the envelope of the measured notch response from the known function $\eta(f)$. More importantly, the interactive length of the electrode in the traveling-wave modulator can be estimated from the transit time τ_L. For the Mach-427 modulator, the interactive length about 32 mm is calculated from the transit time of 0.455 ns. We note that the length of waveguide in this modulator would be much longer than that of the electrode due to a polarizer integrated in the phase modulator [63].

Thus, from the characterization measurements of two phase modulators, the difference between the lumped-type modulator and traveling-wave modulator has been indicated. For the modulator PM-315P, a lumped type, the short length of electrodes has been verified by its high V_π and flat frequency response over the measurement bandwidth. Similarly, for the modulator Mach-427, a traveling-wave type, long electrodes, and waveguide have been verified by its low V_π and the velocity mismatch in frequency response measurement

4.7.3 Comb Spectrum in Active Fiber Ring Resonator Using Phase Modulator

4.7.3.1 Birefringence and Comb Spectrum in the Fiber Ring Using Phase Modulator

Because the integrated EO modulators are normally polarizing elements, the birefringence effect always exists in an active fiber ring laser, even if the birefringence of other components such as optical fibers can be ignored. In fact, a Ti:diffused electro-optic $LiNbO_3$ phase modulator (PM) can support both TE and TM modes propagating at different ordinary and extraordinary effective indices. Although the polarization state is usually controlled by the polarization controller, especially at the input of an integrated EO modulator so as only one polarized state can be preferred through the modulator and hence the forcing of the matching condition of this polarized state, there is still an asymmetrically simultaneous existence of two polarization modes in the ring, especially in the Ti:diffused waveguide. Thus, these modes with different phase delays can couple and interfere at the output of the modulator to form an artificial birefringence filter known as a Lyot filter in the ring cavity. The output spectral response of the ring cavity is similar to that of a Mach–Zehnder interferometer, the all-pass filter with nulls and maxima [64,65]. The output transmittance of the ring cavity using the EO phase modulator is simply given by [64]

$$T = \Gamma_g \cos^2(\Delta\phi/2) \tag{4.18}$$

where Γ_g is the insertion loss of waveguide depending on the input polarization states and the gain bandwidth of the EDFA, $\Delta\phi$ - the effective phase difference between TE and TM modes. Because no polarization maintaining (PM) fiber is used, the effective phase difference in the cavity is dominated by birefringence of the Ti:diffused waveguide. Therefore, the phase difference is given as

$$\Delta\phi = \frac{2\pi l \Delta n_{eff}}{\lambda} \tag{4.19}$$

where l is the waveguide length, and Δn_{eff} is the effective index difference between TE and TM modes. The interference between two polarization modes generates a comb-like spectrum with spacing or free spectral range (FSR):

$$\delta\lambda = \lambda^2 / l \Delta n_{eff} \tag{4.20}$$

On the other hand, the gain spectrum in the fiber cavity is modulated when the waves propagate through the phase modulator.

With two models of phase modulator used in the experiment, the comb-like optical spectra were measured by the appropriate setting to display in an OSA. Figure 4.29 shows the optical spectra in CW operation mode of the ring cavities using models PM-315P and Mach-427, respectively. The emission wavelength is located on one of the maxima of the comb-like response where the gain is maximized. For the ring using model PM-315P, the average FSR is 2.15 nm in a bandwidth of about 12 nm, while the average FSR of the ring using model Mach-4027 is only 0.45 nm in a bandwidth of about 10 nm as observed in Figure 4.29. Based on the characterization and the physical dimensions of two of these models, the waveguide lengths of PM-315P and Mach-427 are assumed to be 16 mm

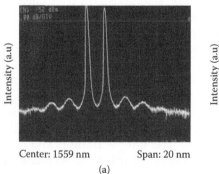
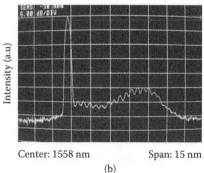

Center: 1559 nm   Span: 20 nm   Center: 1558 nm   Span: 15 nm

(a)   (b)

FIGURE 4.29
The comb-like optical spectra in two setups of the fiber ring laser using (a) model PM-315P and (b) model Mach-427, respectively.

and 64 mm, respectively. Thus, the FSRs of the comb-like spectral response in cases of two models, calculated by Equation (4.20), are 2.0254 nm and 0.4668 nm, respectively, when the effective index difference of TE and TM modes of 0.08 is used [66]. We examined birefringence of the ring cavity in both cases by changing the optical fibers of different lengths. The FSRs remain unchanged in all cases, proving that the birefringence is mainly determined by the Ti:diffused waveguide. These results are reasonable and agree with the results obtained from OSA. The existence of parasitic birefringence in the integrated EO phase modulators forms naturally a comb-like filter that affects mode-locking schemes and characteristics of the mode-locked pulse train.

4.7.3.2 Discrete Wavelength Tuning

One of the influences of the comb-like filtering effect is a discrete wavelength tuning in the FM mode-locked fiber ring laser. In our mode-locked fiber lasers, the mode locking happens only at the peaks of the comb-like optical spectrum where the lightwaves acquire sufficient energy gain to satisfy the mode-locking condition. The wavelength of the sequence of the mode-locked pulses can be tuned only by changing the modulation frequency of the phase modulator. Thus, the wavelength tuning is based on the matching of the dispersed spectrum of the pulse sequence, hence tuning of the pulse central passband [67,68]. The tuning range of the laser depends on the profile of gain spectrum and the total dispersion of the cavity. Because the gain spectrum is modulated by the interference between two polarization modes, the modulation frequency tuning is required by the amount of [67]

$$\delta f_m = -\frac{f_m^2 D L_{cav} \delta \lambda}{m} \tag{4.21}$$

where f_m is the modulation frequency, m is the harmonic order, and D and L_{cav} are the average dispersion and the total length of the ring cavity, respectively. We note that the tuning is only achieved in a specific range of wavelengths to prevent the competition of the CW modes, which suppress the mode locking at other matched wavelengths although the comb-like spectrum can be observed in most of the phase modulators.

In the setup using the phase modulator Mach-4027 with an integrated polarizer at its output, mode locking is achieved at a modulation frequency of about 1 GHz. A 100-m long Corning SMF-28 fiber is inserted after the PM to ensure that the average dispersion in the loop is anomalous. The total loop length is 190 m corresponding to a mode spacing of 1.075 MHz. By tuning the modulation frequency, the lightwaves in the cavity are mode locked to generate a short pulse sequence at the wavelength corresponding to one of the peaks of the comb spectrum as shown in Figure 4.29. A 0.46 nm average spacing between two adjacent mode-locked wavelengths is achieved. By adjusting the PC to optimize the gain spectrum, we can tune over 18

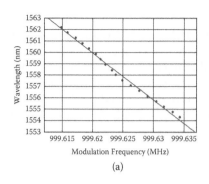
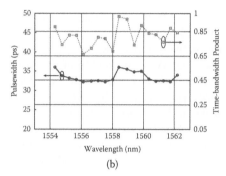

(a)   (b)

FIGURE 4.30
(a) The mode-locked wavelength versus modulation frequency and (b) measured characteristics of mode-locked pulse over the tuning range.

different wavelengths (1554 to 1562 nm) by simply changing the modulation frequency of 1.35 kHz as shown in Figure 4.30a. These measured parameters agree with the theoretically predicted values, which are of 0.4578 nm and 1.27 kHz, respectively. Figure 4.30b shows the pulsewidth and bandwidth-time (BT) product of a mode-locked pulse at 18 wavelengths. The average BT of 0.8 indicates that the output pulses are highly chirped, and they can be compressed by using a suitable dispersive fiber at the output of the laser. In wavelength tunable harmonically mode-locked fiber laser, the supermode noise is an important issue that requires some methods of suppression to improve performance of the generated pulse train [69]. It is worth noting that the mode-locked pulses at all 18 wavelengths are very clean, which is seen from the time trace and its corresponding RF spectrum shown in Figure 4.31. When the f_m is tuned by an amount of $\delta f_m/2$, the mode competition occurs

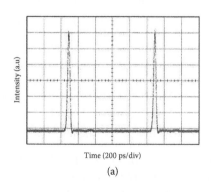
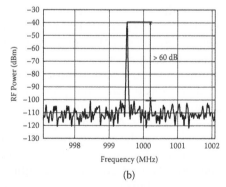

(a)   (b)

FIGURE 4.31
An example of (a) the time trace and (b) radio-frequency spectrum of the mode-locked pulse sequence at one of tuned wavelengths.

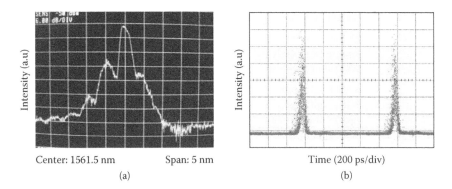

FIGURE 4.32
(a) Optical spectrum and (b) time trace of the output when modulation frequency f_m is tuned by $\pm df_m/2$.

strongly, which degrades the waveform of the output as shown in Figure 4.32. On the other hand, this is the transition state for switching to the adjacent spectral peak, and the mode locking cannot be totally achieved. When the wavelength is tuned to the edge of gain spectrum, multiwavelength operation of the laser can be easily excited, and it also affects the performance of mode-locked pulses as shown in Figure 4.33.

Similarly, a wavelength tuning operation is also observed in the setup using the phase modulator PM-315P, although it is more difficult in adjustment of the polarization controller to obtain a sufficient gain bandwidth covering the tuning range of wavelength because of a wide spacing between. Only three wavelengths at the peaks of the comb-like spectrum are tuned with changing the modulation frequency of 3.1 kHz in this setup to obtain a high-quality pulse train with characteristics summarized in Table 4.2.

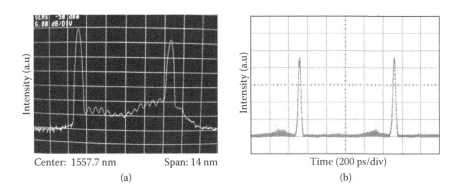

FIGURE 4.33
(a) Optical spectrum and (b) time trace of mode-locked pulse when the wavelength is tuned at the edge of gain spectrum.

TABLE 4.2

Pulse Characteristics at Tuned Wavelengths of the
Frequency Modulation Mode-Locked Laser Using the
Modulator PM-315P

Wavelengths (nm)	1555.98	1558.03	1560.14
Pulse width (ps)	11.9	13	11.3
Spectral width (nm)	0.237	0.24	0.26
BT product	0.3525	0.405	0.3672

4.7.4 Influence of Phase Modulator on Multibound Solitons

4.7.4.1 Formation of Multibound Solitons

The existence of comb-like filtering effects in the cavity using an EO phase
modulator influences remarkably the characteristics of generated pulses. The
artificial filter based on the birefringence of the cavity limits the gain bandwidth and contributes to the mechanism of pulse broadening due to the
limitation of the pulse spectrum. This effect has been experimentally demonstrated by the results of pulse width obtained from two setups of the mode-
locked fiber lasers using the modulators PM-315P and Mach-427. Although the
modulator Mach-427 is driven at a much higher modulation index compared
to the modulator PM-315P because of lower V_π, the pulse generated from the
setup using Mach-427 modulator is more than twice as wide as that generated from the setup using the PM-315P modulator. With a much smaller width
due to broader gain bandwidth, the pulse generated in the case of PM-315P
has a high peak power to create a sufficiently nonlinear phase shift that is
necessary for pulse splitting in the cavity. Moreover, the effective interactions
between pulses in bound states also depend on the pulse width through the
overlap of long pulse tails and the chirping caused by phase modulation. In
the case of PM-315P, the chirp imposed on the generated pulses with sufficient
small width allows an effective attraction to balance the repulsion to generate
a stable bound state. In contrast, a distinct bound state cannot be observed in
the case of Mach-427 with wider-generated pulses as mentioned in Chapter 3.
In this case wider pulses obtain higher chirp from the phase modulator, which
results in a stronger attraction. At once the repulsion force from direct interaction becomes weaker. Therefore, a balanced interaction cannot be achieved in
this case to generate a bound state with obviously resolved pulses although
the cavity is strongly pumped. A state of mode-locked pulse with two-hump
generated from the setup using Mach-427 has been observed as shown in
Figure 4.34. This result may be a new solution of the fiber laser system, yet the
time separation between two humps less than 20 ps may also indicate a strong
attraction between two adjacent pulses.

Furthermore, the polarization sensitivity of the cavity also increases
because of the inherent birefringence of the waveguide. Under a strong

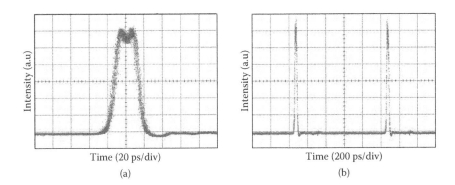

FIGURE 4.34

A mode-locked state generating a two-hump pulse in the cavity using the phase modulator Mach-427. (a) Single pulse trace. (b) Pulse train trace.

pumping scheme, some effects such as spectral hole burning (SHB) and polarization hole burning (PHB) are enhanced in the gain medium, and these make polarization-dependent loss and gain become more complex [70,71]. Inhomogeneous broadening of the gain spectrum associated with SHB and PHB effects [72,73] excites multiwavelength emission in the fiber ring laser, which results in a gain competition between different emission wavelengths. Typically only one of the emission wavelengths satisfies the phase-locking condition in the cavity to generate a pulse train, while other wavelengths may operate in the CW mode or FM mode depending on the polarization setting in the cavity. Thus, if the polarization setting is not optimized, the gain of the mode-locked wavelength can be reduced that results in a switching from higher-order bound solitons to lower-order bound solitons. Figure 4.35 shows a typical example of switching from the

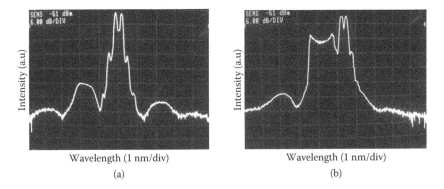

FIGURE 4.35

A switching from the triple-bound soliton into the dual-bound soliton after an adjustment of the polarization controller. (a) Spectrum of triple-bound soliton. (b) Spectrum of dual-bound soliton with an adjacent wavelength in frequency modulation mode.

triple-bound soliton into the dual-bound soliton after the polarization controller is slightly adjusted. The gain reduction of the mode-locked wavelength always accompanies the gain enhancement of adjacent wavelengths as shown in Figure 4.35b. Therefore, it is understandable that the higher-order bound states are more sensitive to polarization settings. However, the polarization-dependent gain effect also provides a flexible mechanism for stabilization of the bound state by adjusting the input polarization state of the phase modulator. If the gain is too high, multibound solitons can fall into unstable states because of the phase fluctuation that leads to a breakdown of interactive balance between pulses. Therefore, a reduction of gain from tuning the polarization state can pull the ring laser back to the stable operation region.

4.7.4.2 *Limitation of Multibound Soliton States*

According to the interaction mechanism of bound solitons in the active mode-locked fiber ring laser, an effective repulsion is induced by linear chirping caused by periodic phase modulation or a balanced interaction between solitons is only achieved around the extreme of modulation cycle where it approximates a parabolic function. However, this approximation is no longer valid if the width of the bunch is so wide. Deviation from linear chirping increases when the number of solitons in bound states increases. Thus, the pulses at the edge of the bunch would experience a nonlinear chirping, rather than linear chirping, which influences the balanced interaction of bound solitons. This explains why the trace of sextuple-bound solitons looks noisy because of oscillation of solitons in the bound state and the spectrum is slightly modulated due to a change in the relative phase difference. It has also been shown that the higher-order multibound solitons are more sensitive to environmental perturbations such as temperature variation. When the fluctuation of cavity length is under a slight detuning, the outer pulses in the multibound state would be most affected by the frequency shifting that breaks the balance in soliton interaction.

The chirping rate determines the separation between bound solitons as described in Chapter 3, and the chirp effect and the number of solitons in the bound states are also correlated. Figure 4.36a depicts the experimental threshold power for the creation of dual- and triple-soliton bound states against the phase modulation index that indicates the increase of required splitting optical power when the chirping is increased. An increase in the chirp rate requires a higher threshold power to maintain a specific number of pulses in the bound state. When the phase modulation index increases,

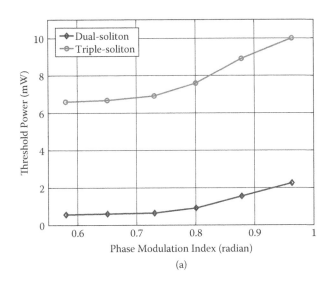

(a)

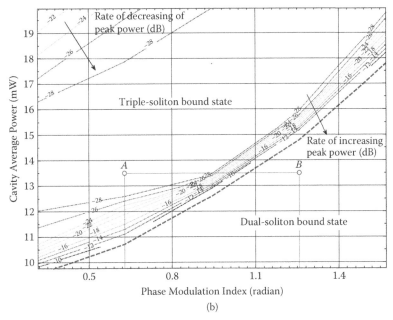

(b)

FIGURE 4.36
(a) Experimentally measured variation of threshold splitting power with the phase modulation index of the bound soliton states. (b) Simulated variation of the peak power of the triple-bound state with the phase modulation index.

the expansion of signal spectrum is enhanced. This is advantageous to not only shortening the pulse width but also more tightly binding the pulses due to the increase in number of phase-locked modes. However, the increase in chirping rate also causes an energy transfer to stronger sidebands that can degrade the soliton content of the pulse sequence. Although the up-chirped pulse is further compressed by anomalous dispersion of the fiber ring at a higher modulation index, the larger frequency chirp at pulse edges also may require higher energy contained in the pulses. This tendency agrees with the theoretical analysis of soliton production from the chirped pulses in [74,75]. At a low phase modulation index, the chirping is almost unaffected by the threshold power due to weak binding of the bound solitons. Consequently, the waveforms of the bound states are noisier and more sensitive to phase fluctuations caused by the ASE noise or the random polarization variation of the guided medium. A simulation result of the power limit of the cavity to maintain the bound state is shown in Figure 4.36b. The rates of increasing and decreasing peak power of bound solitons are indicated showing two distinct regions of operations, the triple and the double bound separated by the dashed bold line. This limit line (dashed bold line) is obtained when a minimum power level circulating in the cavity is required to maintain the bound state that divides the graph into the triple- and double-bound regions. At a specific modulation index, when the average power of the cavity decreases, the variation of the peak power is stronger. When the power is lower than the limit line, the higher-order soliton bound state switches to the lower-order soliton bound state. This tendency also indicates that the increase in chirping diminishes the number of solitons in the bound state at a specific optical power level as shown in Figures 4.37a and 4.37b. These figures show the evolution of the triple-soliton bound state in the cavity at the same average power level but different phase modulation indices that correspond to points A and B, respectively, in Figure 4.36b.

Thus, the phase modulation profile affects significantly the multibound soliton states. In other words, multibound states can be modified by variation of the modulation curve of the phase modulator. Enhancement of higher-order harmonics in the modulation signal is the simplest way to modify the phase modulation profile. Similar to the rational harmonic mode locking, the higher-order harmonics of modulation frequency is strongly enhanced by saturation of the RF power amplifier. When the levels of the second- and the third-harmonics are sufficiently high to distort the phase modulation profile, multiple linear chirps in each modulation cycle can be created. In particular, there are two linear chirps in each modulation cycle created by increase in the input power of the RF amplifier higher than 2 dBm in our setup. In fact a submodulation cycle would appear beside the

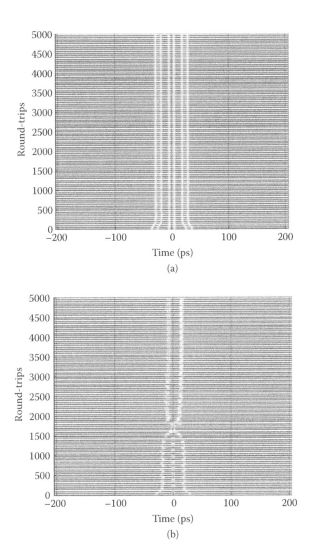

FIGURE 4.37
(a,b) Simulated evolution of the triple-soliton bound state at operation points A and B, respectively, in Figure 4.36b.

main cycle due to the enhancement of the second harmonics in the modulation signal. When the laser is locked in the harmonic mode locking scheme instead of rational harmonic mode locking at the same optical power level of various multibound solitons, the multibound solitons with a smaller number of solitons are formed at two linear chirp positions as shown in

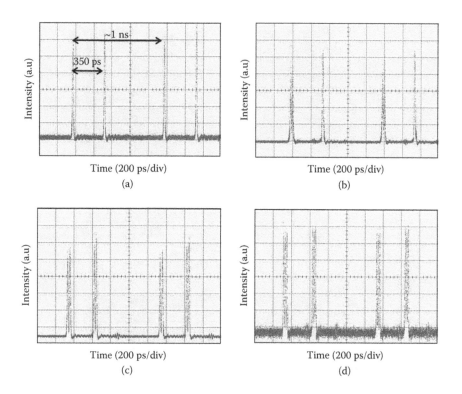

FIGURE 4.38
Splitting of multibound solitons: (a) dual-bound soliton, (b) triple-bound soliton, (c) quadruple-bound soliton, and (d) sextuple-bound soliton, into lower-order bound solitons.

Figure 4.38. The distortion of the phase modulation profile splits higher-order bound solitons into two groups of lower-order bound solitons within the same cycle. Moreover, the change in the higher-order harmonics level also changes the distortion that varies the chirp rate between two groups of solitons. Figure 4.39 shows a variation of time separation between two solitons split from the dual-bound soliton state as a function of the RF input power. Thus, the time separation reduces according to the reduction of the higher harmonic level in the modulation signal. When the RF input power is decreased to 2 dBm, two groups of bound solitons emerge together to form higher-order multibound solitons with the number of solitons equal to the total of solitons in two groups. The emergence of two bound soliton groups is due to the almost disappearance of the submodulation cycle generated by the second harmonic RF signal. It is interesting to show that the change in temporal separation is a complete linear function of the second harmonic RF power as shown in Figure 4.40.

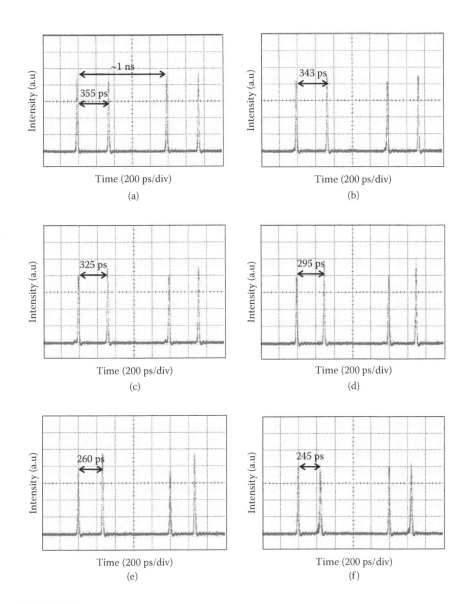

FIGURE 4.39

Time traces shows the variation of time separation between two solitons split from dual-bound soliton versus the change in RF input power: (a) 7 dBm, (b) 6 dBm, (c) 5 dBm, (d) 4 dBm, (e) 3 dBm, and (f) 2.5 dBm.

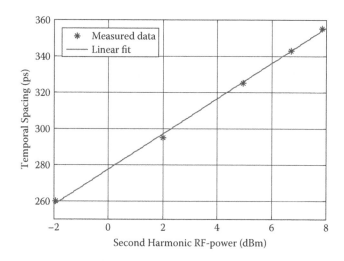

FIGURE 4.40
Correlation between the temporal spacing between two groups of bound soliton and the second harmonic radio-frequency power.

4.8 Remarks

In this section, we demonstrated the influence of the EO phase modulator on multibound solitons. The influence is reflected through two aspects of the phase modulator that consists of the inherent birefringence in the Ti:LiNbO$_3$ waveguide and the phase modulation profile or chirp rate caused by the modulation signal. Two typical phase modulators have been described and characterized to indicate a difference in structure between the lumped type and the traveling-wave type that influences operation modes of the mode-locked fiber laser. Because an artificial comb-like response can be formed in the ring cavity by the birefringence of the integrated phase modulators, a narrow FSR in the response for the phase modulator with a long waveguide can limit the pulse shortening and multibound soliton formation. Moreover, the ability of discrete wavelength tuning can be implemented in the fiber laser through simple tuning of the modulation frequency.

Mode-locking and multibound soliton operation in the active FM-MLFRL is supported by linear chirping that can be approximated by a sinusoidal modulation. Besides the temporal separation of adjacent pulses, the optical power threshold for splitting the solitons and the number of solitons in the bound states are influenced by the chirp rate induced by this phase modulation. When the modulation profile is modified by the enhancement of the higher-order harmonic in the driving modulation signal, higher-order multibound solitons can be split into lower-order multibound solitons with controllable temporal separation between them. More importantly, the experimental results have demonstrated the ability of control through the active device such as an EO phase modulator that is important in potential applications.

References

1. B. A. Malomed, Bound Solitons in the Nonlinear Schrödinger-Ginzburg-Landau equation, *Phys. Rev. A*, 44, 6954, 1991.
2. B. A. Malomed, Bound Solitons in Coupled Nonlinear Schrödinger Equations, *Phys. Rev. A*, 45, R8321, 1992.
3. N. N. Akhmediev et al., Multisoliton Solutions of the Complex Ginzburg-Landau Equation, *Phys. Rev. Lett.*, 79, 4047, 1997.
4. D. Y. Tang et al., Observation of Bound States of Solitons in a Passively Mode-Locked Fiber Laser, *Phys. Rev. A*, 64, 033814, 2001.
5. N. H. Seong and D. Y. Kim, Experimental Observation of Stable Bound Solitons in a Figure-Eight Fiber Laser, *Opt. Lett.*, 27, 1321–1323, 2002.
6. P. Grelu et al., Phase-Locked Soliton Pairs in a Stretched-Pulse Fiber Laser, *Opt. Lett.*, 27, 966–968, 2002.
7. D. Y. Tang et al., Bound-Soliton Fiber Laser, *Phys. Rev. A*, 66, 033806, 2002.
8. Y. D. Gong et al., Close Spaced Ultra-short Bound Solitons from DI-NOLM Figure 8 Fiber Laser, *Opt. Commun.*, 220, 297–302, 2003.
9. P. Grelu et al., Relative Phase Locking of Pulses in a Passively Mode-Locked Fiber Laser, *J. Opt. Soc. Am. B*, 20, 863–870, 2003.
10. D. Y. Tang et al., Compound Pulse Solitons in a Fiber Ring Laser, *Phys. Rev. A*, 68, 013816, 2003.
11. B. Zhao et al., Energy Quantization of Twin-Pulse Solitons in a Passively Mode-Locked Fiber Ring Laser, *Appl. Phys. B: Lasers and Opt.*, 77, 585–588, 2003.
12. Y. D. Gong et al., Regimes of Operation States in Passively Mode-Locked Fiber Soliton Ring Laser, *Opt. and Laser Technol.*, 36, 299–307, 2004.
13. B. Ortac et al., Generation of Bound States of Three Ultrashort Pulses with a Passively Mode-Locked High-Power Yb-Doped Double-Clad Fiber Laser, *Photonics Technol. Lett., IEEE*, 16, 1274–1276, 2004.
14. B. Zhao et al., Passive Harmonic Mode Locking of Twin-Pulse Solitons in an Erbium-Doped Fiber Ring Laser, *Opt. Commun.*, 229, 363–370, 2004.
15. D. Y. Tang et al., Multipulse Bound Solitons with Fixed Pulse Separations Formed by Direct Soliton Interaction, *Appl. Phys. B: Lasers and Opt.*, 80, 239–242, 2005.
16. L. M. Zhao et al., Period-Doubling and Quadrupling of Bound Solitons in a Passively Mode-Locked Fiber Laser, *Opt. Commun.*, 252, 167–172, 2005.
17. M. Grapinet and P. Grelu, Vibrating Soliton Pairs in a Mode-Locked Laser Cavity, *Opt. Lett.*, 31, 2115–2117, 2006.
18. L. M. Zhao et al., Bound States of Dispersion-Managed Solitons in a Fiber Laser at Near Zero Dispersion, *Appl. Opt.*, 46, 4768–4773, 2007.
19. W.-W. Hsiang et al., Stable New Bound Soliton Pairs in a 10 GHz Hybrid Frequency Modulation Mode-Locked Er-Fiber Laser, *Opt. Lett.*, 31, 1627–1629, 2006.

20. L. N. Binh et al., Multi-bound Solitons in a FM Mode-Locked Fiber Laser, in *Conference on Optical Fiber Communication/National Fiber Optic Engineers Conference, 2008. OFC/NFOEC 2008,* 2008, pp. 1–3.
21. G. P. Agrawal, Optical Pulse Propagation in Doped Fiber Amplifiers, *Phys. Rev. A,* 44, 7493, 1991.
22. M. J. Lederer et al., Multipulse Operation of a Ti:Sapphire Laser Mode Locked by an Ion-Implanted Semiconductor Saturable-Absorber Mirror, *J. Opt. Soc. Am. B,* 16, 895–904, 1999.
23. F. X. Kurtner et al., Mode-Locking with Slow and Fast Saturable Absorbers— What's the Difference?, *IEEE J. Selected Topics in Quantum Electron.,* 4, 159–168, 1998.
24. D. Y. Tang et al., Stimulated Soliton Pulse Formation and Its Mechanism in a Passively Mode-Locked Fiber Soliton Laser, *Opt. Commun.,* 165, 189–194, 1999.
25. D. Y. Tang et al., Mechanism of Multisoliton Formation and Soliton Energy Quantization in Passively Mode-Locked Fiber Lasers, *Phys. Rev. A,* 72, 043816, 2005.
26. A. B. Grudinin et al., Energy Quantisation in Figure Eight Fiber Laser, *Electron. Lett.,* 28, 67–68, 1992.
27. G. Martel et al., On the Possibility of Observing Bound Soliton Pairs in a Wave-Breaking-Free Mode-Locked Fiber Laser, *Opt. Lett.,* 32, 343–345, 2007.
28. A. Komarov et al., Ultrahigh-Repetition-Rate Bound-Soliton Harmonic Passive Mode-Locked Fiber Lasers, *Opt. Lett.,* 33, 2254–2256, 2008.
29. J. N. Kutz et al., Stabilized Pulse Spacing in Soliton Lasers due to Gain Depletion and Recovery, *IEEE J. Quantum Electron.,* 34, 1749–1757, 1998.
30. A. N. Pilipetskii et al., Acoustic Effect in Passively Mode-Locked Fiber Ring Lasers, *Opt. Lett.,* 20, pp. 907-909, 1995.
31. W. H. Loh et al., Soliton Interaction in the Presence of a Weak Nonsoliton Component, *Opt. Lett.,* 19, 698–700, 1994.
32. G. P. Agrawal, *Nonlinear Fiber Optics,* 3rd ed. Academic Press, San Diego, 2001.
33. L. Socci and M. Romagnoli, Long-Range Soliton Interactions in Periodically Amplified Fiber Links, *J. Opt. Soc. Am. B,* 16, 12–17, 1999.
34. F. Gutty et al., Stabilisation of Modelocking in Fiber Ring Laser through Pulse Bunching, *Electron. Lett.* 37, 745–746, 2001.
35. D. Y. Tang et al., Soliton Interaction in a Fiber Ring Laser, *Phys. Rev. E,* 72, 016616, 2005.
36. R. P. Davey et al., Interacting Solutions in Erbium Fiber Laser, *Electron. Lett.,* 27, 1257–1259, 1991.
37. D. Krylov et al., Observation of the Breakup of a Prechirped N-Soliton in an Optical Fiber, *Opt. Lett.,* 24, 1191–1193, 1999.
38. S. Longhi, Pulse Dynamics in Actively Mode-Locked Lasers with Frequency Shifting, *Phys. Rev. E,* 66, 056607, 2002.
39. C. R. Doerr et al. Additive-Pulse Limiting, *Opt. Lett.,* 19, 31–33, 1994.
40. S. Wabnitz, Suppression of Soliton Interactions by Phase Modulation, *Electron. Lett.,* 29, 1711–1713, 1993.
41. J. P. Gordon, Interaction forces among solitons in optical fibers, *Opt. Lett.,* 8, 596–598, 1983.
42. Y. Kodama and K. Nozaki, Soliton interaction in optical fibers, *Opt. Lett.,* 12, 1038–1040, 1987.
43. F. M. Mitschke and L. F. Mollenauer, Experimental Observation of Interaction Forces between Solitons in Optical Fibers, *Opt. Lett.,* 12, 355–357, 1987.

44. S. Wabnitz, Suppression of Soliton Interactions by Phase Modulation, *Electron. Lett.*, 29, 1711–1713, 1993.

45. T. Georges and F. Favre, Modulation, Filtering, and Initial Phase Control of Interacting Solitons, *J. Opt. Soc. Am. B*, 10, 1881889, 1993.

46. N. D Nguyen and L. N. Binh, Generation of Bound-Solitons in Actively Phase Modulation Mode-Locked Fiber Ring Resonators, *Opt. Commun.*, 281, 2012–2022, 2008.

47. J. G. Proakis, *Digital Communications*. McGraw-Hill, New York, 2001.

48. L. J. Wang et al., Analysis of Polarization-Dependent Gain in Fiber Amplifiers, *IEEE J. Quantum Electron.*, 34, 413–418, 1998.

49. H. A. Haus et al., Additive-Pulse Modelocking in Fiber Lasers, *IEEE J. Quantum Electron.*, 30, 20208, 1994.

50. T. F. Carruthers et al., Active-Passive Modelocking in a Single Polarization Erbium Fiber Laser, *Electron. Lett.*, 30, 1051–1053, 1994.

51. C. R. Doerr et al., Asynchronous Soliton Mode Locking, *Opt. Lett.*, 19, 1958–1960, 1994.

52. J. J. O'Neil et al., Theory and Simulation of the Dynamics and Stability of Actively Modelocked Lasers, *IEEE J. Quantum Electron.*, 38, 1412–1419, 2002.

53. L. N. Binh, *Photonic Signal Processing: Techniques and Applications*, CRC Press, Boca Raton, FL, 2008.

54. K. Kawano et al., A Wide-Band and Low-Driving-Power Phase Modulator Employing a Ti:LiNbO$_3$ Optical Waveguide at 1.5 mu m, *Photonics Technol. Lett., IEEE*, 1, 33–34, 1989.

55. K. Kawano et al., New Travelling-Wave Electrode Mach-Zehnder Optical Modulator with 20 GHz Bandwidth and 4.7 V Driving Voltage at 1.52 mu m Wavelength, *Electron. Lett.*, 25, 1382–1383, 1989.

56. K. Noguchi et al., A Broadband Ti:LiNbO$_3$ Optical Modulator with a Ridge Structure, *J. Lightwave Technol.*, 13, 1164–1168, 1995.

57. R. Tench et al., Performance Evaluation of Waveguide Phase Modulators for Coherent Systems at 1.3 and 1.5 μm, *J. Lightwave Technol.*, 5, 492–501, 1987.

58. Y. Shi et al., High-Speed Electrooptic Modulator Characterization Using Optical Spectrum Analysis, *J. Lightwave Technol.*, 21, 2358, 2003.

59. R. Regener and W. Sohler, Loss in Low-Finesse Ti:LiNbO$_3$ Optical Waveguide Resonators, *Appl. Phys. B: Lasers and Opt.*, 36, 143–147, 1985.

60. E. Chan and R. A. Minasian, A New Optical Phase Modulator Dynamic Response Measurement Technique, *J. Lightwave Technol.*, 26, 2882–2888, 2008.

61. W. Leeb et al., Measurement of Velocity Mismatch in Traveling-Wave Electrooptic Modulators, *IEEE J. Quantum Electron.*, 18, 14–16, 1982.

62. E. H. W. Chan and R. A. Minasian, Sagnac-Loop-Based Equivalent Negative Tap Photonic Notch Filter, *Photonics Technol. Lett., IEEE*, 17, 1741742, 2005.

63. R. Stubbe et al., Polarization Selective Phase Modulator in LiNbO$_3$, *Photonics Technol. Lett., IEEE*, 2, 187–190, 1990.

64. G. Shabtay et al., Tunable Birefringent Filters—Optimal Iterative Design, *Opt. Express*, 10, 1534–1541, 2002.

65. C. O'Riordan et al., Actively Mode-Locked Multiwavelength Fiber Ring Laser Incorporating a Lyot Filter, Hybrid Gain Medium and Birefringence Compensated LiNbO$_3$ Modulator, in *Ninth International Conference on Transparent Optical Networks, 2007. ICTON '07.* 2007, pp. 248–251.

66. G. J. Sellers and S. Sriram, Manufacturing of Lithium Niobate Integrated Optic Devices, *Opt. News*, 14, 29–31, 1988.

67. S. P. Li and K. T. Chan, Electrical Wavelength Tunable and Multiwavelength Actively Mode-Locked Fiber Ring Laser, *Appl. Phys. Lett.*, 72, 1954–1956, April 1998.

68. Y. Zhao et al., Wavelength Tuning of 1/2-Rational Harmonically Mode-Locked Pulses in a Cavity-Dispersive Fiber Laser, *Appl. Phys. Lett.*, 73, 3483–3485, December 1998.

69. D. Lingze et al., Smoothly Wavelength-Tunable Picosecond Pulse Generation Using a Harmonically Mode-Locked Fiber Ring Laser, *J. Lightwave Technol.*, 21, 93937, 2003.

70. T. Aizawa et al., Effect of Spectral-Hole Burning on Multi-channel EDFA Gain Profile, in *Optical Fiber Communication Conference, 1999, and the International Conference on Integrated Optics and Optical Fiber Communication. OFC/IOOC '99. Technical Digest*, 1999, pp. 102–104, vol. 2.

71. M. Bolshtyansky, Spectral Hole Burning in Erbium-Doped Fiber Amplifiers, *J. Lightwave Technol.*, 21, 1032, 2003.

72. D. Kovsh et al., Gain Reshaping Caused by Spectral Hole Burning in Long EDFA-Based Transmission Links, in *Optical Fiber Communication Conference, 2006 and the 2006 National Fiber Optic Engineers Conference. OFC 2006*, 2006, p. 3.

73. L. Rapp and J. Ferreira, Dynamics of Spectral Hole Burning in EDFAs: Dependency on Pump Wavelength and Pump Power, *Photonics Technol. Lett., IEEE*, 22, 1256–1258, 2010.

74. M. Desaix et al., Propagation Properties of Chirped Soliton Pulses in Optical Nonlinear Kerr Media, *J. Phys. Rev. E*, 65, 2002.

75. J. E. Prilepsky et al., Conversion of a Chirped Gaussian Pulse to a Soliton or a Bound Multisoliton State in Quasi-lossless and Lossy Optical Fiber Spans, *J. Opt. Soc. Am. B*, 24, 1254–1261, 2007.

5

Transmission of Multibound Solitons

Le Nguyen Binh

Hua Wei Technologies, European Research Center, Munich, Germany

CONTENTS

5.1 Introduction

Temporal solitons are attractive for long-haul fiber transmission systems due to their preservation of shape during propagation in nonlinear dispersive medium. In theory, the soliton character only remains unchanged in ideal transmission medium without any perturbation such as loss and noise. Therefore, the understanding of soliton propagation characteristics in real fiber systems is of significance to the design of an optical communication link. Moreover, there is a difference between propagation of solitons in a ring such as a mode-locked fiber laser and a real fiber transmission link. It must be noted that the length of the mode-locked fiber ring is much shorter than the dispersion length of the fiber, L_D, and the pulse sequence is thus operating in the near field region in which the chirp of the lightwave occurs significantly, especially at the edges of the pulses, while the distance of each span in the fiber link is longer than the dispersion length. Therefore, distinct transmission conditions are required to ensure that solitons can be recovered at the end of the link.

Although a number of reports on bound soliton states in mode-locked fiber lasers have been published, to date there is no report on the propagation dynamics of multibound solitons in optical fiber. The difficulty of the generation of a stable bound soliton sequence in a passive mode-locked fiber laser may have prevented the investigation of their propagation and related dynamics. As described in the above chapters, the active frequency modulation (FM) mode-locked fiber laser offers a significant advantage in the generation of an ultrastable multibound soliton sequence at a modulation frequency that would be important for propagation in optical fiber. Dynamics of multibound solitons are an important issue for investigation. The obtained results are significant to possibly potential applications of bound soliton lasers.

5.2 Soliton Propagation in Optical Fibers

5.2.1 Loss Management

Because the group velocity dispersion (GVD) value is unchanged along in optical fibers, the balance between GVD and self-phase modulation (SPM) in soliton transmission would be achieved only if the soliton pulse was not attenuated during propagation. For a lossy fiber in the real transmission system, a reduction of the peak power over distance leads to a soliton broadening. The broadening of soliton can be understood by the reduction of the SPM effect resulting from the reduced peak power that makes the impact of the dispersion effect stronger during propagation. When the fiber loss is assumed as a weak perturbation of the nonlinear Schrödinger equation (NLSE), change in soliton parameters under the influence of the fiber loss has been investigated by using the variation method [1]. Variations of soliton amplitude and phase along the fiber can be given by

$$\eta(\xi) = \exp(-\Gamma\xi) \quad \text{and} \quad \phi(\xi) = \phi(0) + [1 - \exp(-2\Gamma\xi)]/(4\Gamma) \tag{5.1}$$

where $\Gamma = \alpha L_D$ and $\xi = z/L_D$, α is the loss coefficient of the fiber. An exponential decrease in soliton amplitude weakens the SPM effect; hence, the soliton is also broadened with the same manner:

$$T(z) = T_0 \exp(\Gamma\xi) = T_0 \exp(\alpha z) \tag{5.2}$$

where T_0 is the initial width of an unperturbed soliton, T is the width of a soliton at distance z. It notes that a linear increase in pulse width occurs in the linear propagation scheme [2]. When the peak power is considerably attenuated at a long distance, the SPM effect can be negligible and solitons behave like nonsolitary pulses. Therefore, the exponential dependence of soliton width over the distance is only valid at the distance where αz is less than 1.

In order to overcome the broadening problem due to the fiber loss effect, a periodic amplification of soliton is required to recover the soliton energy. A lumped amplification scheme can be used in a fiber link for this purpose [3].

Solitons that propagate in this scheme are called the path-average solitons [4]. Similar to the mode-locked fiber ring system, the adjustment of soliton in the fiber following the amplifier can lead to shedding of a part of soliton energy as dispersive waves that are accumulated to a considerable level over a large number of amplifiers. In order to keep variation of the soliton parameters negligible, two conditions to operate in the average-soliton regime must be satisfied [2,4]:

- The amplifier spacing L_A must be much smaller than the dispersion length L_D ($L_A \ll L_D$). This condition is required to keep radiation of dispersive waves negligibly small because the energy of dispersive waves is inversely proportional to the dispersion length L_D.

- The input peak power of the soliton must be larger than that of the fundamental soliton by a factor

$$\frac{P_s}{P_0} = \frac{G \ln G}{G-1} \tag{5.3}$$

where P_s, P_0 are the peak powers of the path-average soliton and the unperturbed soliton, respectively, G is the amplifier gain factor. This condition is to make certain that the average peak power over the L_A is sufficient to balance the GVD effect.

The soliton energy varies periodically in each span due to the fiber loss, hence other parameters of soliton such as the width and the phase also change accordingly. However, the soliton remains unchanged at ultralong distance if the above conditions are satisfied.

But it is impractical to satisfy the first condition when the existing transmission fibers with L_D of about 10 to 20 km are used for the long-haul transmission system [5,6]. If the L_A is short, there are some disadvantages, such as the high cost and the large accumulated amplification stimulated emission (ASE) noise resulting from a large number of optical amplifiers. To facilitate the first condition, a distributed amplification scheme such as Raman amplification can be employed [7,8]. It is understandable that the variation of peak power is considerably reduced because the amplification process takes place along the transmission fiber. And this scheme allows a transmission with $L_A \gg L_D$. However, an unstable soliton transmission can occur due to the resonance of the dispersive waves and solitons when $L_A/L_D \geq 4\pi$ that need to be carefully considered in the design of soliton transmission system [9].

5.2.2 Dispersion Management

The problem of soliton broadening in lossy fiber can also be overcome by using dispersion-decrease-fiber (DDF) in which its dispersion decreases exponentially corresponding to the reduction of the peak power. From the variation of soliton amplitude in (5.1), the variation of the GVD in DDF follows a function as

$$|\beta_2(z)| = |\beta_2(0)| \exp(-\alpha z) \tag{5.4}$$

Thus, the reduction of the GVD is proportional to the reduction of the SPM effect; hence, solitons can remain unchanged during propagation in the DDF. However, the use of DDF in practical transmission systems is not feasible because of the availability of the DDF as well as the complexity of the system design. Furthermore, the performance would be degraded because the average dispersion along the link is large. Although this scheme is disadvantageous to the system, it is commonly applied for pulse compression based on the soliton effect [10,11].

Therefore, another option is use of dispersion management, which uses alternating positive and negative GVD fibers. This arrangement is commonly employed in high-speed transmission systems today because it offers a relatively improved performance [12,13]. Hence, using a periodic dispersion map along the fiber link has attracted extensive attention for its ability in soliton transmission. Both theoretical and experimental researchers have shown the advantages of dispersion-managed (DM) solitons [12]. Owing to alternating the sign of the fiber dispersion in one map period, the average GVD of the whole link can be kept in a small value while the GVD of each section is large enough to suppress the impairments such as four-wave mixing and third-order dispersion. Two important parameters of the map are the average GVD of the whole link $\bar{\beta}_2$ and the map strength S_m, which are defined as follows [2]:

$$\bar{\beta}_2 = \frac{\beta_{2n}l_n + \beta_{2a}l_a}{l_n + l_a}, \quad S_m = \frac{\beta_{2n}l_n - \beta_{2a}l_a}{t_{FWHM}^2} \tag{5.5}$$

where β_{2n}, β_{2a} are the GVD parameters in the normal and anomalous sections of lengths l_n and l_a, respectively, t_{FWHM} is the full width at half of maximum (FWHM). It has been shown that the shape of DM solitons is closer to a Gaussian pulse rather than a "sech" shape, and their parameters such as the width, peak power, and chirp vary considerably in each map period. Therefore, depending on the map configuration, input parameters of DM solitons should be carefully chosen to ensure that the pulse can recover its state after each map period [14]. In addition, the peak power enhancement of DM soliton compared to the constant-GVD soliton allows an improvement of performance in terms of signal-to-noise ratio while suppressing timing jitter by reducing the average GVD. Interestingly, it has been confirmed that DM solitons exist not only in anomalous average dispersion ($\bar{\beta}_2 < 0$) but also in a normal average dispersion scheme ($\bar{\beta}_2 > 0$) [15]. However, the existence of DM soliton in the map with $\bar{\beta}_2 > 0$ is obtained only when the strength of map S_m is greater than a critical value S_{cr}, which is approximately 4 [16]. Hence, it is not surprising to observe the existence of solitons in the mode-locked fiber laser with normal average dispersion.

Interaction between solitons is an important issue in practice. Because properties of DM solitons are different from the average-GVD solitons, the evolution and dynamics of DM solitons have become more attractive for both theoretical and experimental studies [12,13]. Results of theoretical study have shown that a strong interaction prevents the DM solitons in a positive average GVD map from practical high-speed transmission applications, although they can exist [17]. This also indicates a complexity in the dynamics of DM solitons in propagation, beside a single soliton solution some complex solutions of DM soliton can exist in specific dispersion maps. In one study on a high-order DM soliton it showed a stable evolution of an antisymmetric (antiphase) soliton in the fiber at an ultralong distance [18]. Recently, the dispersion management was proposed to support a transmission of bisoliton, which consists of a couple of DM solitons with a zero-(in-phase) or π-(antiphase) phase difference [19]. By selecting appropriate parameters of the DM map and input pulse, a stable propagation of an in-phase bisoliton for a single channel and an antiphase bisoliton for the multichannel has been confirmed through numerical simulation [19]. In other studies, a structure called soliton molecules consisting of one dark and two bright solitons, which is similar to an asymmetrical soliton, has been numerically and experimentally demonstrated [20–23].

We can see that the soliton complexes in the above studies have a structure similar to pairs of bound solitons generated from the mode-locked fiber lasers. Stable propagation of the soliton complexes can offer new coding schemes for soliton transmission systems. In a conventional coding scheme, data bit "1" or "0" is represented by the presence or absence of a soliton pulse in each time slot. In one proposal, using the soliton complex such as bisoliton for new coding scheme provides code states with a residual bit for error prevention [19]. Therefore, it is necessary to have a new optical source that can generate the complex soliton like a bisoliton in new coding schemes. With the ability of multibound solitons generation, the active FM mode-locked fiber laser offers more states for a coding scheme that allows an improvement of performance and transmission of more than one data bit in each time slot. Hence, propagation characteristics of multibound solitons generated from the actively FM mode-locked fiber laser is of significance for potential applications of multibound soliton lasers in telecommunication systems.

5.3 Transmission of Multibound Solitons

5.3.1 Experimental Setup

To investigate the propagation dynamics of multibound solitons, various multibound solitons with ultrahigh stability from dual to quintuple states are generated by carefully optimizing the locking conditions of the fiber ring.

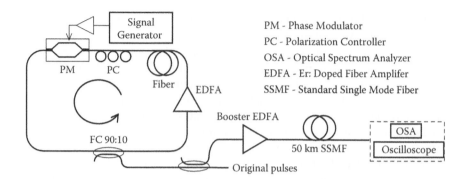

FIGURE 5.1
Experimental setup for propagation of multibound solitons in a standard single fiber.

These states are then propagating through standard single-mode optical fibers (SSMF) in order to investigate their propagation dynamics as shown in Figure 5.1. The length of the fiber is varied to prove the interaction between the soliton pulses. The estimated full width at half of maximum (FWHM) of the individual pulse of original bound soliton pairs (BSPs), triple-bound solitons (TBSs), and quadruple-bound solitons (QBSs) are 7.9, 6.9, and 6.0 ps, respectively. The time separation between two adjacent pulses is about three times of the FWHM pulse width. The repetition rate of these multibound soliton sequences is about 1 GHz.

In this setup, a booster erbium-doped fiber amplifier (EDFA) is used before the transmission section to specify the power of multibound solitons launched into the SSMF. Through adjustment of the launched power, the multibound solitons can propagate under various transmission conditions from linear to nonlinear schemes. Two rolls of standard single-mode fiber with lengths of 1 km and 50 km are used for the investigation. After propagation, the outputs such as time trace and spectrum are monitored by the oscilloscope and the OSA, respectively.

5.3.2 Results and Discussion

It is obvious but worth mentioning that the binding property of the solitons circulating in the fiber ring is different under the case when they are propagating through the optical fiber. In the propagation these individual solitons would be interacting with each other, and naturally they are no longer supported by the periodic phase modulation as when circulating in the ring laser. When propagating through a dispersive fiber, the optical carrier under the envelope of generated multibound solitons is influenced by the chirping effects and then the overlapping between the soliton pulses. When the accumulated phase difference of adjacent pulses is π, the solitons repel each other [24–26]. The solitons in their multibound states acquire the down-chirping effect when propagating in SSMF due to its group velocity dispersion (GVD) induced phase shift. Hence, within the group of multibound solitons the front-end soliton would travel with a higher positive frequency shift, thus a higher group velocity than those at the back end of the multibound group which is influenced with lower negative frequency shift. Therefore, the time separation between adjacent pulses varies with the propagation distance. The variation of the time separation between pulses depends on their relative frequency difference. Figure 5.2 shows the waveforms and their corresponding spectra of

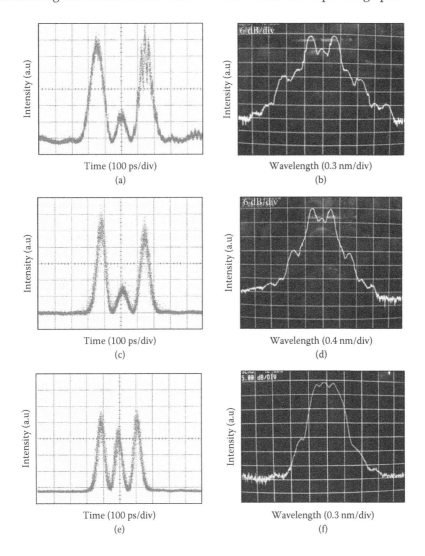

FIGURE 5.2
The time traces and corresponding spectra of the triple-bound soliton at launching powers of (a) through (d) 4.5 dBm, (b) through (e) 10.5 dBm, and © through (f) 14.5 dBm, respectively.

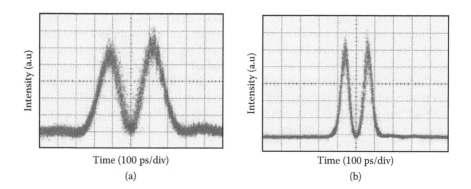

FIGURE 5.3
The time traces of the dual-bound soliton at launching powers of (a) 4.5 dBm, and (b) 17 dBm, respectively, after propagating through 50 km standard single-mode fiber.

triple-bound solitons at the transmitter. Figure 5.3(a) and (b) shows the dual bound solitons after transmission over 50 km of standard single mode fiber with 4.5 and 17 dBm launched power, respectively, after propagating over the fiber.

Figure 5.3 shows the time traces of the dual-bound soliton launched in to the fiber span at launching powers of (a) 4.5 dBm, and (b) 17 dBm, respectively, after propagating through 50 km standard single-mode fiber. We can observe that at higher launched power, the solitons are bound stronger and thus much less broadening of the pulse is achieved. Similarly, a quadruple-bound soliton is launched in the same fiber span of 50 km at launching powers of 5 dBm, and 16.5 dBm. Its envelope is obtained at the span output and displayed in Figure 5.4. A launched power of about 15 dBm would allow a

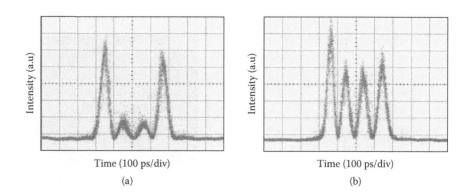

FIGURE 5.4
The time traces of the quadruple-bound soliton at launching powers of (a) 5 dBm, and (b) 16.5 dBm, respectively, after propagating through 50 km SSMF.

near preservation of the quadruple soliton. It is obvious that the higher the order of bound solitons, the higher the optical average power is required so that the binding of individual pulses can be preserved.

In all cases, the time separation significantly increases compared to the initial state due to the repulsion. In the ideal condition of propagation, direct interaction is the only factor that influences the time position and the pulse shape of solitons in bound states. However, propagation in a real optical fiber is considerably influenced by perturbations such as loss of fiber and initial launched powers (P_l) of solitons. The loss of fiber always leads to the broadening of bound solitons. Figures 5.5a and 5.5b show the dependence of pulse width and temporal separation of various multibound solitons with the launched power as a parameter. We observe similar dynamics in the propagation of dual, quadruple, and quintuple groups of solitons. In general, an increase in launched power leads to an enhanced shortening of the pulsewidth or a broadening of bound solitons at a lower rate, and thereby a reduction of the pulse temporal separation due to the enhancement of the nonlinear self-phase-modulation phase shift. Unlike the nearly linear variation of the pulsewidth, variation

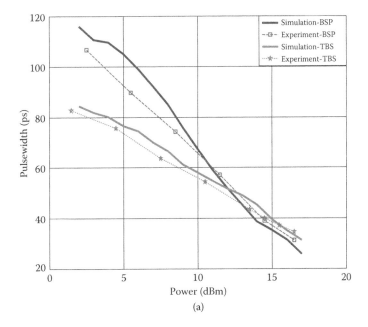

(a)

FIGURE 5.5
The launching power dependent variation of (a) pulsewidth, (b) time separation, and (c) peak power ratio of different multi-soliton bound states after 50 km propagation distance.

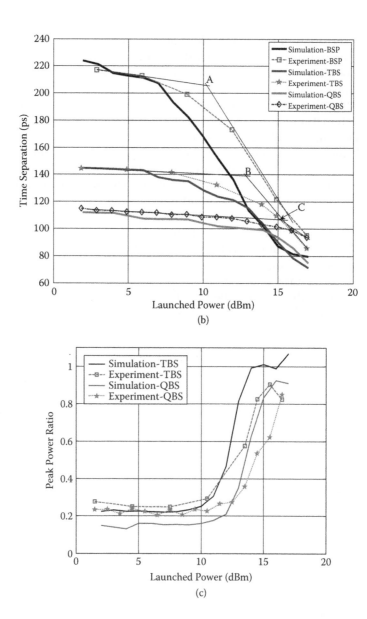

FIGURE 5.5
(Continued)

curves of the temporal separation show nonlinear launched power dependence. As observed in Figure 5.5b, two distinct areas of the curves can be observed as separated at the points A, B, and C. The intersection points A, B, and C of the two tangential curves correspond to the average soliton power (P_{sol}) of the bound state. On the other hand, the points A, B, and C correspond to the minimum powers to launch the bound pulses for soliton propagation. For the transmission over standard single-mode fiber, it is observed that with the experimental parameters of the multibound solitons, the estimated average soliton powers for BSP, TBS, and QBS are 11, 13.5, and 15.2 dBm, respectively.

At P_l lower than P_{sol}, and in addition to the fiber attenuation, the power of multibound soliton is not sufficient to balance the GVD effect. As a result, the pulse widths are rapidly broadened by self-alignment of the multibound solitons due to the perturbations accompanied by partly shading their energy in form of dispersive waves [2]. The broadening of such a dispersive wave is accumulated along the transmission fiber, and oscillations are formed around the multibound solitons as shown by the ripple of the tail of the soliton group in Figure 5.2a. Furthermore, the rate of broadening of the pulses is faster than that of the temporal separation at low launched powers leading to the enhancement of the overlapping between pulses. Therefore, the pulse envelope is consequently modulated by the interference of waves with the phase modulation effect due to the GVD that is the same as the fractional temporal Talbot effect [27,28]. Because of the parabolic symmetry of the anomalous GVD-induced phase shift profile around multibound solitons with a relative phase difference of π, the energy of the inner pulses is shifted to the outer pulses, and hence a decrease of the amplitude of inner pulses. (The frequency components of the inner pulses obtain higher relative velocity or propagate faster than those of the edge pulses that result in the energy transfer from the inner to the outer. This pattern also reconfirms the relative phase difference of π in the bound state.)

At $P_l < P_{sol}$, the multibound states operate in a linear transmission scheme (picture the solitary waves). In contrast, at higher P_l, the SPM phase shift is increased to balance the GVD effect. The pulse width becomes narrower and the ripple of the pulse envelope is lower with higher level of launched power. For the dual-bound soliton, there is no difference in peak power of two pulses in the bound state. For the bound states with the number of solitons greater than two, however, there is significant difference between the inner and the outer pulses because of the energy transfer at $P_l < P_{sol}$. Owing to the soliton content in pulses is enhanced at the average soliton power, there is a jump of peak power ratio between inner and outer pulses as shown in Figure 5.5c.

5.4 Dynamics of Multibound Solitons in Transmission

To verify the experimental results as well as to identify the evolution of multibound solitons propagation in the fiber, we model the multibound solitons as [29]

$$u_{bs} = \sum_{i=1}^{N} u_i(z,t) \tag{5.6}$$

with N as the number of solitons in a bound state, and

$$u_i = A_i \sec h\{[t - iq_0]\} \exp(j\theta_i), \tag{5.7}$$

where A_i and q_0 are amplitude and time separation of solitons, respectively, and the phase difference $\Delta\theta = \theta_{i+1} - \theta_i = \pi$. The propagation of multibound solitons in SSMF is governed by the nonlinear Schrödinger equation with the input parameters as those obtained in the experiment. Shown together with experimental results in Figures 5.5a through 5.5c are the simulation evolution of the parameters (solid curves) over 50 km propagation of various orders of multisoliton bound states. The simulated evolution of the triple-bound soliton is shown in Figure 5.6 over 50 km SSMF, while Figure 5.7 shows the pulse shape and corresponding spectrum of the outputs at various launched power levels. The simulated results generally agree well with those obtained in the experiment (see Figure 5.2). The small residual chirp caused small difference between the experimental and numerical results. Dynamics of multibound solitons consisting of dual bond soliton (DBS), triple bond soliton (TBS), and quadruple bond soliton (QBS) during propagation are shown from Figure 5.8 to Figure 5.10. In each figure, respectively, the evolution of multibound solitons parameters such as the pulse width, the temporal separation between solitons, the ratio between the pulse width and the pulse separation and the peak power of the solitons, is simulated along the transmission fiber with different launched powers. The difference between lower and higher launching powers also obviously exhibited in simulated results. When the launched powers is far from the soliton power P_{sol}, there is oscillation or rapid variation of parameters at initial propagation distance due to the adjustment of multibound solitons to perturbations of propagation conditions as mentioned above. The pulses are compressed at P_is higher than P_{sol}, while they are rapidly broadened at P_is lower than P_{sol}. The slow variation of parameters occurs at P_i close to P_{sol}. However, the time separation of solitons remains unchanged in the

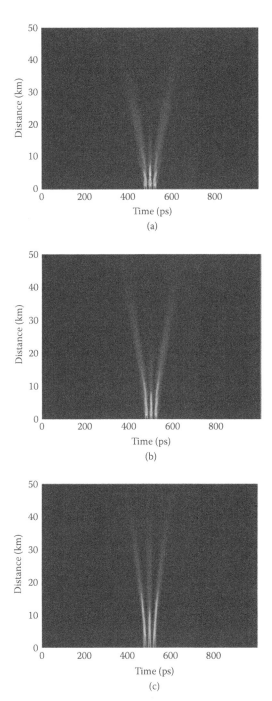

FIGURE 5.6
Numerically simulated evolutions of the triple-bound soliton over 50 km SSMF propagation at
P_1 of respectively (a) 5 dBm, (b) 10 dBm, and (c) 14 dBm.

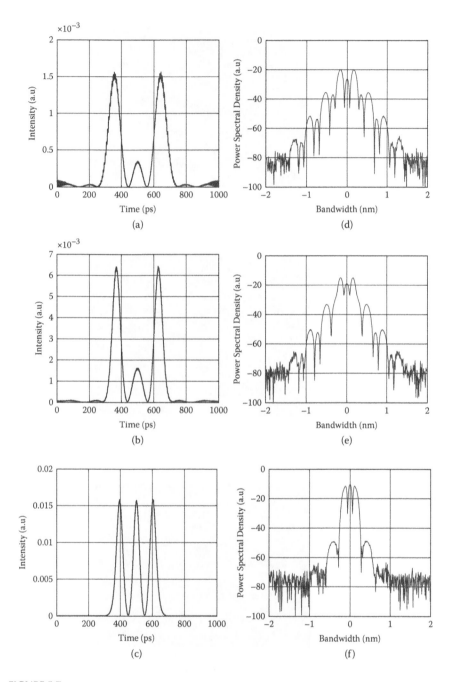

FIGURE 5.7
Row (a) Numerically simulated output pulse shape and Row (c) optical spectrum of triple-soliton bound state over 50 km SSMF propagation at P_1 of respectively 5 dBm, 10 dBm, and 14 dBm, (c), (d), .and (f) are corresponding spectra of bound-solitons depicted in (a), (b), and (c), respectively.

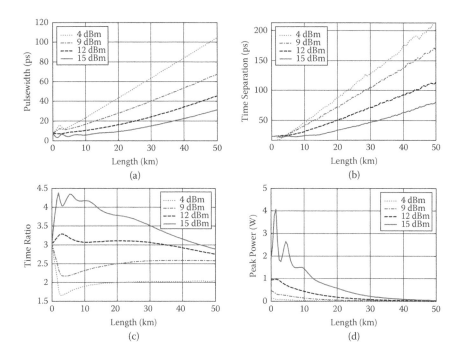

FIGURE 5.8
Evolution of numerically simulated dual-soliton bound state after 50 km propagation distance of SMF-28 fiber: (a) pulse width, (b) temporal separation, (c) ratio between pulse width and separation, and (d) peak power.

propagation distance of one soliton period. We have validated this prediction in our experiment.

Another important property of multibound solitons is the phase difference between pulses that can be determined by the shape of the spectrum. We monitored the optical spectrum of the multibound solitons at both the launched end and the output of the fiber length. The modulated spectrum of multibound solitons is symmetrical with carrier suppression due to a relative phase relationship of π between adjacent solitons. After 50 km propagation at low P_l, both experimental and simulated results show that the spectrum of multibound solitons is nearly the same as that at the launched end (see Figure 5.2d and Figure 5.7d). In a linear-like scheme where the SPM effect is negligible, the GVD only modulates the spectral phase which modifies the temporal profile, but does not modify the spectrum of the bound state. The modulation of the spectrum is modified with an increase in P_l. At sufficient high P_l, the nonlinear phase shift-induced chirp is increased at the edges of pulses. Although the nonlinear phase

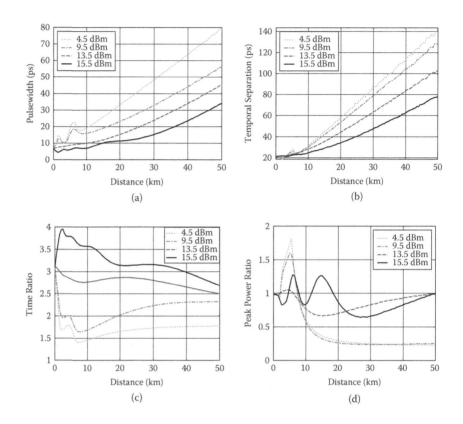

FIGURE 5.9
Evolution of numerically simulated triple-soliton bound state after 50 km propagation distance of SMF-28 fiber: (a) pulse width (b) temporal separation (c) ratio between pulse width and separation (d) peak power ratio between inner pulse and outer pulse.

shift reduces the GVD effect, the phase transition between adjacent pulses is changed due to the direct impact of the nonlinear phase shift. Hence, the small humps in the spectra of multibound states are strengthened, and they may be comparable to main lobes due to enhancement of the far interaction between pulses as shown in Figure 5.2f and Figure 5.7f. Figure 5.11 shows the variation of the relative phase difference between adjacent pulses in various multibound soliton states over 50 km propagation. The simulated results also indicate that the relative phase difference varies differently between two adjacent pulses in propagation. Although the phase difference of π between two central pulses in even-soliton bound states such as DBS and QBS can remain unchanged, it varies in general along propagation distance. At low P_l in all cases, the phase difference is varied to zero or $\pi/2$ value, then recovered to π at the output of the 50 km fiber.

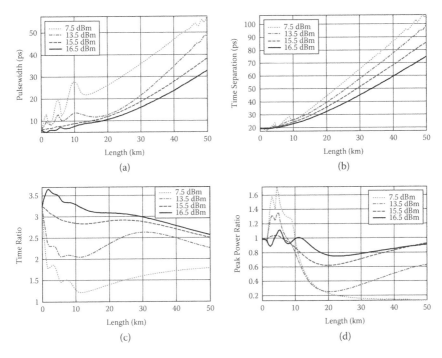

FIGURE 5.10
Evolution of numerically simulated quadruple-soliton bound state after 50 km propagation distance of SMF-28 fiber: (a) pulse width (b) temporal separation (c) ratio between pulse width and separation (d) Peak power ratio between inner pulse and outer pulse.

When the SPM phase shift becomes significant, the phase difference varies in the same manner in the first 40 km propagation, and then it tends to $\pi/2$ that modifies the corresponding spectrum.

5.6 Concluding Remarks

Characteristics of multibound soliton in propagation over 50 km SSMF fiber have been investigated in this chapter. Depending on the launched power level, the variation of multibound states parameters can divide into two propagation schemes: linear and soliton schemes. At low launched power in the linear propagation scheme, the modulated spectrum of multibound states remains unchanged due to the preservation of the phase difference of π, yet the temporal shape is modified so that it can be used to reconfirm the phase difference. At high launched power in a soliton propagation

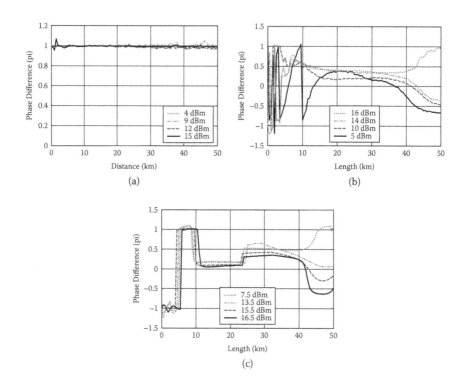

FIGURE 5.11
Evolution of the phase difference between adjacent pulses over 50 km propagation in various bound states: (a) dual-bound soliton, (b) triple-bound soliton, and (c) quadruple-bound soliton.

scheme, the modulated spectrum of multibound states is modified due to the enhancement of the SPM effect; however, the pulses in the bound state behave similar by to the single soliton transmission in perturbed conditions.

References

1. K. J. Blow and N. J. Doran, The Asymptotic Dispersion of Soliton Pulses in Lossy Fibers, _Opt. Commun._, 52, 367–370, 1985.
2. G. P. Agrawal, _Nonlinear Fiber Optics_, 3rd ed. Academic Press, San Diego, 2001.
3. M. Nakazawa et al., Soliton Amplification and Transmission with Er_{3+}-Doped Fiber Repeater Pumped by GaInAsP Diode, _Electron. Lett._, 25, 199–200, 1989.
4, A. Hasegawa and Y. Kodama, Guiding-Center Soliton, _Phys. Rev. Lett._, 66, 161, 1991.
5. Y. Kodama and A. Hasegawa, Amplification and Reshaping of Optical Solitons in Glass Fiber II, _Opt. Lett._, 7, 339–341, 1982.

6. Y. Kodama and A. Hasegawa, Amplification and Reshaping of Optical Solitons in Glass Fiber? III. Amplifiers with Random Gain, *Opt. Lett.*, 8, 342–344, 1983.

7. A. Hasegawa, Amplification and Reshaping of Optical Solitons in a Glass Fiber-IV: Use of the Stimulated Raman Process, *Opt. Lett.*, 8, 650–652, 1983.

8. L. F. Mollenauer and K. Smith, Demonstration of Soliton Transmission over More than 4000 kmin Fiber with Loss Periodically Compensated by Raman Gain, *Opt. Lett.*, 13, 675–677, 1988.

9. L. Mollenauer et al., Soliton Propagation in Long Fibers with Periodically Compensated Loss, *IEEE J. Quantum Electron.*, 22, 157–173, 1986.

10. M. Nakazawa, Tb/s OTDM technology, in *27th European Conference on Optical Communication, Optical Communication, 2001. ECOC '01.* 2001, 184–187, vol. 2.

11. L. Ju Han et al., Wavelength Tunable 10-GHz 3-ps Pulse Source Using a Dispersion Decreasing Fiber-Based Nonlinear Optical Loop Mirror, *IEEE J. Selected Topics in Quantum Electron.*, 10, 181–185, 2004.

12. A. Sano and Y. Miyamoto, Performance Evaluation of Prechirped RZ and CS-RZ Formats in High-Speed Transmission Systems with Dispersion Management, *J. Lightwave Technol.*, 19, 1864–1871, 2001.

13. J. Fatome et al., Practical Design Rules for Single-Channel Ultra High-Speed Dense Dispersion Management Telecommunication Systems, *Opt. Commun.* 282, 1427–1434, 2009.

14. E. Poutrina and G. P. Agrawal, Design Rules for Dispersion-Managed Soliton Systems, *Opt. Commun.*, 206, 193–200, 2002.

15. J. H. B. Nijhof et al., Stable Soliton-Like Propagation in Dispersion Managed Systems with Net Anomalous, zero and Normal Dispersion, *Electron. Lett.*, 33, 1726–1727, 1997.

16. J. H. B. Nijhof et al., Energy Enhancement of Dispersion-Managed Solitons and WDM, *Electron. Lett.*, 34, 481–482, 1998.

17. T. Inoue et al., Interactions between Dispersion-Managed Solitons in Optical-Time-Division-Multiplexed Systems, *Electron. Commun. in Jpn. (Part II: Electron.)*, 84, 24–29, 2001.

18. C. Paré and P. A. Bélanger, Antisymmetric Soliton in a Dispersion-Managed System, *Opt. Commun.*, 168, 103–109, 1999.

19. A. Maruta et al., Bisoliton Propagating in Dispersion-Managed System and Its Application to High-Speed and Long-Haul Optical Transmission, *J. IEEE Selected Topics in Quantum Electron.*, 8, 640–650, 2002.

20. M. Stratmann et al., Dark Solitons Are Stable in Dispersion Maps of Either Sign of Path-Average Dispersion, in *Summaries of Papers Presented at the Quantum Electronics and Laser Science Conference, 2002. QELS '02. Technical Digest*, 2002, p. 226.

21. M. Stratmann and F. Mitschke, Bound States between Dark and Bright Solitons in Dispersion Maps, in *Summaries of Papers Presented at the Quantum Electronics and Laser Science Conference, 2002. QELS '02. Technical Digest.* 2002, pp. 226–227.

22. M. Stratmann et al., Experimental Observation of Temporal Soliton Molecules, *Phys. Rev. Lett.*, 95, 143902, 2005.

23. I. Gabitov et al., Twin Families of Bisolitons in Dispersion-Managed Systems, *Opt. Lett.*, 32, 605–607, 2007.

24. J. P. Gordon, Interaction Forces among Solitons in Optical Fibers, *Opt. Lett.*, 8, 596–598, 1983.

25. Y. Kodama and K. Nozaki, Soliton Interaction in Optical Fibers, *Opt. Lett.*, 12, 1038–1040, 1987.
26. F. M. Mitschke and L. F. Mollenauer, Experimental Observation of Interaction Forces between Solitons in Optical Fibers, *Opt. Lett.*, 12, 355–357, 1987.
27. J. Azana and M. A. Muriel, Temporal Self-Imaging Effects: Theory and Application for Multiplying Pulse Repetition Rates, *IEEE J. Selected Topics in Quantum Electron.*, 7, 728–744, 2001.
28. N. D. Nguyen et al., Temporal Imaging and Optical Repetition Multiplication via Quadratic Phase Modulation, in *Information, Communications and Signal Processing, 2007 Sixth International Conference on*, 2007, pp. 1–5.
29. N. D. Nguyen and L. N. Binh, Generation of Bound-Solitons in Actively Phase Modulation Mode-Locked Fiber Ring Resonators, *Opt. Commun.*, 281, 2012–2022, 2008.

6

Deterministic Dynamics of Solitons in Passive Mode-Locked Fiber Lasers

Tang Ding Yuan and L. M. Zhao

School of Electrical and Electronic Engineering, Nanyang Technological University, Singapore

CONTENTS

6.1 Introduction

Since K. Ikeda first showed that the transmission of a passive nonlinear optical cavity could exhibit multistability and chaos in response to a constant incident light [1], extensive studies on the dynamic property of both the passive and active nonlinear optical cavities have been carried out. It has been shown that period-doubling bifurcation and the period-doubling route to chaos are generic features of light traversing a nonlinear cavity. However, the majority of the researches have been focused on the continuous-wave operation of the cavities [2]. As a matter of fact, apart from the continuous-wave (CW) operation, short-pulse operation of the cavities is also available. Moreover, determined by the cavity parameters, various types of solitons can even be formed in a nonlinear cavity. Solitons as a special nonlinear wave that can propagate long distance in dispersive materials without distorting their shapes have been found in a wide range of physical systems in thermodynamics, plasma physics, condensed matter physics, and optics. As solitons are a nonlinear wave packet intrinsically stable against perturbations, it would be of interest to investigate the dynamical features of the cavities and their manifestations under the ultrashort pulse operation, in particular when the ultrashort pulse is itself an optical soliton.

Soliton operation has been obtained in various ultrashort pulse mode-locked fiber lasers [3–10], as well as in the mode-locked bulk solid-state lasers [11,12]. In the case of a fiber laser, except for the cavity components necessary for achieving mode locking and for laser output, its cavity is mainly made up of optical fibers. Due to the small core size of the single-mode fibers and the long propagating distance, strong nonlinear phase shift can be accumulated when an ultrashort pulse propagates in the fiber laser cavity. Therefore, a mode-locked pulse could be easily shaped into an optical soliton in a fiber laser. It has been shown that despite the actions of the other discrete cavity components, the average dynamics of the formed solitons in a fiber laser is well described by the nonlinear Schrödinger equation (NLSE), a paradigm equation that has been extensively investigated. A soliton formed in a laser is inherently different from those formed in a single-mode fiber as in a laser a soliton also experiences loss and gain. It has been shown both experimentally and numerically that under certain conditions, the solitons formed in a fiber laser could exhibit deterministic dynamics [13–22], such as the soliton period-doubling bifurcations, soliton intermittency, and soliton quasi-periodicity. Moreover, the appearance of the soliton deterministic dynamics is independent of the concrete laser cavity design, showing that it is a general feature of the system.

In this chapter we describe the deterministic dynamics of solitons in fiber lasers. Our discussions are mainly focused on fiber soliton lasers passively mode locked by the nonlinear polarization rotation (NPR) technique. In the fiber lasers the scalar solitons are normally formed. We also investigated the vector soliton dynamics in fiber laser mode locked by the semiconductor saturable absorber mirrors (SESAMs) and the dissipative solitons dynamics in fiber lasers with normal cavity dispersion. The chapter is organized as follows: Section 6.2 presents the general theoretical background for NPR mode locking and soliton generation in fiber lasers. Section 6.3 reviews the deterministic dynamics of various solitons in different fiber lasers, which is followed by the corresponding numerical simulations in Section 6.4. Furthermore, this section describes the cavity-induced modulation instability effect and its relation to the soliton deterministic dynamics. We show that the various forms of the soliton deterministic dynamics observed could be related to the cavity-induced soliton modulation instability effect. Namely, their physical origin could be traced back to the self-induced nonlinear resonant wave coupling in the laser cavity. Section 6.5 presents our conclusion.

6.2 Solution Generation in Fiber Lasers

Solution formation in anomalous dispersion cavity fiber lasers is mainly due to the natural balance between the cavity dispersion and the fiber nonlinear optical Kerr effect. An optical pulse can be routinely generated in a fiber laser by various mode-locking techniques. In terms of the passive mode locking these include the nonlinear optical loop mirror method [23], figure-eight cavity method [24], NPR technique [25], and SESAM method [26]. Among the various passive mode-locking techniques, the NPR technique is widely used. The technique exploits the nonlinear birefringence of the single-mode optical fibers for the generation of an artificial saturable absorber effect in the laser cavity. It is the artificial saturable absorber effect that initiates a self-started mode locking in the laser. To explain the operation principle of the technique, it is to note that generally the polarization state of light passing through a piece of birefringent fiber varies linearly with the fiber birefringence and length. However, when the light intensity is strong, the nonlinear optical Kerr effect of the fiber introduces an extra change to the light polarization. As the extra polarization change is proportional to the light intensity, if a polarizer is placed behind the fiber, then through appropriately selecting the orientation of the polarizer, an effect with the feature that light with higher intensity experiences larger transmission through the polarizer could be obtained. Such an effect is known as an artificial saturable absorber effect. Incorporating such an artificial saturable absorber effect in a fiber laser is equivalent to inserting a saturable absorber in the laser cavity. Under the effect of a saturable absorber, the operation of a laser can automatically become mode locked. In the NPR mode-locked fiber laser, a nonlinear phase shift is required to achieve mode locking. If the nonlinear phase modulation of the pulse could balance the pulse width broadening caused by the cavity dispersion, a soliton is then formed in the laser.

6.2.1 Pulse Propagation in Single-Mode Fibers

When an optical pulse propagates in a birefringent single-mode fiber, not only would the fiber dispersion broaden the pulse width, but also the pulses along different polarizations will have different group velocities, which broadens the pulse width or even splits the pulse. As there exist two polarization components, the nonlinear fiber Kerr effect will not only generate self-phase modulation (SPM) of each component, but also generate cross-phase modulations (XPM) between the two polarization components. The coherent coupling between the two polarization components also generates degenerated four-wave mixing (FWM), whose strength is related to the strength of the linear fiber birefringence [27].

In a fiber laser system, one of the essential components is the active medium, which generally is rare earth ions doped fiber. When an optical pulse propagates in the rare earth ions doped fibers, the effect of light amplification must be considered. A fiber amplifier is characterized by its small signal gain, gain bandwidth, gain saturation, and noise figure. Therefore, to precisely describe pulse evolution in a fiber laser, it is necessary to consider the cavity dispersion, fiber nonlinearity (SPM and XPM), fiber birefringence, FWM, gain and loss, gain dispersion effects, and so on.

Optical pulse propagation in fiber segments in a fiber laser can be described by the extended Ginzburg–Landau equation (GLE) [28]:

$$\frac{\partial A_x}{\partial z} = i\beta A_x - \delta \frac{\partial A_x}{\partial t} - \frac{i}{2}\beta_2 \frac{\partial^2 A_x}{\partial t^2} + i\gamma \left(|A_x|^2 + \frac{2}{3}|A_y|^2 \right) A_x$$

$$+ \frac{i\gamma}{3} A_y^2 A_x^* + \frac{g}{2} A_x + \frac{g}{2\Omega_g^2} \frac{\partial^2 A_x}{\partial t^2}$$

$$\frac{\partial A_y}{\partial z} = -i\beta A_y + \delta \frac{\partial A_y}{\partial t} - \frac{i}{2}\beta_2 \frac{\partial^2 A_y}{\partial t^2} + i\gamma \left(|A_y|^2 + \frac{2}{3}|A_x|^2 \right) A_y$$

$$+ \frac{i\gamma}{3} A_x^2 A_y^* + \frac{g}{2} A_y + \frac{g}{2\Omega_g^2} \frac{\partial^2 A_y}{\partial t^2}$$

(6.1)

where A_x and A_y are the two normalized slowly varying pulse envelopes along the slow and the fast axes, A_x^* and A_y^* are their conjugates, respectively. $2\beta = 2\pi(n_x - n_y)/\lambda = 2\pi/L_b$ is the wave-number difference, and $2\delta = 2\beta\lambda/2\pi c$ is the inverse group-velocity difference. The equations are written in the coordinate system that moves with the average group velocity $\overline{v}_g^{-1} = \overline{\beta}_1 = \frac{\beta_{1x}+\beta_{1y}}{2}$. The GLE is nonintegrable, but numerical simulations are available for studying the pulse dynamics in such fiber lasers.

The last two terms in the right part of Equation (6.1) correspond to the gain and the gain dispersion effects when an optical pulse propagates in the active fiber segments of a fiber laser. g is the peak gain coefficient; Ω_g is the gain bandwidth. When the gain saturation results from light along both polarizations, the saturated gain coefficient is calculated by

$$g = g_0 \exp\left(-\frac{1}{E_s}\int(|A_x|^2 + |A_y|^2)\right) \tag{6.2}$$

where g_0 is the small signal gain, and E_s is the saturation energy. Typical values of E_s for erbium-doped fibers are about 10 μJ. As the pulse energies are normally much smaller than the saturation energy E_s, gain saturation is negligible over the duration of a single pulse. However, in the case of a pulsed fiber laser, the pulse circulates in the laser cavity. The average power of the light will saturate the gain and determine the saturated gain value.

6.2.2 Cavity Transmission of Nonlinear Polarization Rotation Mode-Locked Fiber Lasers

Apart from the pulse propagation in various fiber segments, cavity boundary condition is another intrinsic feature that needs to be considered for soliton generation in fiber lasers. A fiber laser mode locked with the NPR technique can always be equivalently simplified into the three parts as shown in Figure 6.1 where a polarization controller P1 is at the beginning of

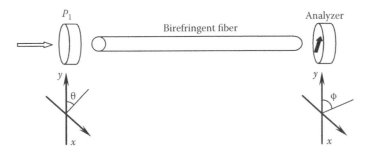

FIGURE 6.1
Equivalent physically simplified laser cavity for fiber lasers using the nonlinear polarization rotation (NPR) mode-locking technique.

the fiber and an analyzer at the end of the fiber, and the fiber is a weakly bire-fringence fiber. The two principal polarization axes of the birefringent fiber are x (horizontal) and y (vertical) axes and we consider that the birefringent axes of the different segments of fibers are the same. The fast axis of P1 and the transmission axis of the analyzer have an angle of θ and φ to the y-axis of the fiber polarization, respectively.

Briefly speaking, the pulse evolution in the fiber laser starts with an arbi-trary weak pulse, and it circulates in the laser cavity until a steady pulse evolution state is established. Whenever the pulse encounters an individual intracavity component except the fiber segments, the Jones matrix of the component is multiplied to the optical field of the pulse.

Starting from a linearly polarized weak pulse F_n, two polarization compo-nents are obtained when F_n travels through P1:

$$\begin{cases} u = F_n \sin\theta \exp(i\Delta\Phi) \\ v = F_n \cos\theta \end{cases} \tag{6.3}$$

where $\Delta\Phi$ is the phase shift between the wave components in the two orthogonal birefringent axes x and y.

Then the two polarization components propagate in the fiber segments, which is governed by the coupled GLEs as shown in previous section.

The light propagates along the laser cavity and finally projects on the transmission axis of the analyzer:

$$F_{n+1} = u' \sin\phi + v' \cos\phi \tag{6.4}$$

where u' and v' are the two orthogonal polarization components of the light after propagation in all the fiber segments.

The intensity transmission T of light through the setup is [29]

$$T = \frac{I_{out}}{I_{in}} = \sin^2\theta\sin^2\varphi + \cos^2\theta\cos^2\varphi + \frac{1}{2}\sin(2\theta)\sin(2\varphi)\cos(\Delta\Phi_F) \tag{6.5}$$

where Φ_F is the phase delay generated between the two light polarization components when they traverse the birefringent fiber. The phase delay actu-ally consists of two parts: a linear phase delay $\Delta\Phi_l$ raised because of the lin-ear birefringence of the fiber, this part always exists; and a nonlinear part $\Delta\Phi_{nl}$ generated by the nonlinear effects of the fiber. If the light intensity is weak, the nonlinear part becomes zero. Hence, one can divide the light trans-mission into the linear transmission and the nonlinear transmission. The linear intensity transmission can be written as

$$T_l = \sin^2\theta\sin^2\varphi + \cos^2\theta\cos^2\varphi + \frac{1}{2}\sin(2\theta)\sin(2\varphi)\cos(\Delta\Phi_l) \tag{6.6}$$

Once the orientations of the polarizer and analyzer with respect to the fiber polarization principal axes are fixed, the linear transmission of the setup is a sinusoidal function of the linear phase delay between the two polarization components. Although the relative orientation between the polarizer and the analyzer could be arbitrarily selected, in order to possibly achieve a stable mode-locked operation, it is preferred to choose the relative orientation between the polarizer and the analyzer as 90 degrees, as under such a selection the linear transmission is possible to set to zero.

To illustrate the effect of the nonlinear polarization rotation on the intensity transmission, we assume that the nonlinear fiber birefringence only introduces a small extra phase delay $\Delta\Phi_{nl}$. Then the intensity transmission can be written as

$$T = T_l - \frac{1}{2}\sin(2\theta)\sin(2\varphi)\sin(\Delta\Phi_l)\Delta\Phi_{nl} \qquad (6.7)$$

The nonlinear phase delay introduces a transmission change relative to the value of the linear transmission T_l. The change is not only the nonlinear phase delay $\Delta\Phi_{nl}$; therefore, the light intensity dependent, but also the orientation of the polarizer, as well as the linear phase delay are dependent. As a result, in order to generate an artificial saturable absorber effect by the nonlinear fiber birefringent effect, it is necessary to appropriately select the combination of all these parameters.

It is instructive to consider a special case to illustrate it. Consider that the polarizer and the P1 are set orthogonally. In this case the linear intensity transmission is

$$T_l = \sin^2 2\theta \frac{[1 - \cos(\Delta\Phi_l)]}{2} \qquad (6.8)$$

Therefore, the maximum linear intensity transmission is limited by $\sin^2 2\theta$. The intensity transmission under the existence of a nonlinear phase delay $\Delta\Phi_{nl}$ is

$$T = T_l + \frac{1}{2}\sin^2(2\theta)\sin(\Delta\Phi_l)\Delta\Phi_{nl} \qquad (6.9)$$

The orientation of the polarizer determines the projection of light on the two polarization axes of the fiber and, therefore, determines whether a positive $\Delta\Phi_{nl}$ or a negative $\Delta\Phi_{nl}$ would be generated. Assuming that a negative $\Delta\Phi_{nl}$ is generated, then further depending on the linear phase delay, either an increase or decrease in the intensity transmission could be caused by the nonlinear polarization rotation. The magnitude of $\Delta\Phi_{nl}$ is always proportional to the intensity of light. This means that with the increase of the light intensity, either an increase or a decrease in the transmission could be generated, which is purely determined by the linear phase delay selection.

Strictly speaking, the nonlinear phase delay between the two polarization components of an optical pulse traversing a piece of birefringent fiber can only be determined by numerically solving the coupled GLEs (Equation 6.1). However, in the case of CW light propagation and ignoring the effect of energy exchange between the two polarization components, this can be calculated explicitly as [28]

$$\Delta\Phi_{nl} = \frac{\gamma P_0 L}{3} \cos(2\theta) \tag{6.10}$$

where P_0 is the power of the light, and L is the length of the fiber. Using the nonlinear phase delay, the intensity transmission can be written as [30]

$$T = T_l + \kappa |E_{in}|^2 \tag{6.11}$$

where $\kappa = -\frac{\gamma L}{12} \sin(4\theta) \sin(2\theta) \sin(\Delta\Phi_l)$. As far as $\kappa > 0$, a saturable absorber effect can be obtained. Based on the formula one could estimate the optimum selection for the θ value so that the strongest saturable absorber effect could be achieved [31]. In this case it occurs at $\theta = 27.7°$.

In practice it is difficult to set the linear phase delay just within the range where the saturable absorption effect is achieved. Therefore, a polarization controller is normally inserted in the setup to efficiently control the value of the linear phase delay $\Delta\Phi_l$. Mathematically, this is equivalent to adding a variable linear phase delay bias term in the intensity transmission formula. Hence, the intensity transmission can be further written as

$$T = \frac{I_{out}}{I_{in}} = \sin^2\theta \sin^2\varphi + \cos^2\theta\cos^2\varphi + \frac{1}{2}\sin(2\theta)\sin(2\varphi)\cos(\Delta\Phi_{PC} + \Delta\Phi_F) \tag{6.12}$$

where Φ_{pc} is the phase delay bias introduced by the polarization controller, and it is continuously tunable. With the insertion of a polarization controller in the setup, it is always possible to achieve an artificial saturable absorber effect through changing the linear birefringence of the fiber.

In summary, the soliton generation in fiber lasers should be the stable pulse propagation that can both fulfill pulse propagation in fiber segments as described by Equation (6.1) and the cavity transmission as indicated by Equation (6.12).

6.3 Deterministic Dynamics of Solitons in Fiber Lasers

6.3.1 Experimental Configuration

We have experimentally investigated deterministic soliton dynamics in passive mode-locked fiber lasers with various cavity design and cavity parameters. Figure 6.2 shows a typical fiber laser we used. The fiber laser has a ring

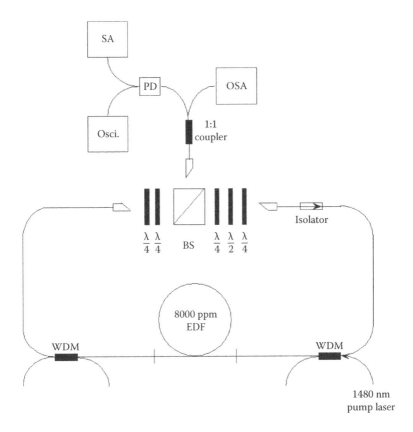

FIGURE 6.2
The experimental setup. $\lambda/4$: quarter-wave plate; $\lambda/2$: half-wave plate; BS: beam splitter; WDM: wavelength-division multiplexer; EDF: erbium-doped fiber; OSA: optical spectrum analyzer; PD: photodetector; Osci.: oscilloscope; SA: rf spectrum analyzer.

cavity composed of a segment of the erbium-doped fiber used as the laser gain medium. The erbium fiber is pumped by a high-power Fiber Raman Laser source (BWC-FL-1480-1) of wavelength 1480 nm. The pump light is coupled into the cavity through a fiber wavelength division multiplexer (WDM). A fiber pigtailed isolator is inserted in the cavity to force the unidirectional operation of the cavity. We used the NPR technique for the self-started mode locking of the laser. To this end, two polarization controllers, one consisting of two quarter-wave plates and the other two quarter-wave plates and one half-wave plate, are used to adjust the polarization of the light. A cubic polarization beam splitter is used to output the laser pulses and set the polarization at the position of the cavity. The polarization controllers and the beam splitter are mounted on a 7-cm-long fiber bench. The soliton output of the laser is monitored with an optical spectrum analyzer (Ando AQ-6315B), a 26.5 GHz radio-frequency (RF) spectrum analyzer (Agilent E4407B ESA-E series), and a 350 MHz oscilloscope (Agilent 54641A) together with a 5 GHz photodetector. A commercial optical autocorrelator (Autocorrelator Pulsescope) was used to measure the soliton pulse width.

In our experiments we used optical fibers with various parameters, such as different fiber group velocity dispersion (GVD), different lengths, and different doping concentrations of the gain fiber. The purpose of using fibers of different properties and lengths is to change the laser cavity parameters, so possibly different soliton dynamics in the cavity could be observed. Experimentally we found that with the appropriate selection on the orientations of the waveplates, self-started mode locking of the fiber lasers could always be achieved, and as far as the cavity nonlinearity is not too strong, stable soliton operation of the fiber lasers can always be observed. Such stable soliton operation of the fiber lasers has been extensively investigated and reported previously [29]. However, we also found that when the peak power of the formed solitons reached a certain high level in a fiber laser, deterministic dynamics of the optical solitons could also be observed.

6.3.2 Soliton Deterministic Dynamics

Suggested by the average soliton theory of lasers [32], it is generally believed that the output of the fiber soliton lasers is a uniform soliton pulse train. However, Kim et al. found theoretically that depending on the strength of the fiber birefringence and the alignment of the polarizer with the fast- and slow-polarization axes of the fiber, the output pulse train of a fiber laser mode locked with the NPR technique could exhibit periodic fluctuations in pulse intensity and polarization [33,34]. Nevertheless, the nonuniformity of the pulse train could be diminished by aligning the polarizer with either the fast or slow axis of the fiber. We had also experimentally investigated the output property of a fiber soliton ring laser passively mode locked by the NPR technique [35] and found that the soliton pulse nonuniformity is in fact an intrinsic feature of the laser, whose appearance is independent on the orientation of the polarizer in the cavity but closely related to the pump power. Based on numerical simulations we showed that depending on the linear cavity phase delay bias, the nonlinear polarization switching effect could play an important role on the soliton dynamics of the lasers. When the linear cavity phase delay bias is set close to the nonlinear polarization switching point and the pump power is strong, the soliton pulse peak intensity could increase to so high that the generated NPR cross over the nonlinear polarization switching point and, consequently, drive the laser cavity from the positive feedback regime to the negative feedback regime. Eventually the competition between the soliton pulses with the linear waves in the cavity such as the dispersive waves or CW laser emission then causes the amplitude of the soliton pulses to vary periodically. There are two methods to suppress the periodical intensity fluctuations: one is to reduce the pump power so that the peak intensity of the solitons is below the polarization switch threshold; the other is to increase the polarization switching power of a laser. However, the latter method needs to appropriately adjust the linear phase delay bias of the cavity.

Experimentally, lasers of the same cavity configuration but with different cavity lengths, different fiber birefringence, and different fiber dispersion properties have been set up and investigated [13,15,17,19,20]. In all lasers if the pump power is set beyond the mode-locking threshold, and the linear cavity phase delay is appropriately selected, self-started mode locking can always be obtained. However, depending on the laser parameter selection, deterministic dynamics of the solitons may not necessarily appear. In the following we summarize the typical soliton deterministic dynamics of the lasers observed.

6.3.3 Period-Doubling Bifurcation and Period-Doubling Route to Chaos of Single-Pulse Solitons

Provided that the orientations of the polarization controllers are appropriately set, self-started soliton operation of the lasers is automatically obtained by simply increasing the pump power beyond the mode-locking threshold. Multiple solitons are initially obtained. However, by decreasing the pump power, the state with only one soliton existing in the cavity can always be achieved. Starting from a stable soliton operation state, experimentally it was noticed that turning the orientation of one of the quarter-wave plates to one direction, which theoretically corresponds to shifting the linear cavity phase delay bias away from the nonlinear polarization switching point, the peak power of the soliton pulse formed in the cavity increases. Consequently, the strength of nonlinear interaction of the soliton pulses with the cavity components such as the optical fiber and the gain medium increases. To a certain level of the nonlinear interaction, it was observed that the output soliton intensity pattern of the laser experiences period-doubling bifurcations and a period-doubling route to chaos. Figure 6.3 shows as an example an experimentally observed period-doubling route to chaos. The results were obtained with fixed linear cavity phase delay bias but increasing pump power. At a relatively weak pump power, a stable soliton pulse train with uniform pulse intensity was obtained. The pulses repeat themselves with the cavity fundamental repetition rate (Figure 6.3a). We experimentally measured the laser output power when it is operating in such a state. With a pump power of about 26 mW, an average output power of about 140 μW was obtained, which illustrates that the single soliton pulse energy is about 8.05 pJ. Carefully increasing the pump power further, the intensity of the soliton pulse becomes no longer uniform, but alternates between two values (Figure 6.3b). Although the round-trip time of the solitons circulating in the cavity is still the same, the pulse energy returns only every two round-trips,

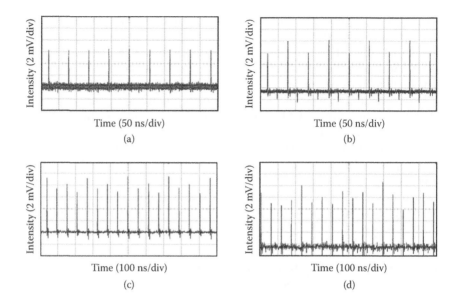

FIGURE 6.3

Period-doubling bifurcation to chaos of the soliton trains. (a) Period 1 state; (b) Period 2 state; (c) Period 4 state; (d) chaotic state. From (a) to (d) the pump intensity is increased.

forming a period-doubled state as compared with that of Figure 6.3a. Further slightly increasing the pump power, a period-quadrupled state then appears (Figure 6.3c). Eventually the process ends up with a chaotic soliton pulse energy variation state (Figure 6.3d).

With fixed pump power all the states shown are stable. Provided there are no great disturbances, they can last for several hours. It was also confirmed experimentally by combined use of the autocorrelator (PulseScope, scan range varies from 50 fs to 50 ps) and a high-speed oscilloscope (Agilent 86100A 50 GHz) that there is only one soliton in the cavity. Limited by the resolution of our autocorrelator, the measured autocorrelation traces show no unusual features of the soliton pulses and thus give no evidence of the behavior of period-doubling bifurcations, which is similar to that observed by G. Sucha et al. [36]. In all states the average soliton duration measured was about 316 ± 10 fs. The experimental results demonstrate that contrary to the general understanding to the mode-locked lasers, after one round-trip the mode-locked pulse does not return to its original value but does it in every two or four cavity round-trips in the stable periodic states. Depending on the strength of the nonlinear interaction between the pulse and the cavity components, the pulse could never return back to its original state in the chaotic state. To exclude any possibility of artificial digital sampling effect of the oscilloscope, we checked the pulse intensity alternation of the various periodic states by using a high-speed sampling oscilloscope (Agilent 86100A 50 GHz). Figure 6.4

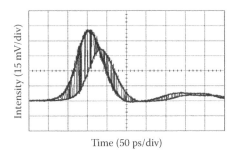

Time (50 ps/div)

FIGURE 6.4
Oscilloscope traces of a period-two state of the laser emission.

shows the result corresponding to a Period 2 state. In obtaining the figure we used the soliton pulse as the trigger for the oscilloscope and a high oscilloscope resolution of 50 ps/division. Due to the high scan speed of the oscilloscope we could clearly see that the individual soliton pulse trace on the screen now becomes broader. Therefore, no sampling problem exists. Triggered by different pulses the oscilloscope traces formed have two distinct peak intensities, indicating that the solitons in the laser output have two different pulse energies.

The optical spectra of the solitons corresponding to the Period 1 and Period 2 states are shown in Figures 6.5a and 6.5b, respectively. While the spectral curve shown in Figure 6.5a is smooth, the spectral curve shown in Figure 6.5b exhibits clear modulations. The spectral curve shown in Figure 6.5a possesses typical features of the soliton spectra of the passive mode-locked fiber lasers, characterized by the existence of sidebands superposing on the soliton spectrum. As in a Period 1 state solitons are identical in the soliton train, the spectrum shown in Figure 6.5a is also the optical spectrum of each individual soliton. In contrast, the optical spectrum shown in Figure 6.5b is an average of the spectra of two different solitons, each with different energy and frequency chirps. Based on Figure 6.5b it is to conclude that after a period-doubling bifurcation, the solitons possess different frequency properties as that of the solitons before bifurcation.

We also measured the intensity modulation of the soliton train with an RF spectrum analyzer. If period doubling does occur, there should appear a new frequency component in the RF spectrum locating exactly at the half of the fundamental cavity repetition rate position. Figure 6.6 shows the RF spectra of the laser output measured. As expected, after a period-doubling bifurcation a new frequency component of about 8.7 MHz appears in the spectrum. The amplitude of the new frequency component is quite strong compared to the fundamental frequency component, which vividly suggests that the soliton peak intensity alternates between two values with large difference. When period quadrupling occurs, in the RF spectrum we found that the amplitude of the new frequency component decreased; however, the

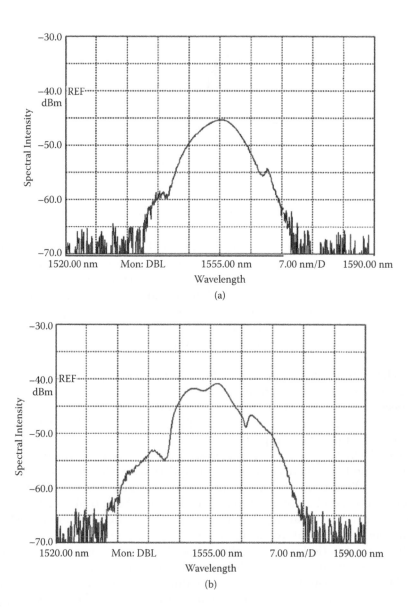

FIGURE 6.5
Optical spectra of the laser measured in the states of (a) Period 1 and (b) Period 2.

frequency components corresponding to the period quadrupling were too weak to be clearly distinguished from the background noise.

It is to be noted that if the linear cavity phase delay bias is selected close to the nonlinear polarization switching point of the cavity, although the single soliton operation of the laser can still be obtained, because the peak intensity of the soliton pulses is limited by the nonlinear polarization switching

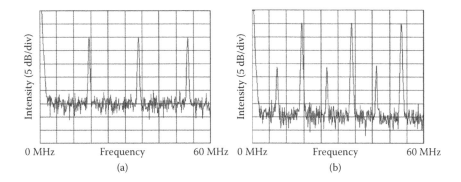

FIGURE 6.6
Radio-frequency spectra corresponding to (a) Period 1 and (b) Period 2 states.

power, which is weak under the linear cavity phase delay bias selection, no matter how strong the pump power is, no period-doubling bifurcation could be observed. There exists a threshold for the occurrence of the period-doubling bifurcation. Only when the linear cavity phase delay bias, which determines the stable soliton peak intensity, is appropriately set so that the stable soliton peak intensity exceeds a certain value, can period-doubling bifurcation be achieved. The experimental result further confirms that the appearance of the period-doubling bifurcations and period-doubling route to chaos is soliton pulse intensity dependent, and it is a nonlinear dynamic feature of the laser. One thing confirmed in our experiment is that for the occurrence of the effect, the soliton pulse energy or peak power must be strong. It is to imagine that in this case the nonlinear interaction between light and the gain medium, light and the nonlinear laser cavity will also become strong. It is well known that as a result of strong nonlinear interaction between light and gain medium in the laser cavity, a laser operating in the CW or Q-switched mode can exhibit a period-doubling route to chaos. Our experimental result now further demonstrated that this phenomenon could even appear in a mode-locked soliton laser. Finally, we point out that Côté and van Driel reported period doubling of a femtosecond Ti:sapphire laser by total mode locking of the TEM_{00} and TEM_{01} modes in an effective confocal cavity [37]. They believe that the gain saturation is a likely mechanism to support the transverse mode locking and the period doubling. However, in our laser there only exists one transverse mode of the fiber due to the characteristics of the single-mode fibers.

6.3.4 Period Doubling and Quadrupling of Bound Solitons

The full route of soliton period-doubling bifurcation to chaos was only observed under appropriately selected laser cavity parameters. In most cases only one or two soliton period-doubling bifurcations could be obtained.

A common feature of all soliton fiber lasers is the multiple soliton generation under strong pumping. And in the steady states all the solitons generated have identical properties, a result of the soliton energy quantization effect [38]. Interaction between the multiple solitons has been extensively investigated previously. It was found that various modes of the multiple soliton operation could be formed [39]. A special situation also experimentally observed is that under appropriate laser cavity conditions, the solitons in the cavity could automatically bind together and form states of bound solitons [40,41]. Depending on the strength of the soliton binding, a certain state of the bound solitons can even become the only stable state in a laser. In such a state the bound solitons as an entity exhibit features that are in close similarity to those of the single pulse solitons [40,42,43]. It was speculated that such a state of bound solitons could even be regarded as a new form of multipulse solitons in the lasers [44]. We found experimentally that period-doubling bifurcations could also occur on these bound solitons. Like the conventional single-pulse solitons, the bound solitons as an entity can exhibit complicated deterministic dynamics.

Figures 6.7a, 6.7b, and 6.7c show, for example, an experimentally observed period-doubling bifurcation route of a state of bound solitons. In the current case several solitons are tightly bound and move together at the cavity fundamental repetition frequency. When the peak intensity of the bound solitons is strong, the total intensity of the bound solitons exhibits the period-doubling bifurcation. The response time of the photodetector used in our experiment is about 140 ps, which can clearly resolve the pulse train but not the intensity profile of the bound solitons. Therefore, the measured oscilloscope traces shown in Figure 6.7 have no distinctions to those of the period-doubling bifurcations of the single-pulse solitons [13]. Figures 6.7d, 6.7e, and 6.7f are the simultaneously measured optical spectra corresponding to Figures 6.7a through 6.7c. Clear optical spectral modulations exist on the spectra, which show that more than one soliton are actually in the cavity and they are closely spaced. The oscillating peaks of the optical spectra have a peak separation of 0.7 nm. Assuming that the central wavelength of the soliton pulse is 1550 nm, based on the Fourier transformation, the solitons have a peak separation of about 11.5 ps in the time domain. The bound soliton nature of the state is also confirmed by the simultaneous autocorrelation trace measurement as shown in Figure 6.8. Limited by our autocorrelator resolution, the measured autocorrelation traces show no distinguishable features among different period-doubled states. The measured single soliton duration of the state is about 326 ± 12 fs, and the soliton separation between neighboring solitons is about 11.8 ps, which agrees with the 0.70 nm period of optical spectral modulation within the experimental errors. The soliton separations under different period-doubled states are also the same. From the autocorrelation traces it is shown that there are at least three solitons with the same soliton separations binding together in the bound-soliton state. Due to the limited scan range of our autocorrelator we could not identify

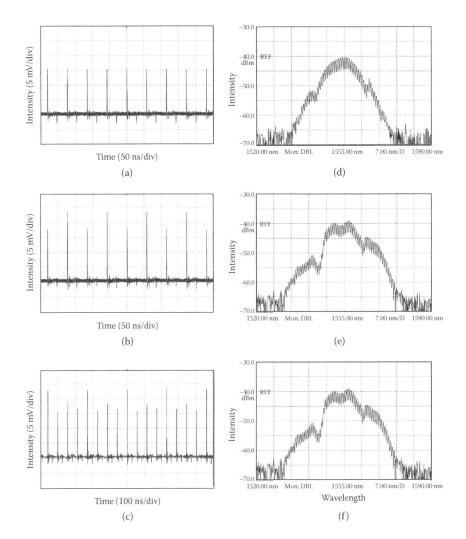

FIGURE 6.7
Oscilloscope traces and corresponding spectra of period-doubling bifurcations of a bound-soliton pulse train. (a,d) Period 1 state; (b,e) period-doubled state; and (c,f) period-quadrupled state.

how many solitons are actually in the state. Nevertheless, it is confirmed that all the solitons in the state are equally spaced, and the bound soliton as an entity exhibits period-doubling bifurcations. We note that the state of bound solitons shown in Figure 6.7 is the only pulse pattern existing in the cavity, which is confirmed by the combined monitoring of the autocorrelation trace and a high-speed oscilloscope trace.

Figures 6.7a, 6.7b, and 6.7c are obtained with continuously increased pump power while keeping all the other cavity parameters fixed. The chaotic state of the bound solitons could be obtained with further increasing pump power

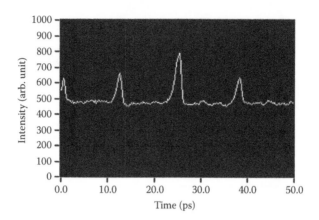

FIGURE 6.8
A typical measured autocorrelation trace of the bound soliton.

from the state in Figure 6.7c. However, the chaotic state was not stable. It quickly evolved into a chaotic state of the single-pulse soliton. The change of the laser emission from a chaotic state of bound solitons to that of single-pulse solitons was experimentally identified by the change of the optical spectra. It was observed that whenever the bound-soliton became chaotic, the modulations on the optical spectrum, which is a direct indication of close soliton separation in the time domain, disappeared, and subsequently the optical spectrum had exactly the same profile as that of the chaotic state of the single-pulse solitons. Considering that in a chaotic state the energy of solitons varies randomly, which may affect the binding energy between the solitons and destroy their binding, this result seems also plausible. The states of period-doubled and quadrupled-bound solitons are stable. Nevertheless, compared with those states of the single-pulse solitons [13], they are more sensitive to the environment perturbations. Analyzing the optical spectra of the period-doubled bound solitons, it is found that the overall spectral profile in each state is similar to those of the single-pulse soliton undergoing the period doubling, except that it is now modulated [13]. The feature of the optical spectra indicates that the individual solitons within a bound-soliton are still the same as the single-pulse soliton of the laser.

RF spectra of the bound-soliton pulse train also disclose different period-doubled states. The emergence of a new frequency component at the position of half of the fundamental cavity repetition frequency clearly shows that the repetition rate of the bound solitons is doubled. The amplitude of the new frequency component is nearly half of that of the fundamental frequency component, which vividly suggests that the total peak intensity of the bound solitons alternates between two values with a large difference. The same as observed in [13], the new frequency component corresponding to the period-quadrupling is not distinguishable from the background noise. Limited by the resolution of our measurement system, the detailed intensity variations of each of the solitons under period-doubling bifurcations could not be resolved. There are two possible ways for a two-pulse bound soliton exhibiting a period-doubling intensity pattern. Either the two solitons simultaneously experience the period doubling, or only one soliton experiences the period doubling while the other remains stable. With more solitons binding together the process could become more complicated.

Experimentally, we also observed period-doubling bifurcations of bound solitons with different soliton separations. Bound solitons with different pulse separations can be easily distinguished by their optical spectra, as different spectral modulation periods correspond to different pulse separations in the time domain. This experimental result suggests that the appearance of the phenomenon should be a generic feature of the laser, which is independent of the concrete property of the optical pulses. The period-doubling route to chaos is a well-known nonlinear dynamic phenomenon widely investigated. The period-doubling route to chaos of the CW and the Q-switched lasers as a result of strong nonlinear interaction between the light field and the gain medium have already been reported [45,46]. Except that the laser modes are phase locked, physically the interaction between the light and the gain medium in a mode-locked laser is still the same. Therefore, it is not surprising that under the existence of strong mode-locked pulses, the period-doubling route to chaos on the pulse repetition rate could still be obtained. Nevertheless, it was a little bit unexpected that a bound-soliton can exhibit period-doubling bifurcations as intuitively the dynamic bifurcation of the laser state could easily damage the binding between the solitons. In our experiment a dispersion-managed laser cavity was used. The purpose of using a dispersion-managed cavity is to possibly make the energy of the formed solitons strong, so an average strong nonlinear interaction between the pulses with the cavity components could be achieved.

6.3.5 Period Doubling of Multiple Solitons

It is well known that a passive mode-locked fiber laser can operate with multiple solitons in a cavity, and depending on the soliton interaction, the multiple solitons can either form a soliton bunch or randomly distribute with stable relative soliton separations [47]. With more than two solitons coexisting in a cavity, soliton interaction cannot be ignored. In the previous section we reported the period-doubling bifurcation of bound solitons. In that case two or more solitons coexist in the cavity. However, because the solitons are strongly coupled, the formed state of bound solitons actually behaves like a single pulse soliton. No difference in their period-doubling bifurcations was observed to those of the single pulse soliton. Different from the case of the bound solitons, the solitons are now distributed far apart in the cavity with stable relative separations. We show that even with multiple solitons coexisting in a fiber cavity, and there is obviously gain-competition between them,

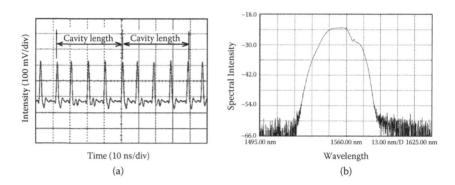

FIGURE 6.9
(a) Oscilloscope trace and (b) optical spectrum of a period-one state of the randomly distributed multiple solitons (four solitons in the cavity).

period-doubling bifurcation can still occur in the laser, and specifically, each soliton in the cavity experiences period-doubling bifurcation. However, the intensity variation of the individual solitons is not necessarily synchronized.

Figure 6.9 shows, for example, a typical state of the multiple soliton operation observed in a fiber laser. The laser has a cavity round-trip time of about 38 ns. It is to see that four solitons coexist in the cavity and locate far apart from each other. In the case of current soliton operation, all solitons have the same pulse height in the oscilloscope trace, indicating that after every cavity round-trip the energy of each soliton returns to its previous value. Such a multiple soliton operation has also been reported by other authors, which is a typical case of the conventional soliton fiber laser operation [6]. Changing the orientation of one of the waveplates in the laser cavity, which corresponds to shifting the linear cavity phase delay bias away from the cavity polarization switching point, and therefore increasing the formed soliton peak power [29], period-doubling bifurcation occurs in the laser. Figure 6.10 shows a case of the period-doubled state of the

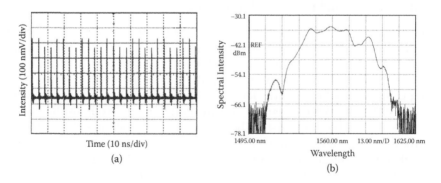

FIGURE 6.10
(a) Oscilloscope trace and (b) optical spectrum of a period-doubled multiple-soliton state (two solitons in the cavity).

laser with two solitons coexisting in the cavity. Comparing with the oscilloscope trace shown in Figure 6.9, the pulse height of each soliton in the oscilloscope returns to its value after every two cavity round-trips. It is obvious that each individual soliton in the cavity experiences the period-doubling bifurcation, suggesting that each of them has the same bifurcation threshold and behaves identically in the cavity. For the sake of completeness, we have also shown in Figures 6.9b and 6.10b the corresponding optical spectrum of the solitons. As the solitons have large separations in the cavity, the spectrum of the multiple solitons is the same as that of the single soliton. Again it is to see that after the period-doubling bifurcation, extra spectral structures appear on the soliton spectrum, indicating the existence of dynamical sideband generation [18].

As there are only two solitons in the cavity, it is difficult to judge from the oscilloscope trace whether the intensity variations of the solitons are synchronized or not. To clarify, we show in Figure 6.11a another period-doubled state of the multiple solitons, where six pulses randomly scattered in the cavity. Again after the period-doubling bifurcation, the period of each pulse in the cavity becomes doubled, but it is now clear to see that the intensity variation of the pulses is not all synchronized. In the same cavity round-trip some pulses are in their high-power state while the others are in their low-power state. Note that one pulse in the oscilloscope trace has significantly larger pulse height than the others. It is actually due to that two solitons are too close in the cavity that our detector cannot resolve them. Therefore, in the oscilloscope trace they appear as one pulse. It is to point out that due to the opposite intensity variation of the two solitons in the current state, the total pulse height of them in the oscilloscope exhibits no change. It is only an experimental artificial appearance. In the experiment we also obtained a state as shown in Figure 6.11b, where the strong pulse in the oscilloscope trace also exhibits period doubling, indicating that the two solitons that form the pulse have synchronized intensity variation. Using the commercial autocorrelator, we measured the soliton pulse width of our laser, which is about 1.54 ps if a Gaussian pulse shape is assumed.

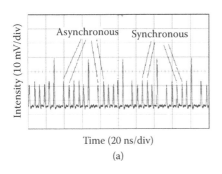
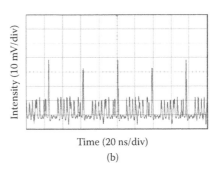

FIGURE 6.11
Oscilloscope traces of period-doubled multiple-soliton states. (The higher pulses with nearly twice intensity are caused by two closely spaced solitons.): (a) period doubling under moderate intensity excitation, (b) period doubling with high intensity.

It was shown that even under the existence of multiple solitons in the cavity the laser still can experience period-doubling bifurcation. In particular, each soliton in the cavity exhibits the same period-doubling bifurcation. However, the detailed intensity variation of the solitons could be unsynchronized, indicating that their period doubling is actually not related.

6.3.6 Period Doubling of Dispersion-Managed Solitons at around Zero Cavity Dispersion

Although a dispersion-managed cavity consists of fibers with either positive or negative dispersion, and on average the pulse peak power is lower than that of the pulses formed in the equivalent uniform dispersion cavity, as far as the net cavity GVD is negative and large, conventional solitary waves (solitons with clear sidebands) could still be formed in the lasers. Furthermore, in the regime of near-zero net cavity GVD a different type of solitary wave known as the dispersion-managed solitons [48] could also be formed. Comparing with the conventional solitons formed in a laser, the dispersion-managed solitons are characterized by that their optical spectra have a Gaussian profile without spectral sidebands. Dispersion-managed solitons in the fiber transmission lines have been extensively investigated [48–53]. It was shown that such a soliton could even exist in a system with positive near-zero net GVD [49–52].

It is well known that cavity dispersion is wavelength related, and the central wavelength of the mode-locked pulses shifts with the experimental conditions. It is difficult to accurately determine the net cavity dispersion under various laser mode-locking states. However, one can roughly estimate the dispersion of a laser cavity by simply accumulating the dispersion of each cavity component, and then fine-tune it through cutting back the single mode fiber (SMF) length.

We also investigated the period-doubling bifurcation of the dispersion-managed solitons. After having obtained a uniform dispersion-managed soliton pulse train as shown in Figure 6.12a, we then carefully tuned one of the waveplates to the direction that causes the pulse peak power to increase

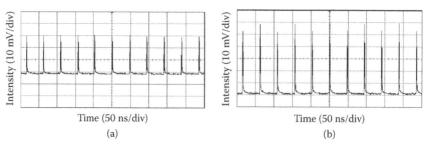

(a)   (b)

FIGURE 6.12
Oscilloscope traces of the dispersion-managed solitons: (a) Period 1 state and (b) period-doubled state.

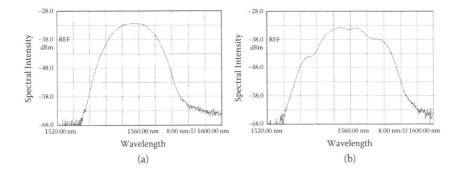

FIGURE 6.13

Dispersion-managed solitons of the laser: (a) optical spectrum of the Period 1 state and (b) optical spectrum of the period-doubled state.

while keeping all other laser parameters unchanged. Physically this action corresponds to shift the linear cavity phase delay bias to the direction that causes the peak power of the optical pulse to be clamped at a higher level by the cavity [14]. To a certain level of the pulse peak power, it was observed that the dispersion-managed solitons exhibited period-doubling bifurcation as shown in Figure 6.12b. Obviously, in a Period 2 state the pulse intensity returned to its original value by every two cavity round-trips. Figure 6.13b shows the optical spectrum of the dispersion-managed solitons corresponding to Figure 6.12b. The fundamental repetition rate of the laser is 21.3 MHz. Limited by the resolution of our autocorrelator, the autocorrelation trace measured under a period-doubled state has no obvious difference to that of the uniform soliton pulse train. However, the optical spectrum of the state shows clear modulation as compared with that of the Period 1 state (Figure 6.13a). The measured optical spectrum is an average of the pulse spectra. As in the period-doubled state the laser emits alternately between two mode-locked pulse states, it is expected that the resultant spectrum is different from that of the Period 1 state. Monitored by a high-speed sampling oscilloscope, we confirmed experimentally that in the state shown there is only one soliton pulse propagating in the cavity.

Figure 6.14 further shows the RF spectra of the laser outputs. Obviously, in the period-doubled state a new frequency component appears at half of the cavity repetition rate, while there is no such frequency component in the case of the Period 1 state.

6.3.7 Period Doubling of Gain-Guided Solitons in Fiber Lasers with Large Net Normal Dispersion

Although dynamics of the lasers with large positive cavity dispersion is still determined by the extended GLE, the formed soliton pulses in these lasers have distinct features from those of the solitons formed in lasers with

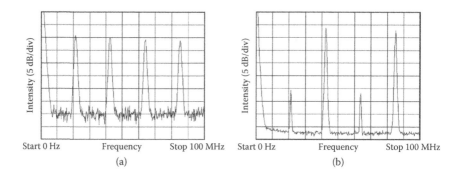

FIGURE 6.14
Radio-frequency spectrum of the dispersion-managed solitons: (a) Period 1 state and (b) period-doubled state.

negative cavity dispersion. While the solitons formed in the latter are dominantly a result of the balanced interaction between the cavity negative dispersion and the fiber nonlinear Kerr effect, the solitons formed in lasers with positive cavity dispersion are due to the spectral filtering of the limited gain bandwidth and the cavity nonlinearity. To highlight their differences solitons formed in fiber lasers with positive cavity dispersion were also called gain-guided solitons (GGSs) [54]. The GGSs belong to the family of dissipative solitons [55]. They are a localized, stable chirped nonlinear wave. It was found that the GGSs could also exhibit deterministic dynamics despite the fact that GGSs have large chirp and broad pulse width.

GGSs are characterized by their steep spectral edges and pump-power-dependent spectral bandwidth. They could be easily obtained in a mode-locked fiber laser with large positive cavity dispersion. To observe period-doubling bifurcation of GGSs, one starts from a stable GGS operation state. If the saturable absorption strength of the cavity is gradually increased, this could be done through shifting the linear cavity phase delay bias away from the polarization switching point [29] of the laser, the peak power of the mode-locked pulses increases. Note that accompanying the increase of the saturable absorption strength, the pump power should also be carefully increased in order to maintain the stable GGS operation. To a certain point of the pulse peak power it could be observed that some spectral spikes suddenly appear on the long wavelength side of the soliton spectrum as shown in Figure 6.15a. However, the appearance of the soliton spikes does not affect the stable GGS operation. Figures 6.15b and 6.15c show the oscilloscope trace and the RF spectrum measured immediately after the appearance of the spectral spikes. Obviously, the laser still emitted uniform pulses. The laser cavity length was about 4.5 m, which matched the soliton repetition rate of 44.8 MHz shown in Figure 6.15c. Based on autocorrelation measurements, the GGSs of the laser have pulse duration of about 3.18 ps if a Gaussian pulse profile is assumed. After the spectral spikes

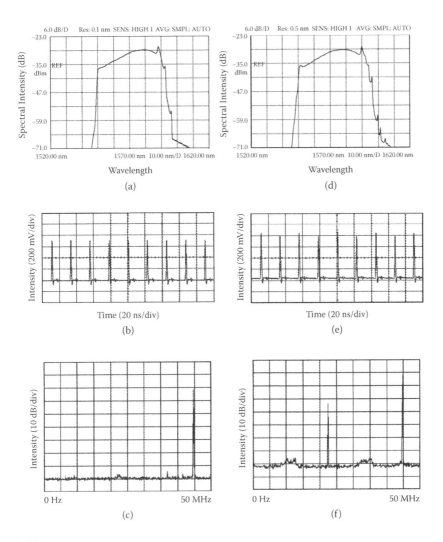

FIGURE 6.15
Period 1 doubling of gain-guided solitons in the laser: (a,d) optical spectrum, (b,e) oscilloscope trace, and (c,f) radio frequency (RF) spectrum.

are obtained, if the pump power is further increased but with all other laser operation conditions fixed, a period-doubling bifurcation of the GGS is then observed, as shown in Figures 6.15d through 6.15f. Associated with the period-doubling bifurcation, more spectral spikes appear on the soliton spectrum (Figure 6.15d). The period doubling of the soliton is clearly visible on the oscilloscope trace, where after every two cavity round-trips the pulse energy returned, and on the RF spectrum, where a new spectral component appears at the position of half-cavity fundamental repetition frequency. The above process is stable and repeatable in a laser. Nevertheless, in our experiments no further period-doubling bifurcation but a noise-like state was observed when the pump power was further increased.

6.3.8 Period Doubling of Vector Solitons in a Fiber Laser

A characteristic of the NPR mode-locked fiber lasers is that a polarizer is inserted in the laser cavity for the generation of an artificial saturable absorption effect in the cavity [56]. As the intracavity polarizer fixes light polarization at the cavity position, solitons formed in the lasers are considered as scalar solitons.

Due to technical limitation in fabricating perfectly circular cross-sectional core fibers and random mechanical stresses, in practice a single-mode fiber always supports two polarization eigenmodes, or in other words, possesses weak birefringence. It has been shown that without a polarizer in the cavity, solitons formed in a fiber laser could exhibit complicated polarization dynamics. S. T. Cundiff et al. demonstrated the formation of vector solitons in a fiber laser mode locked with a semiconductor saturable absorber mirror (SESAM) [57]. A vector soliton differs from a scalar soliton in that it consists of two orthogonal polarization components, and the two polarization components are nonlinearly coupled. Depending on the features of their coupling there are different types of vector solitons, the Polarization–locked vector solitons (PLVSs), the group velocity locked vector solitons (GVLVSs), and the polarization rotating vector solitons.

A vector soliton fiber laser can also exhibit deterministic dynamics. Using a similar cavity configuration as reported in [58], we have first experimentally observed period-doubling bifurcation of a vector soliton fiber laser. Briefly, the vector soliton fiber laser has a ring cavity with a length of about 9.40 m, which consists of 2.63 m erbium-doped fiber (StockerYale EDF-1480-T6) and all other fibers used are the standard single-mode fibers (SMFs). With the help of a three-port polarization independent circulator, an SESAM is introduced in the ring cavity of the laser. A 1480 nm Raman fiber laser with maximum output of 220 mW is used to pump the laser. The backward pump scheme is adopted to avoid the CW overdriving of the SESAM by the residual pump strength. The laser outputs through a 10% fiber coupler. A fiber-based polarization controller is inserted in the cavity to control the cavity birefringence. The SESAM used has a saturable absorption of 8%, and a recovery time of 2 ps.

As no explicit polarization discrimination components are used in the cavity, and all the fibers used have weak birefringence, vector solitons are easily obtained in the laser by simply increasing the pump power above the mode-locking threshold. Determined by the detailed laser operation conditions, various types of vector solitons, such as the PLVS, incoherently coupled vector soliton, and polarization rotating vector solitons are obtained in the fiber laser. In particular, it is found that the polarization rotation of

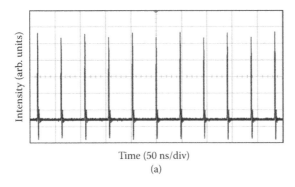

Time (50 ns/div)

(a)

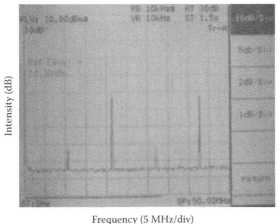

Frequency (5 MHz/div)

(b)

FIGURE 6.16
The (a) oscilloscope trace and (b) corresponding radio-frequency spectrum of direct intensity period doubling.

the formed polarization rotating vector solitons could be locked to the laser cavity round-trip time or a multiple of it [58].

In the parameter regime of a polarization rotation locked vector soliton, a state of the period-doubling of the vector solitons, as shown in Figure 6.16a, is also observed. The vector soliton output of the fiber laser is directly monitored by a photodetector with 2 GHz bandwidth. Although the pulse intensity difference between two adjacent cavity round-trips is weak, intensity period-doubling of the soliton pulse train is evident. The RF spectrum of the laser emission is also measured, as shown in Figure 6.16b. A weak but clearly visible frequency component appears at the position of half of the cavity fundamental repetition frequency. Measured after passing through an external polarizer, a polarization rotation state with polarization rotation locked to twice of the cavity round-trip time would have the same result as

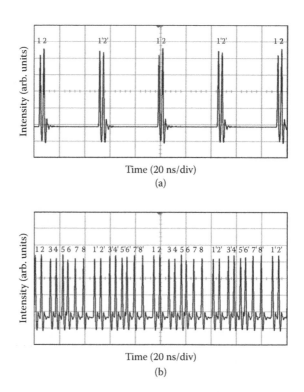

FIGURE 6.17
Period doubling of multiple vector solitons: (a) two vector solitons and (b) eight vector solitons.

shown in Figure 6.16. However, such a state is not a period-doubled state as the pulse intensity alternation observed on the oscilloscope trace is due to the polarization rotation of the vector soliton. The period doubling shown in Figure 6.16 is formed due to the intrinsic dynamic feature of the laser.

After the period-doubling state is achieved, keeping all other laser operation parameters fixed but increasing the pump power the number of vector solitons circulating in the cavity increases. The new vector solitons generated exhibit the same period-doubling feature. Figures 6.17a and 6.17b show, for example, the cases where two vector solitons and eight vector solitons coexist in the cavity, respectively. Period-doubling of the vector solitons can be clearly identified. Decreasing pump power reduces the number of vector solitons in the cavity. Varying pump power could also change the soliton pulse intensity within a small range. However, no Period 1 state could be obtained by simply decreasing the pump power. Period 1 state could be obtained only if the cavity birefringence is changed. Similarly, after a Period 1 state is obtained, the period-doubling state could not be obtained by simply increasing the pump strength. Experimentally it is found that in a period-doubled state the polarization rotation of the vector solitons is locked to twice that of the cavity round-trip time.

6.4 Numerical Simulations on the Soliton Deterministic Dynamics in Fiber Lasers

6.4.1 Round-Trip Model of the Soliton Fiber Lasers

The soliton propagation in a fiber laser is characterized by the soliton circulation in the nonlinear ring cavity and the periodical interaction with the cavity components. Therefore, in order to simulate the soliton dynamics, in particular to gain an insight into the deterministic dynamics of the various types of solitons in the mode-locked fiber lasers, a round-trip model is used to simulate the soliton evolution in the lasers. Concretely, we start a simulation with an arbitrarily small light pulse and let it circulate in the cavity. We follow the pulse circulation in the laser cavity. Whenever the pulse encounters a discrete cavity component, for example, the cavity output coupler, the action of the cavity component on the light pulse is considered by multiplying the transfer-matrix of the discrete cavity component to the pulse. The pulse propagation in the fiber segments of the cavity is described by Equation (6.1). After one round of circulation in the cavity, the result of the previous round of calculation is then used as the input of the next round of calculation until a steady state is reached, which is denoted as a stable soliton state of the laser.

The round-trip model has several advantages. The calculation is made following the pulse propagation in the cavity. Therefore, the detailed pulse evolution within one cavity round-trip can be studied. Within each step of calculation the pulse's variation is always small, even if the change of a pulse within one cavity round-trip is big. As there is no limitation on the pulse change within one cavity round-trip, dynamical process of soliton evolution, such as the process of new soliton generation in the cavity, soliton interaction, soliton collapse, and so forth can be investigated. In addition, the effect of discrete cavity components on the soliton, the influence of the dispersive waves, and the different order of the cavity components on the soliton property are automatically included in the calculation.

The soliton operation of a fiber laser is simulated using exactly the laser cavity parameters whenever they are known. Numerically we found that independent of the concrete laser cavity design, as far as the intensity of the formed soliton is strong, soliton quasi-periodicity and period-doubling bifurcations could always be obtained. Under certain laser designs, soliton intermittency could also be numerically observed. Nevertheless, in order to obtain a full period-doubling route to chaos, the cavity parameters such as the fiber dispersion and fiber lengths must be appropriately selected.

6.4.2 Deterministic Dynamics of Solitons in Different Fiber Lasers

Based on the round-trip model the soliton operation of the passive mode-locked fiber ring lasers mode locked with the nonlinear polarization rotation technique were numerically simulated. Properties of the soliton pulses formed in the lasers, influence of different laser cavity parameters on the soliton properties, as well as the intrinsic laser cavity effect on the soliton operation were numerically studied. In all of the numerical simulations, if it is not explicitly pointed out, the same simulation parameters as the following were used. The fiber nonlinearity: $\gamma = 3$ W^{-1}m^{-1}; fiber dispersion: -12.8 ps^2/km for the erbium-doped fiber, -23 ps^2/km for the single-mode fiber, -2.6 ps^2/km for the dispersion-shifted fiber; fiber beat length: $L_b = L/2$; the orientation of the intracavity polarizer to the fiber fast birefringent axis: $\Psi = 0.152\pi$; gain saturation energy: $E_{sat} = 300$ pJ; gain bandwidth: $\Omega_g = 16$ nm; laser cavity length: $L = 6$ m, and the orientations of the polarizer and the analyzer with the fast axis of the birefringent fiber: $\theta = \pi/8$ and $\varphi = \pi/2 + \pi/8$.

6.4.3 Period-Doubling Route to Chaos of the Single-Pulse Solitons

Through properly choosing the linear cavity phase delay bias, which corresponds in the experiment to appropriately selecting the orientations of the polarization controllers, soliton operation can always be obtained in our simulations. With a fixed linear cavity phase delay bias but different values of gain, as long as the generated peak power of the soliton pulse is weaker than that of the polarization switching power of the cavity [29], a stable uniform soliton pulse train can always be obtained. Figure 6.18a shows the soliton profiles of the laser under different gain coefficients when the linear cavity phase delay bias is fixed at $\delta\phi = 1.2\pi$. Figure 6.18b is the corresponding soliton spectra. The exact soliton parameters, such as the pulse width and peak power, are determined by the laser parameter settings and the laser operation condition such as the gain value. Under a larger pump power the solitons generated have higher peak power and narrower pulse width. With the current linear cavity phase delay bias selection, the nonlinear polarization

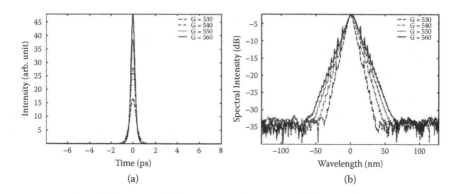

FIGURE 6.18
Soliton profiles (a) and the corresponding optical spectra (b) when the linear cavity phase delay bias is set as $\delta\phi = 1.2\pi$.

switching (NPS) threshold of the cavity is low. Therefore, the maximum peak power of the solitons reachable is clamped by the nonlinear polarization switching effect. Once the soliton peak is clamped, further increasing the gain, instead of further increasing the soliton peak power, a new soliton is generated, and consequently a multiple soliton operation state of the laser can be obtained. As a laser operating in the state has weak linear cavity loss, mode locking can be easily achieved. Therefore, in practice a laser will always start the soliton operation from the state. However, our numerical simulations show that no soliton period-doubling bifurcation could occur at such a linear cavity phase delay bias as the soliton peak power is weak.

When the linear cavity phase delay bias is chosen as $\delta\phi = 1.6\pi$, which corresponds to lifting the nonlinear polarization switching threshold of the cavity higher so that the soliton pulse formed could have higher peak power, stable soliton operation can still be obtained. As now the soliton peak power can reach a very high value under strong pumping, it is observed that when the peak power of the soliton increases to a certain value, period-doubling bifurcations and period-doubling route to chaos of the solitons as observed in the experiments automatically appear. Figure 6.19 shows, for example, a numerically calculated period-doubling route to chaos of a laser. When the gain coefficient was set at $G = 800$, a stable and uniform high-intensity soliton train is obtained (Figure 6.19a). Increasing the value of G and keeping all the other parameters fixed, the soliton repetition period in the cavity is then doubled at $G = 850$ (Figure 6.19b). At $G = 902$ it doubles again (Figure 6.19c) and further doubles at $G = 908$ (Figure 6.19d). Eventually the soliton repetition in the laser becomes chaotic (Figure 6.19e). Figures 6.19f through 6.19j show the corresponding optical spectra of the solitons in Figures 6.19a through 6.19e. Associated with the soliton intensity variation the soliton spectrum also exhibits period-doubling changes. Figure 6.20 further shows the soliton spectral variation within one period of the Period 4 state. In this state after every four cavity round-trips the soliton intensity and profile return to the original value and shape. From Figure 6.20, see that the soliton spectrum in each round-trip is different. Corresponding to the soliton of the highest peak power, the soliton spectrum (Figure 6.20a) also exhibits the strongest sidebands and spectral modulations, indicating the existence of strong nonlinear self-phase modulation (SPM) on the pulse. The change from one soliton operation state to the other is abrupt. At the bifurcation point when the gain coefficient is slightly increased, the soliton quickly jumps to another state with the doubled periodicity, exhibiting the universal characteristic of the period-doubling bifurcation and route to chaos of the nonlinear dynamics systems.

The appearance of the period-doubling bifurcation is independent of the concrete laser cavity design. Under different cavity parameter settings, as far as the soliton peak power is unlimited by the nonlinear polarization switching effect of the cavity, we could always obtain the phenomenon in our simulations. Even the period-doubling route to chaos of a Period 3 state, which in terms of the nonlinear dynamics theory is known as a periodic window

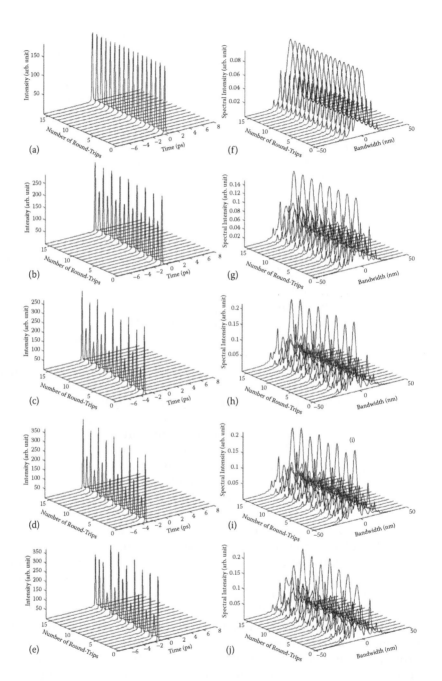

FIGURE 6.19
Soliton pulse evolution and the corresponding optical spectra numerically calculated under different pump strengths. The linear cavity phase delay bias is set as df = 1.6π. (a,f) Period 1 soliton state, $G = 800$; (b,g) Period 2 soliton state, $G = 850$; (c,h) Period 4 soliton state, $G = 902$; (d,i) Period 8 soliton state, $G = 908$; (e,j) chaotic soliton state, $G = 915$.

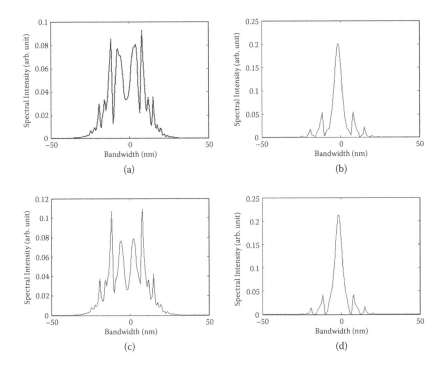

FIGURE 6.20
Soliton spectral variation within one Period 4 state [soliton evolves from (a) to (d)].

within the chaotic regime, has also been numerically revealed. However, it is to point out that due to the coexistence of other effects in the laser, for example, the soliton peak nonuniformity [35] and soliton collapse [59], in order to obtain a full period-doubling route to chaos in a laser, the cavity parameters must be appropriately selected. In some of our numerical simulations frequently only certain period-doubling bifurcations, for example, the Period 1 to Period 2 bifurcation and then to chaos, or the Period 1 to chaotic state, could be obtained. If the linear cavity phase delay bias is not set large enough, the nonlinear polarization switching effect could also limit the peak power of the solitons. Consequently, only bifurcations to a certain periodic state, for example, the Period 4 state, could be reached. Further increase of the gain would cause the generation of a new soliton rather than the further bifurcation to chaos, which clearly shows the direct relation of the period-doubling bifurcation to the soliton peak power.

Previous studies on the synchronously pumped passive ring cavities have also revealed a period-doubling cascade to chaos in the sequence of pulses emerging from the cavities [60,61]. It was shown that the bifurcations and route to chaos of the system were caused by the repetitive interference between the input pulse and the pulse that has completed a round-trip in the cavity. As the pulse traveling in the cavity suffers nonlinear phase shift, which itself is pulse intensity dependent, the transmission of the cavity is a nonlinear function of the pulse intensity. It is an intrinsic property of such a nonlinear cavity that under larger nonlinear phase shift of the pulse, its output exhibits the period-doubling route to chaos [2]. We note that a similar repetitive interference process exists in the fiber soliton lasers. Due to the birefringence of the laser cavity, the pulse propagation in the laser actually includes two orthogonal polarization components. Although there are non-linear couplings between the two orthogonally polarized pulses as can be seen from Equation (6.1), after one round-trip they experience different linear and nonlinear phase shifts. The interference between them at the intracav-ity polarizer results in that the effective cavity transmission is a nonlinear function of the soliton intensity. Based on the studies on the synchronously pumped passive ring cavities, it is therefore to imagine that under strong soliton peak intensity, the period-doubling bifurcation and route to chaos could also appear in the lasers. Our numerical simulation shows that the effect could only occur in the fiber lasers when the linear cavity phase delay bias is set away from the nonlinear polarization switching point. In this case the soliton peak power is unclamped by the cavity and can increase to a very high value with the increase of the pump strength. A high soliton peak power generates a large nonlinear phase shift difference between the two polarization components, which causes the intrinsic instability of the system.

In order to obtain a full route to chaos, the cavity parameters such as the fiber dispersion and fiber lengths must be appropriately selected. In one of our numerical simulations even the period-doubling route to chaos of a Period 3 state, which in the nonlinear dynamics theory is known as a peri-odic window within the chaotic regime, was also obtained as shown in Figure 6.21.

6.4.4 Period-Doubling Route to Chaos of the Bound Solitons

Period-doubling bifurcations and route to chaos have also been numerically revealed for the bound solitons. Figure 6.22 shows, for example, one of such results obtained. In calculating the state we used the same laser parame-ters as those for obtaining the single-pulse soliton period-doubling route to chaos, only the pump strength and initial state are different. Figure 6.22a shows that two solitons coexist in the cavity and bind together. Note that due to the close separation between the solitons, their optical spectra have strong intensity modulations, but from round to round the modulation pat-terns do not change, which indicates that the phase difference between the solitons is fixed as well. The binding nature of the solitons is represented by the fixed soliton separation and phase difference even under the existence of soliton interaction between them. Increasing pump strength from the state of Figure 6.22a, both solitons experience simultaneously period-doubling bifurcations and route to chaos. Associated with the soliton period doubling, the soliton separation between the bound solitons also changes slightly.

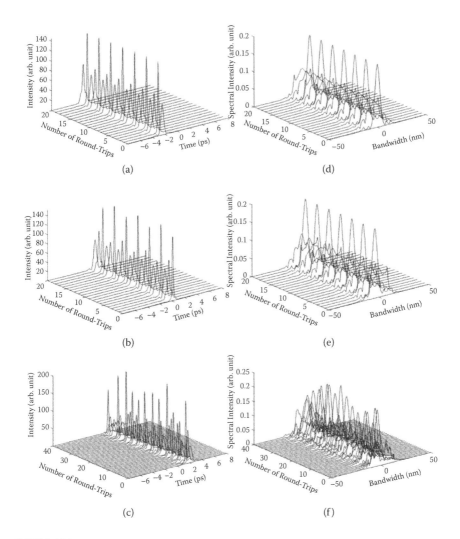

FIGURE 6.21
Soliton profiles and corresponding optical spectra numerically calculated. (a,d) State of Period 3, $G = 730$; (b,e) state of Period 6, $G = 735$; (c,f) chaotic state, $G = 750$.

However, after the period-doubling bifurcation the soliton separation then remains constant again. This soliton separation change suggests that the dynamic bifurcation of the system could affect the soliton interaction.

6.4.5 Period Doubling of Multiple Solitons

As shown previously, multiple soliton formation in the fiber lasers is a result of the cavity pulse peak clamping effect [29]. Therefore, through appropriately setting the linear cavity phase delay bias and the pumping strength,

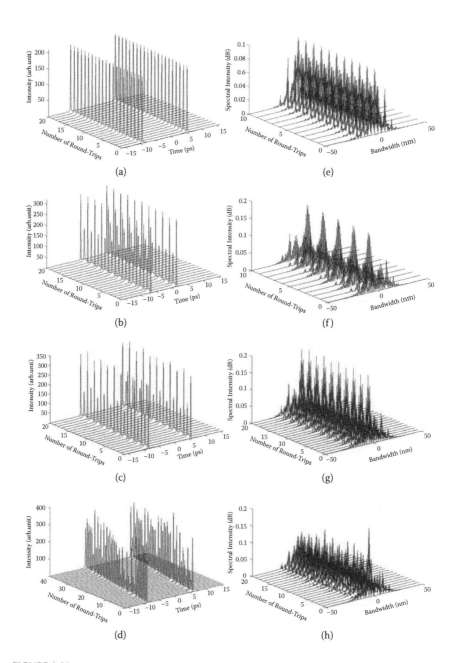

FIGURE 6.22
Period-doubling route to chaos of the bound solitons. (a,e) State of stable bound solitons, G = 1149; (b,f) state of Period 2 of the bound solitons, G = 1300; (c,g) state of Period 4 of the bound solitons, G = 1353; and (d,h) chaotic state of the bound solitons, G = 1358.

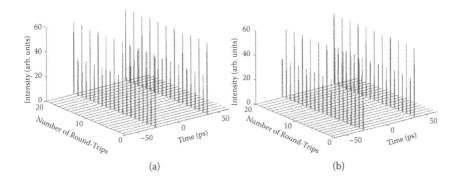

FIGURE 6.23
Calculated pulse intensity evolution with the cavity round-trips of the period-doubled solitons: (a) synchronous evolution and (b) asynchronous evolution.

randomly distributed multiple solitons can be numerically easily obtained. The maximum achievable soliton pulse peak power is also determined by the setting of the linear cavity phase delay bias. If the linear cavity phase delay bias is set too close to the cavity polarization switching point, only low peak power of the solitons could be obtained. In this case increasing the pumping strength except that the number of the solitons in the cavity will increase, no period-doubling bifurcation of the solitons could be observed. Therefore, to obtain the soliton period-doubling bifurcation it is necessary to set the linear cavity phase delay bias away from the polarization switching point. Numerically we find that with current simulation parameters only when the linear cavity phase delay bias is larger than 1.4π, the phenomenon of soliton period doubling could be achieved. In our simulation we fixed it as $\delta\varphi = 1.5\pi$. Depending on the initial state and pump strength, either synchronous or asynchronous period doubled evolutions are obtained. Figures 6.23a and 6.23b show the typical numerical results, where two solitons have a separation of 64 ps in cavity. Figure 6.23a shows that both solitons are period-doubled and have exactly the same intensity evolution with the cavity round-trips, while Figure 6.23b shows a case where the solitons have unsynchronized pulse intensity evolution with the cavity round-trips. Numerically we also obtained states of period-doubling of multiple solitons with more than two solitons and confirmed all of the experimental observations.

6.4.6 Period Doubling of Dispersion-Managed Solitons

The cavity of a mode-locked fiber laser is a nonlinear cavity. Therefore, it is expected that when the intensity of the pulse circulating in the fiber ring laser has strong peak power, the laser output could exhibit similar nonlinear dynamical behaviors. Furthermore, as the occurrence of the period doubling is a property of the nonlinear laser cavity, no matter whether the pulse circulating in it is a conventional soliton or a dispersion-managed soliton, the same cavity dynamics should be observed.

Figure 6.24 shows, for example, the numerically calculated dispersion-managed soliton for the positive near-zero cavity dispersion case. The optical spectrum of the dispersion-managed soliton shown in Figure 6.24a has a Gaussian-like spectral profile and no obvious sidebands, which agrees with the experimental observation. Figure 6.24b shows the corresponding temporal pulse profile. Again it is closer to the Gaussian shape than the sech^2 form. Our numerical results clearly show that the dispersion-managed soliton could be formed in lasers with positive near-zero cavity dispersion just as experimentally observed.

After a Gaussian-like stable uniform pulse train is obtained, with all the other simulation parameters fixed but the small signal gain coefficient is increased, a period-doubled state as shown in Figure 6.24c is then observed, which shows that, like the conventional solitons, the dispersion-managed solitons can experience the period-doubling bifurcation as well. With our current laser cavity design only a Period 2 state could be obtained. If the pump strength is further increased, a new dispersion-managed soliton is generated instead of a further period-doubling of the pulse [62]. Similar results were also obtained when the net cavity dispersion is set negative near zero or zero.

6.4.7 Period Doubling of the Gain-Guided Solitons

The period-doubling bifurcation of the GGSs could also be numerically simulated. Figure 6.25 shows, for example, a numerically calculated period doubling of a GGS when the pump strength was selected as $G = 4300$. Figure 6.25a shows the evolution of the calculated GGS with the cavity round-trips. Period doubling of the pulse is evidenced by the pulse returning to its previous parameters at every two cavity round-trips. Figure 6.25b shows the optical spectra of the soliton in two adjacent round-trips and the averaged one. Clear differences between them are visible. However, no obvious spectral spikes were obtained. We believe the absence of the spectral spikes could be caused by the parabolic gain profile approximation used in our simulations. From the experimental results the spikes appeared only on the edges of the spectrum, but where the parabolic gain profile artificially introduced large losses. Nevertheless, the numerical simulations have reasonably reproduced the essential features of the soliton period-doubling bifurcation (e.g., the simulated spectra have a flat and smooth top, and the spectral variations between the Period 1 and period-doubled states only occur on the edges of the spectrum, which are in agreement with the experimental observations and different from those of the soliton period doubling observed in fiber lasers of negative cavity dispersion) [13].

Figure 6.26 shows a comparison between the evolutions of the GGSs in cavity before and after the period-doubling bifurcation was numerically

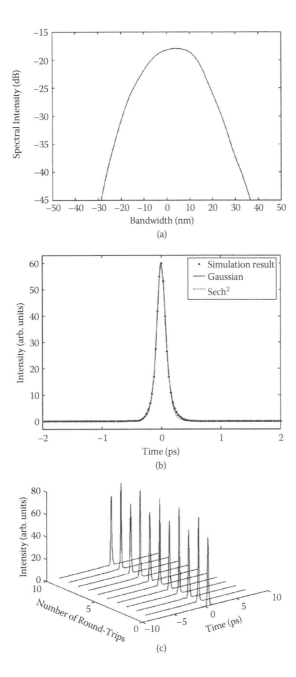

FIGURE 6.24

Numerical simulations of the dispersion-managed soliton in the positive near-zero net cavity group velocity dispersion regime: (a) optical spectrum and (b) temporal profile of a dispersion-managed soliton: small signal gain coefficient $g_0 = 520$; (c) a period-doubled state: small signal gain coefficient $g_0 = 550$.

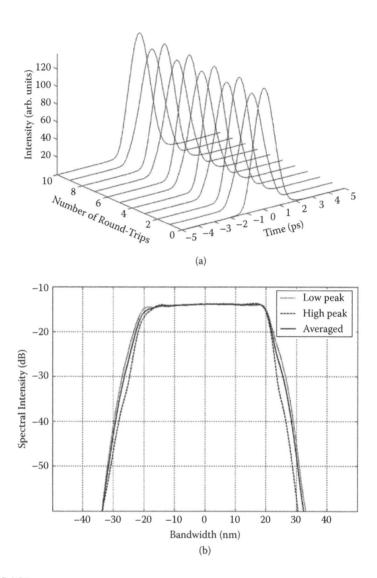

FIGURE 6.25
Period doubling of the gain-guided solitons numerically calculated. (a) Soliton evolution with the cavity round-trips. (b) Optical spectra of the soliton in two adjacent round-trips and their average.

calculated. Figures 6.26a and 6.26b show the pulse evolution in the time domain along the cavity; Figures 6.26c and 6.26d show the corresponding pulse width variation along the cavity; Figures 6.26e and 6.26f show the evolution of the pulse spectrum corresponding to Figures 6.26a and 6.26b, respectively. We note that Figure 6.26 has shown pulse evolution in two cavity round-trips in order to display the period-doubling effect. With all other parameters fixed, the Period 1 state could be obtained in a

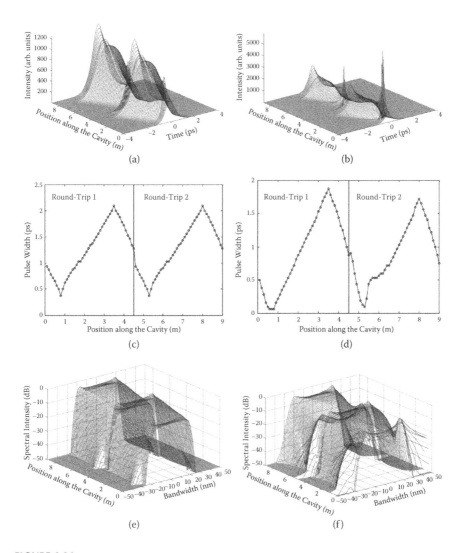

FIGURE 6.26
Comparison between the soliton evolutions in cavity before and after the period-doubling bifurcation. (a,b) Pulse evolution in time domain; (c,d) pulse width evolution; and (e,f) optical spectrum evolution in double cavity length.

large pump strength range. In a Period 1 state the chirped GGS is compressed in the SMF segments. The stronger the pump strength, the larger is the chirp accumulated in the EDF, and consequently, narrower pulse and higher pulse peak power is obtained in the SMF. After that the pulse peak power is beyond a certain value, period doubling of the pulse occurs. Numerically we found that in a period-doubled state, the GGS may be dechirped to a transform-limited pulse in the SMF, and remain the shortest pulse width until entering the EDF as shown in Figure 6.26d. However, a pedestal also associates with the compressed pulse, which could be understood as a result of the pulse breaking.

6.4.8 Period Doubling of Vector Solitons

Due to the existence of a polarizer in the cavity, which fixes the polarization of light in the position of the cavity, solitons formed in fiber lasers mode locked with the NPR technique are considered as scalar solitons. In fiber lasers if there are no polarization-dependent components in the cavity, vector solitons could be formed. The formed vector solitons have two coupled orthogonal polarization components. Numerical simulations have shown that the vector solitons can also exhibit period-doubling bifurcations as observed in the experiments.

Similarly, in the experiments where in order to obtain vector soliton operation of a fiber laser instead of the NPR mode locking but a polarization-insensitive SESAM was used, to numerically simulate the features of vector solitons formed in a fiber laser, the effect of a real saturable absorber is considered. Figure 6.27 shows the results of simulations obtained under different net cavity birefringence but the same other laser parameters. Numerically it is found that when the cavity birefringence is selected as $L_b = L/0.4$, where L is the cavity length, a Period 1 vector soliton state is obtained. When the cavity birefringence is selected as $L_b = L/0.09$, a period-doubled vector soliton is then obtained. Figures 6.27a and 6.27b show the calculated vector soliton evolution with the cavity round-trips. The soliton period-doubling of the state shown in Figure 6.27a in comparison with that shown in Figure 6.27b is evident. Figure 6.27c/Figure 6.27e, and Figure 6.27d/ Figure 6.27f show further the evolution of the vector soliton components, of the period-doubled vector soliton and the period-one vector soliton, respectively. Each component of the period-doubled vector soliton also exhibits period doubling. However, the pulse intensity variation between the two orthogonal polarization components is antiphase. It is because of this antiphase pulse intensity evolution of the period-doubled vector soliton, only a weak period-doubling could be observed on the vector soliton intensity evolution, as shown in Figure 6.27a. The result agrees with our experimental observations of Figure 6.16a.

6.4.9 Cavity-Induced Soliton Modulation Instability Effect

Modulation instability (MI) is a well-known phenomenon that destabilizes strong CW propagation in dispersive media. In optics, Tai et al. first reported the experimental observation of MI in light propagation in single-mode fibers [63]. They found that MI is physically a special case of four-wave mixing where the phase-matching condition is self-generated by the nonlinear refractive index change and the anomalous dispersion of the fibers.

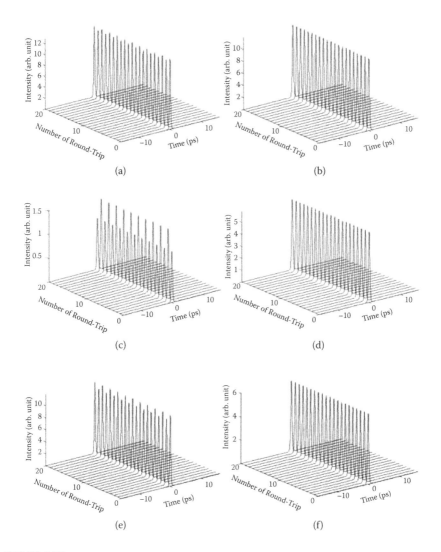

FIGURE 6.27
Numerically simulated results of period-doubling/Period 1 of (a,b) the vector soliton, (c,d) the vector soliton along the horizontal birefringent axis, and (e,f) the vector soliton along the vertical birefringent axis with the cavity round-trips.

Soliton propagation in single-mode fibers is intrinsically stable against MI. However, it was theoretically shown that if dispersion of a fiber is periodically varied or the intensity of a solitary wave is periodically modulated, MI of solitons could still occur [64,65]. MI establishes resonant energy exchange between the soliton and dispersive waves, which results in a new type of spectral sideband generation on the soliton spectrum. Soliton propagation in a laser cavity experiences periodically fiber dispersion and soliton energy

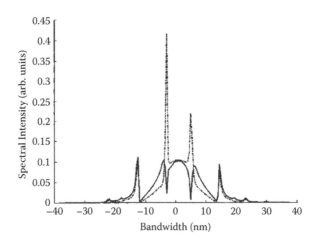

FIGURE 6.28
Comparison between soliton spectra of adjacent round-trips in the Period 2 state.

variations. It is to be expected that under certain conditions such a soliton MI could appear.

We point out that MI is an intrinsic feature of the solitons formed in a laser. The development of MI leads to strong resonant wave coupling between the soliton and dispersive waves, which further leads to the dynamic instability of the soliton. We believe that it is the dynamic instability of a soliton that ultimately limits the performance of a soliton laser. MI of a soliton in lasers is characterized by the formation of a new set of spectral sidebands on the soliton spectrum.

Figure 6.28 shows a case of the soliton spectra numerically calculated for a Period 2 soliton state. In the soliton Period 1 state the soliton spectrum has only the Kelly sidebands. The Kelly sidebands are generated by the constructive interference between the soliton and the dispersion waves in the cavity, whose positions are almost fixed for a fiber laser once its cavity parameters are fixed. We note that the formation of Kelly sidebands is a linear effect. However, after the soliton experienced a period-doubling bifurcation, it is observed that apart from the Kelly sidebands a new set of spectral sidebands also appears on the soliton spectrum. Positions of the new spectral sidebands vary with the laser cavity detuning. At certain particular cavity detuning settings, or when the new sidebands appear at certain positions on the soliton spectrum, the soliton period-doubling bifurcation occurs, while at the other positions the soliton quasi-periodicity effect is observed. Carefully checking the soliton spectral evolution within one cavity round-trip, it is further identified that in a soliton Period 2 state, while in one cavity round-trip the new sideband is a spectral spike, in the subsequent cavity round-trip the new spectrum will become a spectral dip, and vice versa, which indicates that the new spectral sidebands are formed due to a parametric wave interaction process.

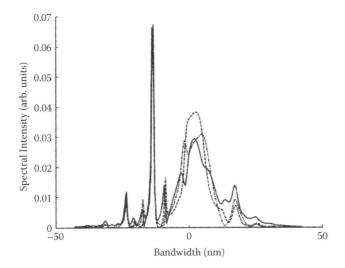

FIGURE 6.29
Comparison between soliton spectra of adjacent round-trips in the Period 3 state.

Figure 6.29 shows a comparison between the soliton spectra of the adjacent cavity round-trips of a Period 3 state. Due to that the soliton period in the state is three cavity lengths the variation of the extra sidebands between two adjacent round-trips is not exactly opposite, but $2\pi/3$ out of phase.

Based on the new spectral sideband formation and their relation to the observed soliton dynamics, we speculate that the appearance of the deterministic soliton dynamics is related to the cavity-induced soliton modulation instability. Independent of the cavity dispersion, soliton propagation in the cavity experiences a periodic intensity variation, which causes a phase matching of the soliton with certain particular frequencies of weak light or dispersive waves. Energy exchange between the soliton pulse and the phase-matched dispersive waves could be built up. When the phase difference between the soliton and the dispersive waves within one cavity round-trip is 2π, then a soliton Period 2 state is formed. As energy exchange between the soliton and weak light depends on their phase difference, within half of the period energy flows from the soliton to the dispersive waves, while in the other half period it then flows reversely, resulting in a soliton Period 2 state. In the case that the phase difference between the soliton and dispersive waves is not a multiple of 2π but an arbitrary value, the quasi-periodicity state occurs. This could be why the quasi-periodic soliton operation state could be experimentally easily observed, while the period-doubling and period-doubling route to chaos could only be observed under certain special laser conditions.

6.5 Conclusions

In conclusion, we investigated both experimentally and numerically the deterministic dynamics of solitons formed in passive mode-locked fiber lasers. We have shown experimentally that under strong soliton strength, deterministic soliton dynamics, these are the soliton period-doubling bifurcations and soliton period-doubling route to chaos, could appear. In particular, the deterministic soliton dynamics is independent of the number of solitons in the cavity and the types of soliton formed. We observed the period-doubling route to chaos on the soliton repetition rate of both the single pulse soliton and the bound solitons; on the conventional solitons formed in anomalous dispersion cavity fiber lasers, or on the dispersion-managed solitons generated in around net zero dispersion fiber lasers, or on the gain-guided solitons generated in normal dispersion fiber lasers. Period doubling of vector solitons was also experimentally observed. Based on the coupled extended GLEs and the round-trip model, we numerically simulated the soliton operation of various fiber lasers and successfully reproduced all the observed soliton deterministic dynamics experimentally observed. Moreover, based on results of both experimental studies and numerical simulations, we found that the appearance of these soliton dynamics could be related to the cavity-induced soliton modulation instability effect. We speculate that the cavity-induced modulation instability is also responsible for the deterministic dynamics of the nonlinear cavity under CW operation.

References

1. K. Ikeda, Multiple-Value Stationary State and Its Instability of the Transmitted Light by a Ring Cavity System, *Opt. Commun.*, 30, 257–261, 1979.
2. K. Ikeda, H. Daido, and O. Akimoto, Optical Turbulence: Chaotic Behavior of Transmitted Light from a Ring Cavity, *Phys. Rev. Lett.*, 45, 709–712, 1980.
3. D. U. Noske, N. Pandit, and J. R. Taylor, Subpicosecond Soliton Pulse Formation from Self-Mode-Locked Erbium Fiber Laser Using Intensity Dependent Polarization Rotation, *Electron. Lett.*, 28, 1391–1393, 1992.
4. V. J. Matsas, T. P. Newson, D. J. Richardson, and D. N. Payne, Self-Starting Passive Mode-Locked Fiber Ring Soliton Laser Exploiting Nonlinear Polarization Rotation, *Electron. Lett.*, 28, 1391–1393, 1992.
5. I. N. Duling III, All-Fiber Ring Soliton Laser Mode Locked with a Nonlinear Mirror, *Opt. Lett.*, 16, 539–541, 1991.
6. D. J. Richardson, R. I. Laming, D. N. Payne, M. W. Phillips, and V. J. Matsas, 320fs Soliton Generation with Passively Mode-Locked Erbium Fiber Laser, *Electron. Lett.*, 27, 730–732, 1991.

7. B. C. Collings, K. Bergman, S. T. Cundiff, S. Tsuda, J. N. Kutz, J. E. Cunningham, W. Y. Jan, M. Koch, and W. H. Knox, Short Cavity Erbium/Ytterbium Fiber Lasers Mode-Locked with a Saturable Bragg Reflector, *IEEE J. Select Topics in Quantum Electron.*, 3, 1065–1075, 1997.

8. S. Gray and A. B. Grudinin, Soliton Fiber Laser with a Hybrid Saturable Absorber, *Opt. Lett.*, 21, 207–209, 1996.

9. D. J. Jones, H. A. Haus, and E. P. Ippen, Subpicosecond Solitons in an Actively Mode-Locked Fiber Laser, *Opt. Lett.*, 21, 1818–1820, 1996.

10. R. P. Davey, N. Langford, and A. I. Ferguson, Interacting Solitons in Erbium Fiber Laser, *Electron. Lett.*, 27, 1257–1259, 1991.

11. Y. P. Tong, P. M. W. French, J. R. Taylor, and J. O. Fujimoto, All-Solid-State Femtosecond Source in the Near Infrared, *Opt. Commun.*, 136, 235–238, 1997.

12. A. M. Kowalevicz, Jr., A. Tucay Zare, F. Kartner, J. G. Fujimoto, S. Dewald, U. Morgner, V. Scheuer, and G. Angelow, Generation of 150-nJ Pulses from a Multiple-Pass Cavity Kerr-Lens Mode-Locked Ti:Al$_2$O$_3$ Oscillator, *Opt. Lett.*, 28, 1597–1599, 2003.

13. L. M. Zhao, D. Y. Tang, F. Lin, and B. Zhao, Observation of Period-Doubling Bifurcations in a Femtosecond Fiber Soliton Laser with Dispersion Management Cavity, *Opt. Express*, 12, 4573–4578, 2004.

14. D. Y. Tang, L. M. Zhao, and F. Lin, Numerical Studies of Routes to Chaos in Passively Mode Locked Fiber Soliton Ring Lasers with Dispersion-Managed Cavity, *Europhys. Lett.*, 71(1), 56–62, 2005.

15. L. M. Zhao, D. Y. Tang, and B. Zhao, Period-Doubling and Quadrupling of Bound Solitons in a Passively Mode-Locked Fiber Laser, *Opt. Commun.*, 252, 167–172, 2005.

16. L. M. Zhao, D. Y. Tang, and A. Q. Liu, Chaotic Dynamics of a Passively Mode-Locked Soliton Fiber Ring Laser, *Chaos*, 16, 013128, 2006.

17. L. M. Zhao, D. Y. Tang, T. H. Cheng, and C. Lu, Period-Doubling of Multiple Solitons in a Passively Mode-Locked Fiber Laser, *Opt. Commun.*, 273, 554–559, 2007.

18. D. Y. Tang, J. Wu, L. M. Zhao, and L. J. Qian, Dynamic Sideband Generation in Soliton Fiber Lasers, *Opt. Commun.*, 275, 213–216, 2007.

19. L. M. Zhao, D. Y. Tang, T. H. Cheng, H. Y. Tam, C. Lu, and S. C. Wen, Period-Doubling of Dispersion-Managed Solitons in an Erbium-Doped Fiber Laser at around Zero Dispersion, *Opt. Commun.*, 278, 428–433, 2007.

20. L. M. Zhao, D. Y. Tang, Wu, H. Zhang, Period-Doubling of Gain-Guided Solitons in Fiber Lasers of Large Net Normal Dispersion, *Opt. Commun.*, 281, 3557–3560, 2008.

21. L. M. Zhao, D. Y. Tang, H. Zhang, X Wu, C. Lu, and H. Y. Tam, Period-Doubling of Vector Solitons in a Ring Fiber Laser, *Opt. Commun.*, 281, 5614–5617, 2008.

22. D. Y. Tang, L. M. Zhao, Wu, and H. Zhang, Soliton Modulation Instability in Fiber Lasers, *Phys. Rev. A*, 80, 023806, 2009.

23. F. Ö. Ilday, F. W. Wise, and T. Sosnowski, High-Energy Femtosecond Stretched-Pulse Fiber Laser with a Nonlinear Optical Loop Mirror, *Opt. Lett.*, 27, 1531–1533, 2002.

24. M. J. Guy, D. U. Noske, and J. R. Taylor, Generation of Femtosecond Soliton Pulses by Passive Mode Locking of an Ytterbium-Erbium Figure-of-Eight Fiber Laser, *Opt. Lett.*, 18, 1447–1449, 1993.

25. D. Y. Tang, W. S. Man, H. Y. Tam, and M. S. Demokan, Modulational Instability in a Fiber Soliton Ring Laser Induced by Periodic Dispersion Variation, *Phys. Rev. A*, 61, 023804, 2000.

26. L. M. Zhao, D. Y. Tang, H. Zhang, Wu, and N. Xiang, Soliton Trapping in Fiber Lasers, *Opt. Express*, 16, 9528–9533, 2008.

27. C. R. Menyuk, Nonlinear Pulse Propagation in Birefringent Optical Fibers *IEEE Journal of Quantum Electronics*, 23, 174–176, 1987.

28. G. P. Agrawal, *Nonlinear Fiber Optics*, 4th ed., Academic Press, Boston, 2007.

29. D. Y. Tang, L. M. Zhao, B. Zhao, and A. Q. Liu, Mechanism of Multisoliton Formation and Soliton Energy Quantization in Passively Mode-Locked Fiber Lasers, *Phys. Rev. A*, 72, 043816, 2005.

30. M. Hofer, M. H. Ober, F. Haberl, and M. E. Fermann, Characterization of Ultrashort Pulse Formation in Passively Mode-Locked Fiber Lasers, *IEEE J. Quantum Electron.*, QE-28, 720–728, 1992.

31. R. P. Davey, N. Langford, and A. I. Ferguson, Role of Polarization Rotation in the Modelocking of an Erbium Fiber Laser, *Electron. Lett.*, 29, 758–760, 1993.

32. S. M. J. Kelly, K. Smith, K. J. Blow, and N. J. Doran, Average Soliton Dynamics of a High-Gain Erbium Fiber Laser, *Opt. Lett.*, 16, 1337–1339, 1991.

33. A. D. Kim, J. N. Kutz, and D. J. Muraki, Pulse-Train Uniformity in Optical Fiber Lasers Passively Mode-Locked by Nonlinear Polarization Rotation, *IEEE J. Quantum Electron.*, 36, 465–471, 2000.

34. K. M. Spaulding, D. H. Yong, A. D. Kim, and J. N. Kutz, Nonlinear Dynamics of Mode-Locking Optical Fiber Ring Lasers, *J. Opt. Soc. Am. B*, 19, 1045–1054, 2002.

35. B. Zhao, D. Y. Tang, L. M. Zhao, and P. Shum, Pulse-Train Nonuniformity in a Fiber Soliton Ring Laser Mode-Locked by Using the Nonlinear Polarization Rotation Technique, *Phys. Rev. A*, 69, 043808, 2004.

36. G. Sucha, S. R. Bolton, S. Weiss, and D. S. Chemla, Period Doubling and Quasi-Periodicity in Additive-Pulse Mode-Locked Lasers, *Opt. Lett.*, 20, 1794–1796, 1995.

37. D. Côté and H. M. van Driel, Period Doubling of a Femtosecond Ti:Sapphire Laser by Total Mode Locking, *Opt. Lett.*, 23, 715–717, 1998.

38. A. B. Grudinin, D. J. Richardson, and D. N. Payne, Energy Quantization in Figure Eight Fiber Laser, *Electron. Lett.*, 28, 67–68, 1992.

39. D. J. Richardson, R. I. Laming, D. N. Payne, V. J. Matsas, and M. W. Phillips, Characterization of a Self-Starting, Passively Mode-Locked Fiber Ring Laser That Exploits Nonlinear Polarization Evolution, *Electron. Lett.*, 27, 1451, 1991.

40. D. Y. Tang, W. S. Man, H. Y. Tam, and P. D. Drummond, Observation of Bound States of Solitons in a Passively Mode-Locked Fiber Laser, *Phys. Rev. A*, 64, 033814, 2001.

41. Ph. Grelu, F. Belhache, F. Gutty, and J.-M. Soto-Crespo, Soliton Pairs in a Fiber Laser: from Anomalous to Normal Average Dispersion Regime, *Opt. Lett.*, 27, 966, 2002.

42. B. Zhao, D. Y. Tang, P. Shum, Y. D. Gong, C. Lu, W. S. Man, and H. Y. Tam, Energy Quantization of Twin-Pulse Solitons in a Passively Mode-Locked Fiber Ring Laser, *Appl. Phys. B*, 77, 585, 2003.

43. B. Zhao, D. Y. Tang, P. Shum, W. S. Man, H. Y. Tam, Y. D. Gong, and C. Lu, Passive Harmonic Mode Locking of Twin-Pulse Solitons in an Erbium-Doped Fiber Ring Laser, *Opt. Commun.*, 229, 363, 2004.

44. D. Y. Tang, B. Zhao, D. Y. Shen, C. Lu, W. S. Man, and H. Y. Tam, Compound Pulse Solitons in a Fiber Ring Laser, *Phys. Rev. A*, 68, 013816, 2003.

45. C. O. Weiss and H. King, Oscillation Period Doubling Chaos in a Laser, 59-61, *Opt. Commun.*, 44, 59, 1982.

46. D. Y. Tang, S. P. Ng, L. J. Qin, and L. Meng, Deterministic Chaos in a Diode-Pumped NdYAG Laser Passively Q Switched by a Cr 4+: YAG Crystal, *Opt. Lett.*, 28, 325, 2003.

47. D. Y. Tang, B. Zhao, L. M. Zhao, and H. Y. Tam, Soliton Interaction in a Fiber Ring Laser, *Phys. Rev. E*, 72, 016616, 2005.

48. N. J. Smith, F. M. Knox, N. J. Doran, K. J. Blow, and I. Bennion, Enhanced Power Solitons in Optical Fibres with Periodic Dispersion Management, *Electron. Lett.*, 32, 54–55, 1996.

49. J. H. B. Nijhof, N. J. Doran, W. Forysiak, and F. M. Knox, Stable Soliton-Like Propagation in Dispersion Managed Systems with Net Anomalous, Zero and Normal Dispersion, *Electron. Lett.*, 33, 1726–1727, 1997.

50. J. N. Kutz and S. G. Evangelides, Jr., Dispersion-Managed Breathers with Average Normal Dispersion, *Opt. Lett.*, 23, 685–687, 1998.

51. J. H. B. Nijhof, W. Forysiak, and N. J. Doran, Dispersion-Managed Solitons in the Normal Dispersion Regime: A Physical Interpretation, *Opt. Lett.*, 23, 1674–1676, 1998.

52. Y. Chen and H. A. Haus, Dispersion-Managed Solitons in the Net Positive Dispersion Regime, *J. Opt. Soc. Am. B*, 16, 24–30, 1999.

53. V. S. Grigoryan, R. -M. Mu, G. M. Carter, and C. R. Menyuk, Experimental Demonstration of Long-Distance Dispersion-Managed Soliton Propagation at Zero Average Dispersion, *IEEE Photonics Technol. Lett.*, 12, 45–46, 2000.

54. L. M. Zhao, D. Y. Tang, T. H. Cheng, and C. Lu, Gain-Guided Solitons in Dispersion-Managed Fiber Lasers with Large Net Cavity Dispersion, *Opt. Lett.*, 31, 2957–2959, 2006.

55. N. Akhmediev and A. Ankiewicz (Eds.), *Dissipative Solitons: Lecture Notes in Physics, V. 661*, Springer, Heidelberg, 2005.

56. H. A. Haus, J. G. Fujimoto, and E. P. Ippen, Analytic Theory of Additive Pulse and Kerr Lens Mode Locking, *IEEE J. Quantum Electron.*, 28, 2086–2096, 1992.

57. S. T. Cundiff, B. C. Collings, and W. H. Knox, Polarization Locking in an Isotropic, Modelocked Soliton Er/Yb Fiber Laser, *Opt. Express*, 1, 12–20, 1997.

58. L. M. Zhao, D. Y. Tang, H. Zhang, and H. Wu, Polarization Rotation Locking of Vector Solitons in a Fiber Ring Laser, *Optics Express*, 16 (4), 10053–10058, 2008.

59. A. I. Chernykh and S. K. Turitsyn, Soliton and Collapse Regimes of Pulse Generation in Passively Mode-Locking Laser Systems, *Opt. Lett.*, 12, 1011, 1987.

60. H. Nakatsuka, S. Asaka, H. Itoh, K. Ikeda, and M. Matsuoka, Observation to Chaos in All-Optical Bistabe System, *Phys. Rev. Lett.*, 50, 109, 1983.

61. G. Steinmeyer, A. Buchholz, M. Hansel, M. Heuer, A. Schwache, and F. Mitschke, Dynamical Pulse Shaping in a Nonlinear Resonator, *Phys. Rev. A*, 52, 830, 1995.

62. F. Ilday, J. Buckley, and F. Wise, Period-Doubling Route to Multiple-Pulsing in Femtosecond Fiber Lasers, *Proc. of Conf. on Nonlinear Guided Waves and Their Applications*, Section Optical Amplifiers (MD), Paper MD9, Canada, March 28, 2004.

63. K. Tai, A. Hasegawa, and A. Tomita, Observation of Modulational Instability in Optical Fibers, *Phys. Rev. Lett.*, 56, 135–138, 1986.

64. F. Matera, A. Mecozzi, M. Romagnoli, and M. Settembre, Sideband Instability Induced by Periodic Power Variation in Long-Distance Fiber Links, *Opt. Lett.*, 18, 1499–1501, 1993.

65. N. J. Smith and N. J. Doran, Modulation Instabilities in Fibers with Periodic Dispersion Management, *Opt. Lett.*, 21, 570–572, 1996.

7

Bistability, Bifurcation, and Chaos in Nonlinear Loop Fiber Lasers

Le Nguyen Binh

Hua Wei Technologies, European Research Center, Munich, Germany

CONTENTS

7.1 Overview

All real physical systems are nonlinear in nature as briefly declared in Chapter 2. Apart from systems designed for linear signal processing, many systems have to be nonlinear by assumption, for instance flip-flops, modulators, demodulators, amplifiers, and so forth. In this section, we focus on the optical bistability, bifurcation, and chaos of a nonlinear system.

Bidirectional lightwaves propagation in an erbium-doped fiber ring laser and the behavior of optical bistability are studied. We exploit this commonly known undesirable bidirectional propagation of lightwaves for constructing fiber laser configurations based on nonlinear optical loop mirror (NOLM) and nonlinear amplifying loop mirror (NALM) structure. The lasers are operated in different regions. The switching capability of the laser based on its bistability characteristics is identified. Chaotic phenomena are also observed in these fiber lasers.

7.2 Introduction

The nonlinear phenomena of optical bistability have been studied in nonlinear resonators since 1976 by placing the nonlinear medium inside a cavity formed by using multiple mirrors [1]. As for the fiber-based devices, single-mode fiber was used as the nonlinear medium inside a ring cavity in 1983 [2]. Since then, the study of nonlinear phenomena in fiber resonators has remained a topic of considerable interest.

Fiber ring laser is a rich and active research field in optical communications. Many fiber ring laser configurations have been proposed and constructed to achieve different objectives. It can be designed for continuous-wave (CW) or pulse operation, linear or nonlinear operation, fast or slow repetition rate, narrow or broad pulse width, and so forth, for various kinds of photonic applications.

The simplest fiber laser structure is an optical closed loop with a gain medium and some associated optical components such as optical couplers. The gain medium used can be any rare earth element doped fiber amplifier, such as erbium and ytterbium, semiconductor optical amplifier, parametric amplifier, and so forth, as long as it provides the gain requirement for lasing. Without any mode-locking mechanism, the laser will operate in the CW regime. By inserting an active mode-locker into the laser cavity (i.e., either amplitude or phase modulator), the resulting output will be an optical pulse train operating at the modulating frequency, when the phase conditions is matched, as discussed in Chapter 3. This often results in a high-speed optical pulse train, however with broad pulse width. There is another kind of fiber laser that uses the nonlinear effect in generating the optical pulses and is known as the passive mode-locking technique as described in this chapter. Saturable absorber, stretched pulse mode-locking [3], nonlinear polarization rotation [4], figure-eight fiber ring lasers [5], and so forth can be grouped under this category. This type of laser generates shorter optical pulse sequences in exchange for high repetition frequency. Hence, this is the trade-off between the active and passive mode-locked systems.

Although there are different types of fiber lasers, with different operating regimes, one common criterion is the unidirectionality, besides the gain and phase-matching conditions. Unidirectional propagation has been proven to offer better lasing efficiency, less sensitivity to back reflections, and good potential for single longitudinal mode operation [6], and can be achieved by incorporating an optical isolator within the laser cavity. Shi et al. [7] has demonstrated a unidirectional inverted S-type erbium-doped fiber ring laser without the use of optical isolator, but with optical couplers. In their laser system, the lightwaves are passed through in one direction and suppressed in the other using certain coupling ratios and, hence, achieving unidirectional operation. However, only the power difference between the two CW lightwaves was studied, and the works did not extend further into unconventional and nonlinear regions of operation.

In this chapter, we describe the bidirectional lightwaves propagation in an erbium-doped fiber ring laser and its behavior due to optical bistability that has been well reported [8–10]. We exploit this commonly known process in constructing a kind of fiber laser configuration based on nonlinear optical loop mirror (NOLM) and nonlinear amplifying loop mirror (NALM) structures. This laser configuration is similar to the one demonstrated by Shi et al. [11], but operating the laser in different regimes. We investigate the switching capability of the laser based on its bistability behavior. Furthermore, we focus on the nonlinear dynamics of the fiber laser and describe the optical bifurcation phenomenon.

7.3 Optical Bistability, Bifurcation, and Chaos

An optical bistable device is a device with two possible operation points. It will remain stable in any of the two optical states, one of high transmission and the other of low transmission, depending upon the intensity of the light passing through it. In this section, we will discuss the effect of optical nonlinearity together with the proper feedback, which can give rise to optical bistability and hysteresis. This is expected as these two effects are also observed in nonlinear electronic circuits with feedback, such as the Schmitt trigger, as well as hybrid optical devices, such as an acousto-optic device with feedback [12].

A typical bistability curve is shown in Figure 7.1, with Yi and Yo the input and output parameters, respectively. For an input value between Y1 and Y2,

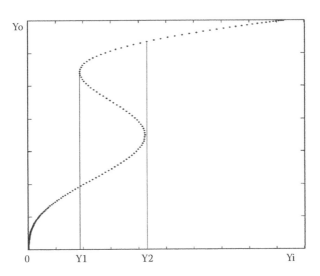

FIGURE 7.1
Bistability curve.

there are three possible output values. The middle segment, with negative slope, is known to be always unstable. Therefore, the output will eventually have two stable values. When two outputs are possible, which one of the outputs is eventually realized depends on the history of how the input is reached, and hence the hysteresis phenomenon. As the input value is increased from zero, the output will follow the lower branch of the curve until the input value reaches Y2. Then it will jump up and follow the upper branch. However, if the input value is decreased from some points after the jump, the output will remain on the upper branch until the input value hits Y1, then the output will jump down and follow the lower branch. Hence, the bistability region observed is from Y1 to Y2.

In an optical ring cavity, the lightwaves can be split into two counterpropagating components that can mutually interact leading to optical gain competition. Hence, in a ring laser, the lightwave may propagate in one direction or another depending on the initial configuration and is thus a running wave in general. To eliminate this randomness, a device such as optical isolator can be inserted into the ring cavity to block the unwanted wave.

There are two different types of optical bistability, namely, absorptive bistability and dispersive bistability in a nonlinear ring cavity composed of a two-level gain medium as nonlinear medium. Absorptive bistability is the case when the incident optical frequency is close to or equal to the transition frequency of the atoms from one level to another. In other words, the system is in perfect resonance condition. In this case, the absorption coefficient becomes a nonlinear function of the incident frequency of the lightwaves.

On the other hand, if the frequencies are far apart, the gain medium behaves like a Kerr-type material, and the system exhibits what is called dispersive bistability. A nonzero atomic detuning would introduce saturable dispersion in response to the medium. In this case the material can be modeled by an effective nonlinear refractive index (the Kerr's effects), that is its refractive index varies nonlinearly with respect to its intensity [13].

For optical bistability in a ring cavity, there is a possible instability due to the counterpropagating wave. Hence, it is interesting to know the lightwave behavior for both co- and counterpropagating components. The bidirectional operation of an optical bistable ring system has not yet been studied thoroughly. Although some studies have been done relating to the bistability properties of the ring cavities, they focused on forcing the system to support only unidirectional operation [14]. The restriction to unidirectional propagation has been quite consistent with the experimental results; however, the exceptions were noted [15]. Unfortunately, no further investigation has been carried out since then.

It was discovered that the nonlinear response of a ring resonator could initiate a period doubling (bifurcation) route to optical chaos [16]. In a dynamical system, a bifurcation is a period doubling, quadrupling, and so forth, that accompanies the onset of chaos. It represents a sudden appearance of a qualitatively different solution for a nonlinear system as some parameters

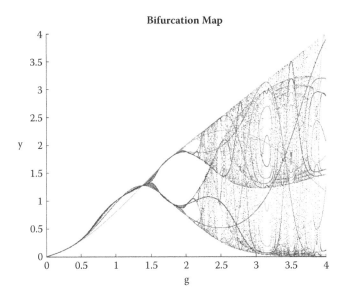

FIGURE 7.2
Typical bifurcation map.

are varied. A typical bifurcation map is shown in Figure 7.2, with g and y the input and output system parameters, respectively. Period doubling action starts at $g = $ ~1.5, and period quadrupling at $g = $ ~2. The region beyond $g = $ ~2.3 is known as chaos (see Figure 7.3).

Each of the local bifurcations may give rise to a distinct route to chaos if the bifurcations appear repeatedly when changing the bifurcation parameter. These routes are important because it is difficult to conclude from experimental data alone whether irregular behavior is due to measurement noise or chaos. Recognition of one of the typical routes to chaos in experiments is a good indication that the dynamics may be chaotic [17]:

- Period doubling route to chaos—When a cascade of successive period-doubling bifurcations occurs when changing the value of the bifurcation parameter it is often the case that finally the system has reached chaos.

- Intermittency route to chaos—The route to chaos caused by saddle-node bifurcations. The common feature of which is a direct transition from regular motion to chaos.

- Torus breakdown route to chaos—The quasi-periodic route to chaos results from a sequence of Hopf bifurcations.

It is well known that self-pulsing often leads to optical chaos in the laser output, following a period doubling or a quasi-periodic route. The basic idea

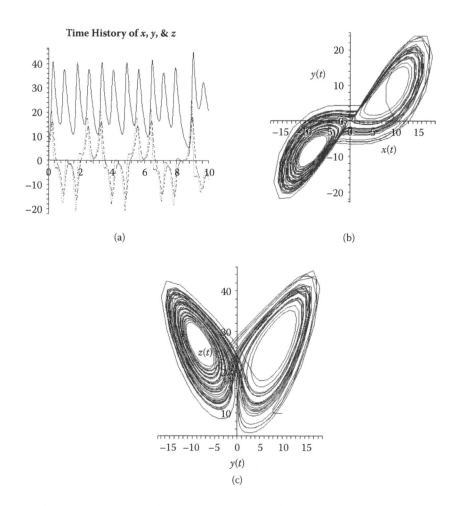

FIGURE 7.3
System response of Lorenz attractor. (a) Time traces of x, dash; y, dotted; and z, solid. (b) x Trajectory. (c) z-x Trajectory.

is that the dynamics of the intracavity field is different from one round-trip to another in a nonlinear fashion. The characteristics of a chaotic system [18] are

- Sensitive dependence on initial conditions—It gives rise to an apparent randomness in the output of the system and the long-term unpredictability of the state. Because the chaotic system is deterministic, two trajectories that start from identical initial states will follow precisely the same paths through the state space.

- Randomness in the time domain—In contrast to the periodic waveforms, chaotic waveform is quite irregular and does not appear to repeat itself in any observation period of finite length.

- Broadband power spectrum—Every periodic signal may be decomposed into a Fourier series, a weighted sum of sinusoids at integers multiples of a fundamental frequency. Thus, a periodic signal appears in the frequency domain as a set of spikes at the fundamental frequency and its harmonics. The chaotic signal is qualitatively different from the periodic signal. The aperiodic nature of its time-domain waveform is reflected in the broadband noise like a power spectrum. This broadband structure of the power spectrum persists even if the spectral resolution is increased to a higher frequency.

A typical example of the chaotic system is the Lorenz attractor, or more commonly known as the butterfly attractor, with the following system description, with $a = 10$, $b = 28$, and $c = 8/3$:

$$\frac{dx}{dt} = a(y - x) \tag{7.1}$$

$$\frac{dy}{dt} = x(b - z) - y \tag{7.2}$$

$$\frac{dz}{dt} = xy - cz \tag{7.3}$$

7.4 Nonlinear Optical Loop Mirror

The concept of the nonlinear optical loop mirror (NOLM) was proposed by Doran and Wood [19]. It is basically a fiber-based Sagnac interferometer that uses the nonlinear phase shift of optical fiber for optical switching. This configuration is inherently stable because the two arms of the structure reside in the same fiber and same optical path lengths for the signals propagating in both arms, however in the opposite direction. There is no feedback mechanism in this structure because all lightwaves entering the input port exit from the loop after a single round-trip. The NOLM in its simplest form contains a fiber coupler, with two of its output ports connected together, as shown in Figure 7.4, with κ the coupling ratio of the coupler, E_1, E_2, E_3, and E_4 are the fields at ports 1, 2, 3, and 4, respectively.

Under a lossless condition, the input-output relationships of a coupler with an intensity coupling ratio κ are

$$E_3 = \sqrt{\kappa}E_1 + j\sqrt{1-\kappa}E_2$$

$$E_4 = \sqrt{\kappa}E_2 + j\sqrt{1-\kappa}E_1 \tag{7.4}$$

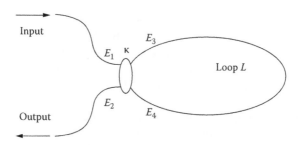

FIGURE 7.4
Nonlinear optical loop mirror.

where $j^2 = -1$. If low-intensity light is fed into Port 1 (i.e., no nonlinear effect), the transmission T of the device is

$$T = \frac{P_{out}}{P_{in}} = 1 - 4\kappa(1 - \kappa) \tag{7.5}$$

If $\kappa = 0.5$, all light will be reflected back to the input; therefore, the name loop mirror, or else the counterpropagating lightwaves in the loop will have different intensities, thus leading to different nonlinear phase shifts. This device can be designed to transmit a high-power signal while reflecting it at low power levels, thus acting as an all-optical switch.

After L propagation, the output signals E_{3L} and E_{4L} become as follows with nonlinear phase shifts taken into account:

$$E_{3L} = E_3 \exp\left(j\frac{2\pi n_2 L}{\lambda}|E_3|^2 \right)$$

$$E_{4L} = E_4 \exp\left(j\frac{2\pi n_2 L}{\lambda}|E_4|^2 \right) \tag{7.6}$$

where $|E_3|^2$ and $|E_4|^2$ are the light intensities of the two arms, n_2 is the nonlinear coefficient, L is the loop length, λ is the operating wavelength of the signal, and $E_1 = E_{01}\exp(j\omega_1 t)$ and $E_2 = E_{02}\exp(j\omega_2 t)$:

$$|E_{3L}|^2 = |E_3|^2$$
$$= \kappa|E_1|^2 + (1-\kappa)|E_2|^2 + j\sqrt{\kappa(1-\kappa)}(E_1^*E_2 - E_1E_2^*)$$
$$= \kappa|E_{01}|^2 + (1-\kappa)|E_{02}|^2 - 2\sqrt{\kappa(1-\kappa)}E_{01}E_{02}\sin[(\omega_2 - \omega_1)t]$$

$$|E_{4L}|^2 = \|E_4\|^2 \tag{7.7}$$
$$= \kappa|E_2|^2 + (1-\kappa)|E_1|^2 + j\sqrt{\kappa(1-\kappa)}(E_2^*E_1 - E_2E_1^*)$$
$$= \kappa|E_{02}|^2 + (1-\kappa)|E_{01}|^2 + 2\sqrt{\kappa(1-\kappa)}E_{01}E_{02}\sin[(\omega_2 - \omega_1)t]$$

The last term of (7.7) represents the interference pattern between E_1 and E_2. Most of the literature considers only a simple case with single input at Port 1 (i.e., $E_2 = 0$) and hence reduce the interference effect between the signals, and the outputs at Ports 1 and 2 are given as follows:

$$|E_{o1}|^2 = |E_{01}|^2 \left\{ 2\kappa(1-\kappa) \left[1 + \cos\left((1-2\kappa) \frac{2\pi n_2 |E_{01}|^2 L}{\lambda} \right) \right] \right\}$$

$$|E_{o2}|^2 = |E_{01}|^2 \left\{ 1 - 2\kappa(1-\kappa) \left[1 + \cos\left((1-2\kappa) \frac{2\pi n_2 |E_{01}|^2 L}{\lambda} \right) \right] \right\}$$

(7.8)

However, for more complete studies, we consider inputs at both ports; and the outputs can be expressed as

$$
\begin{aligned}
|E_{o1}|^2 &= \kappa |E_{4L}|^2 + (1-\kappa)|E_{3L}|^2 + j\sqrt{\kappa(1-\kappa)}(E_{4L}^* E_{3L} - E_{4L}E_{3L}^*) \\
&= [\kappa^2 + (1-\kappa)^2]|E_{02}|^2 + 2\kappa(1-\kappa)|E_{01}|^2 - 2(1-2\kappa)\sqrt{\kappa(1-\kappa)}E_{01}E_{02} \\
&\quad \times \sin[(\omega_2 - \omega_1)t] + j\sqrt{\kappa(1-\kappa)}(E_3 E_4^* \exp(j\Delta\theta) - E_3^* E_4 \exp(-j\Delta\theta)) \\
&= |E_{01}|^2 [2\kappa(1-\kappa)[1+\cos(\Delta\theta)]] + |E_{02}|^2 [1-2\kappa(1-\kappa)[1+\cos(\Delta\theta)]] \\
&\quad - 2\sqrt{\kappa(1-\kappa)}E_{01}E_{02}[(1-2\kappa)\sin[(\omega_2 - \omega_1)t][1+\cos(\Delta\theta)] \\
&\quad + \cos[(\omega_2 - \omega_1)t]\sin(\Delta\theta)]
\end{aligned}
$$

(7.9)

$$
\begin{aligned}
|E_{o2}|^2 &= \kappa |E_{3L}|^2 + (1-\kappa)|E_{4L}|^2 + j\sqrt{\kappa(1-\kappa)}(E_{3L}^* E_{4L} - E_{3L}E_{4L}^*) \\
&= [\kappa^2 + (1-\kappa)^2]|E_{01}|^2 + 2\kappa(1-\kappa)|E_{02}|^2 + 2(1-2\kappa)\sqrt{\kappa(1-\kappa)}E_{01}E_{02} \\
&\quad \times \sin[(\omega_2 - \omega_1)t] + j\sqrt{\kappa(1-\kappa)}(E_4 E_3^* \exp(j\Delta\theta) - E_4^* E_3 \exp(-j\Delta\theta)) \\
&= |E_{02}|^2 [2\kappa(1-\kappa)[1+\cos(\Delta\theta)]] + |E_{01}|^2 [1-2\kappa(1-\kappa)[1+\cos(\Delta\theta)]] \\
&\quad + 2\sqrt{\kappa(1-\kappa)}E_{01}E_{02}[(1-2\kappa)\sin[(\omega_2 - \omega_1)t][1+\cos(\Delta\theta)] \\
&\quad + \cos[(\omega_2 - \omega_1)t]\sin(\Delta\theta)]
\end{aligned}
$$

(7.10)

where

$$\Delta\theta = \theta_3 - \theta_4 = 2\pi n_2 L \left\{ (1-2\kappa) \left[\frac{|E_{02}|^2}{\lambda_2} - \frac{|E_{01}|^2}{\lambda_1} \right] - \frac{4\sqrt{\kappa(1-\kappa)}E_{01}E_{02}\sin[(\omega_2 - \omega_1)t]}{\sqrt{\lambda_1\lambda_2}} \right\}$$

(7.11)

and λ_1 and λ_2 are the operating wavelengths of E_1 and E_2, respectively.

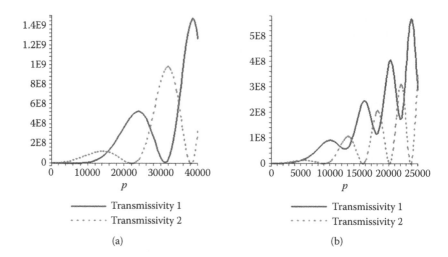

FIGURE 7.5
Switching characteristics for nonlinear optical loop mirror with (a) $\kappa = 0.45$ and (b) $\kappa = 0.2$.

For $\lambda_1 = \lambda_2$ and $E_2 = pE_1$ with p as a constant, the switching characteristics for (a) $\kappa = 0.45$ and (b) $\kappa = 0.2$ are shown in Figure 7.5. Switching occurs only for a large energy difference between the two ports (i.e., $p \gg 1$). Also, as can be seen from the figure, the switching behavior is better for a coupling ratio close to 0.5.

7.5 Nonlinear Amplifying Loop Mirror

The structure of the nonlinear amplifying loop mirror (NALM) (as shown in Figure 7.6) is somehow similar to the NOLM structure, and it is an improved exploitation of NOLM. For NALM configuration, a gain medium with gain coefficient, G is added to increase the asymmetric nonlinearity within the loop [20,21]. The amplifier is placed at one end of the loop, closer to Port 3 of the coupler, and is assumed short relative to the total loop length, as shown in Figure 7.8. One lightwave is amplified at the entrance to the loop, while the other experiences amplification just before exiting the loop. Because the intensities of the two lightwaves differ by a large amount throughout the loop, the differential phase shift can be quite large.

By following the analysis procedure stated in the previous section, we arrive at the input-output relationships as follows:

$$|E_{o1}|^2 = G\left\{\begin{array}{l} |E_{01}|^2 \, [2\kappa(1-\kappa)[1+\cos(\Delta\theta)]] + |E_{02}|^2 \, [1-2\kappa(1-\kappa)[1+\cos(\Delta\theta)]] \\ -2\sqrt{\kappa(1-\kappa)}E_{01}E_{02}[(1-2\kappa)\sin[(\omega_2-\omega_1)t][1+\cos(\Delta\theta)] \\ +\cos[(\omega_2-\omega_1)t]\sin(\Delta\theta)] \end{array}\right\} \qquad (7.12)$$

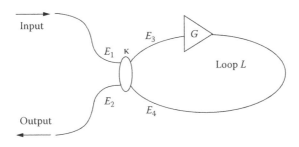

FIGURE 7.6
Nonlinear amplifying loop mirror.

$$|E_{02}|^2 = G \left\{ \begin{matrix} |E_{02}|^2 \, [2\kappa(1-\kappa)[1+\cos(\Delta\theta)]]+|E_{01}|^2 \, [1-2\kappa(1-\kappa)[1+\cos(\Delta\theta)]] \\ +2\sqrt{\kappa(1-\kappa)}E_{01}E_{02}[(1-2\kappa)\sin[(\omega_2-\omega_1)t][1+\cos(\Delta\theta)] \\ +\cos[(\omega_2-\omega_1)t]\sin(\Delta\theta)] \end{matrix} \right\} \quad (7.13)$$

$$\Delta\theta = 2\pi n_2 L \left\{ \begin{matrix} (1-\kappa-G\kappa)\left[\dfrac{|E_{02}|^2}{\lambda_2} - \dfrac{|E_{01}|^2}{\lambda_1} \right] \\ -\dfrac{2(G+1)\sqrt{\kappa(1-\kappa)}E_{01}E_{02}\sin[(\omega_2-\omega_1)t]}{\sqrt{\lambda_1\lambda_2}} \end{matrix} \right\} \quad (7.14)$$

7.6 Nonlinear Optical Loop Mirror/Nonlinear Amplifying Loop Mirror (NOLM-NALM) Fiber Ring Lasers

The configuration of the NOLM-NALM fiber ring laser is shown in Figure 7.7. It is simply coupled loop mirrors, with NOLM (A-B-C-E-A) on one side and NALM (C-D-A-E-C) on the other side of the laser, with a common path in the middle section (A-E-C). One interesting feature about this configuration is the feedback mechanism of the fiber ring: one is acting as the feedback path to another (i.e., NOLM is feeding-back part of the NALM's signal and vice versa).

The complex systems tend to encounter bifurcations that when amplified can lead to either order or chaos. The system can transit into chaos, through period doubling, or order through a series of feedback loops. Hence, bifurcations can be considered as critical points in this system transition.

In the formularization process, we ignore the nonlinear phase shift due to laser pulsation. The assumption is valid because of the saturation effect of gain medium, as well as the energy stabilization provided by the filter.

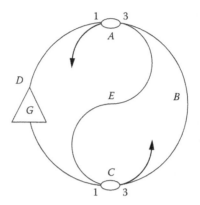

FIGURE 7.7

A nonlinear optical loop mirror/nonlinear amplifying loop mirror (NOLM-NALM) fiber ring laser.

Simulation of the laser behavior is conducted by combining the effects described in the previous section, and bifurcation maps are obtained based on the CW operation as shown in Figures 7.8a and 7.8b, with P_{o1} and P_{o2} the output powers at points A and C, respectively; $\kappa_1 = 0.55$ and $\kappa_2 = 0.65$; L_a, L_b, and L_c are 20 m, 100 m, and 20 m, respectively; nonlinear coefficient of 3.2 $\times$ 10^{-20} m^2/W; gain coefficient of 0.4/m at 1550 nm with erbium-doped fiber (EDF) length of 10 m and a saturation power of 25 dBm; and fiber effective area of 50 mm^2. κ_1 and κ_2 are the coupling ratios of couplers from Port 1 to Port 3 at points A and C, respectively; L_a, L_b, and L_c are the fiber lengths of the left, mid, and right arm of the laser cavity. The maps are obtained with 200 iterations. Figure 7.8 shows the bifurcation maps with the contribution of

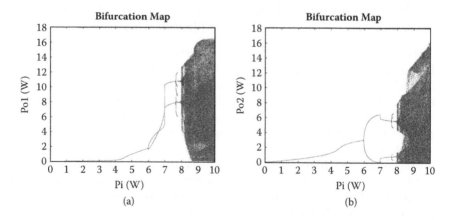

FIGURE 7.8

Bifurcation maps for (a) P_{o1} and (b) P_{o2} with $\kappa_1 = 0.55$ and $\kappa_2 = 0.65$ with self-phase modulation consideration only.

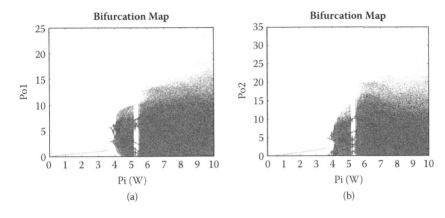

FIGURE 7.9
Bifurcation maps for (a) P_{o1} and (b) P_{o2} with $\kappa_1 = 0.55$ and $\kappa_2 = 0.65$ with both self-phase modulation and cross-phase modulation considerations.

self phase modulation (SPM) effect only, whereas in Figure 7.9 both SPM and cross phase modulation (XPM) effects are included.

In constructing the bifurcation maps of the system, we separate the lightwaves into clockwise and counterclockwise directions; similarly for the couplers involved in the system (i.e., four couplers are used in simulating this bifurcation behavior). Initial conditions are set to be the pump power of the EDFA. We then propagate the lightwaves in both directions within the fiber ring, with various component effects taken into account, such as coupling ratios, gain, and SPM and XPM effects. The outputs of the system are then served as the input to the system for the next iteration. The process is repeated a number of times (i.e., 200 times). Each iterated set of outputs is then combined together in constructing the bifurcation maps.

There are some similarities between Figure 7.8 and Figure 7.9. Both indicate different operating regions at different power levels. With both nonlinear modulation effects (SPM and XPM) are included (Figure 7.9), the transition from one operation state to another is faster, leading to earlier chaotic behavior with lower pump power because the strength of the XPM effect is twice of that of the SPM effect. For simplicity, we discuss only the bifurcation behavior of Figure 7.8 in order to illustrate the transition behavior from one state to another.

From the bifurcation maps depicted in Figure 7.8, under this system setting, there are three operating regions within this NOLM-NALM fiber ring laser. The first is the linear operation region ($0 < P < 6$ W), where there are single-value outputs, and the output power increased with the input power. The period-doubling effect starts to appear when the input power reaches ~6 W. This is the second operating region ($6 \leq P < 8$ W) of the laser, where double-periodic and quasi-periodic signals can be found. When the input power level reaches beyond 8 W, the laser enters into the chaotic state operation.

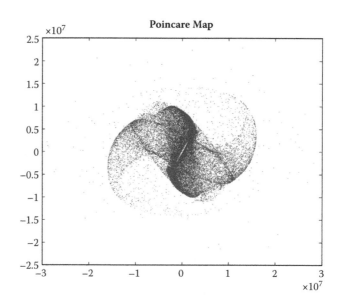

FIGURE 7.10
Poincaré map of the system under high pump power with $\kappa_1 = 0.55$ and $\kappa_2 = 0.65$.

The Poincaré map of the above system configuration with high pump power is shown in Figure 7.10. It shows the pattern of attractor of the system when the laser is operating in the chaotic region. The powers required for the operations can be reduced by increasing the lengths of the fibers, L_a, L_b, and L_c.

By setting $\kappa_2 = 0$, the laser will behave like a NOLM. With an input pulse to Port 1 of Coupler A and we observe pulse compression at its Port 2, as shown in Figures 7.11 and 7.12. Figure 7.12 presents the transmission capability of the setup at Port 2 for various κ_1 values, when the input is injected to Port 1

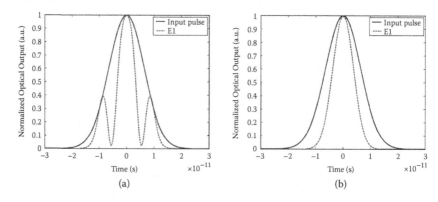

FIGURE 7.11
Input and output pulse at Port 1 and Port 2 comparison with $\kappa_2 = 0$ and (a) $\kappa_1 = 0.41$ and (b) $\kappa_1 = 0.49$ (solid line, input pulse; dotted line, output pulse at Port 1).

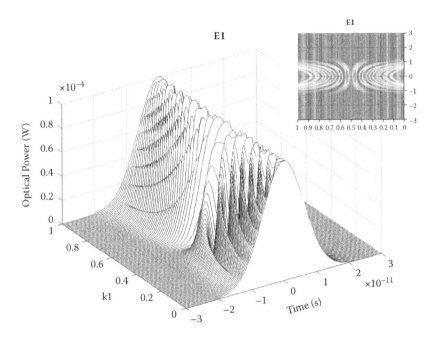

FIGURE 7.12
Output pulse behavior at various coupling coefficients κ_1 and when $\kappa_2 = 0$ for small input power (Inset: top view of the pulse behavior).

and $\kappa_2 = 0$. When $\kappa_1 = 0.5$, we observe no transmission at Port 2, as the entire injected signal has been reflected back to Port 1, where the mirror effect takes place. By changing the coupling ratio of κ_2, we are able to change the zero transmission point away from $\kappa_1 = 0.5$. Figure 7.13 shows the transmissivities of P_{o1} and P_{o2} for various sets of κ_1 and κ_2, The figure shows the complex

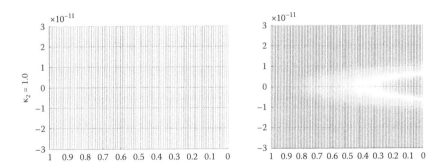

FIGURE 7.13
Intensity variation as a function of P_{o1} and P_{o2} for different κ_2 values. (*y*-time axis, *x*-axis are the values of κ_1.)

switching dynamics of the laser for different sets of κ_1,κ_2, also with the pulse compression capability.

7.7 Experiment Setups and Analyses

7.7.1 Bidirectional Erbium-Doped Fiber Ring Laser

We start with a simple erbium-doped fiber ring laser; an optical closed loop with EDFA and some fiber couplers. It is used to study the bidirectional lightwave propagation behavior of the laser. The EDFA is made of a 20 m erbium-doped fiber (EDF) and is dual pumped by 980 and 1480 nm diode lasers, with the saturation power of about 15 dBm. The slope efficiencies of the pump lasers are 0.45 mW/mA and 0.22 W/A, respectively. No isolator and filter are used in the setup to eliminate the direction and spectra constraints. Two outputs are taken for examination (i.e., Output 1 in counterclockwise direction, and Output 2 in clockwise direction). All connections within the fiber ring are spliced together to reduce the possible reflections within the system. Output 1 (counterclockwise, ccw) and Output 2 (clockwise, cw) as shown in Figure 7.14 are taken for investigations.

The amplification stimulated emission (ASE) spectrum of the laser covers the range from 1530 nm to 1570 nm. By increasing both the 980 nm and 1480 nm pumping currents to their maximum allowable values, we obtained bidirectional lasing, which is shown in Figure 7.15. The upper plot is the lightwave propagating in the clockwise direction while the other one is in the opposite direction. To obtain bidirectional lasing, the losses and the gain must be balanced for the two lightwaves that propagate around the cavity in opposite directions [22]. The pump lasers used in the setup are not identical (in terms of power and pumping wavelength), and this gives rise to different

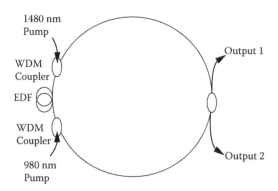

FIGURE 7.14
Bidirectional Erbium-doped fiber ring laser.

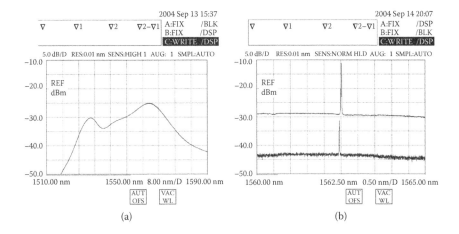

FIGURE 7.15
(a) ASE spectrum of the laser. (b) Lasing characteristics in both directions (upper trace: Output 2, lower trace: Output 1).

lasing behaviors for both clockwise and counterclockwise directions. Due to the laser diode controllers' limitation, the maximum pumping currents for both 980 nm and 1480 nm laser diodes are capped at 300 mA and 500 mA, respectively, which correspond to 135 mW and 110 mW in clockwise pump and counterclockwise pump directions. This explains the domineering clockwise lasing as shown in the figure. One thing to note is that the lasings of the laser in both directions are not very stable due to the disturbance from the opposite propagating lightwave and the ASE noise contribution due to the absence of filter. The Output 1 is mainly contributed by the back reflections from the fiber ends and connectors, as well as some back-scattered noise. Because all the connections within the fiber ring are spliced together, the reflection due to the connections is lowest. However, there are still some unavoidable reflections from the fiber ends, which contribute to the lightwaves in the opposite directions. Besides that, back-scattered noise adds to the lightwaves in the opposite directions.

We maintain 980 nm pump current at a certain value, and adjust 1480 nm pump current upwards and then downwards to examine the bistability behavior of the laser. The bistable characteristics at a lasing wavelength of about 1562.2 nm for both clockwise (cw) and counterclockwise (ccw) directions are shown in Figures 7.16 and 7.17, with log and linear scales for the vertical axes of Figures 7.16a and 7.17a and Figures 7.16b and 7.17b, respectively. We obtain about 15 dBm difference between the two propagating lightwaves. The bistable region is obtained at a ~30 mA of pump current of 1480 nm source and at a fixed value 100 mA for 980 nm pump laser. No lasing is observed in the ccw lightwave propagation. This bistable region can be further enhanced by increasing the 980 nm pump power to a higher level. By maintaining the

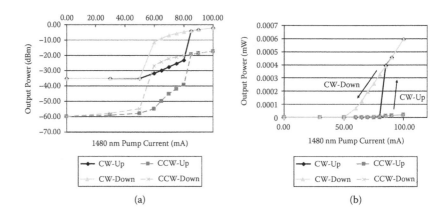

(a)   (b)

FIGURE 7.16
Hysteresis loops of EDFRL at 980 nm pump laser at a driving current of 100 mA (a) log scale, (b) linear scale.

980 nm pump current at ~175 mA, a bistable region can be observed as wide as at ~70 mA of the 1480 nm pump current. When the 980 nm pump current is maintaining at a higher level (>175 mA), the lasing of Output2 remains, even when the 1480 nm is switched off, as shown in Figure 7.17. This bistable behavior is mainly due to the saturable absorption of the Er-doped fiber section.

7.7.2 NOLM-NALM Fiber Ring Laser

The experimental setup of the NOLM-NALM fiber ring laser is shown in Figure 7.18. It is simply a combination of NOLM on one side and NALM on

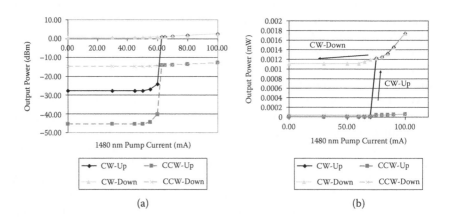

(a)   (b)

FIGURE 7.17
Hysteresis loops obtained from the EDFRL for 980 nm pump current = 200 mA (a) log scale, (b) linear scale.

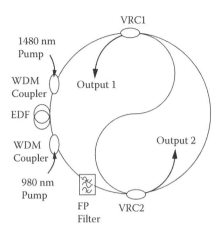

FIGURE 7.18
Experiment schematic of nonlinear optical loop mirror/nonlinear amplifying loop mirror (NOLM-NALM) fiber ring laser.

the other side of the laser, with a common path in the middle section. The principal element of the laser is an optical closed loop with an optical gain medium, two variable ratio couplers (VRCs), an optical bandpass filter (BPF), optical couplers, and other associated optics. The gain medium used in our fiber laser system is the amplifier used in the preceding experiment. The two VRCs, with coupling ratios ranging from 20% to 80% and insertion loss of about 0.2 dB are added into the cavity at positions shown in the figure to adjust the coupling power within the laser. They are interconnected in such a way that the output of one VRC is the input of the other. A tunable bandpass filter with 3 dB bandwidth of 2 nm at 1560 nm is inserted into the cavity to select the operating wavelength of the generated signal and to reduce the noise in the system. It is noted that the lightwaves are traveling simultaneously in both directions, as there is no isolator used in the laser. Output 1 and Output 2 as shown in the figure as are the outputs of the laser.

One interesting phenomenon observed before the bandpass filter is inserted into the cavity is the wavelength tunability. The lasing wavelength is tunable from 1530 nm to 1560 nm (almost the entire EDFA C-band) by changing the coupling ratios of the VRCs. We believe that this wavelength tunability is due to the change in the traveling lightwaves' intensities, which contributes to the nonlinear refractive index change, and in turn modifies the dispersion relations of the system, and hence the lasing wavelength. Therefore, the VRCs within the cavity determine not only the directionality of the lightwave propagation but also the lasing wavelength.

For a conventional erbium-doped fiber ring laser, bistable stability is not observable when the pump current is far above the threshold value, where saturation starts to take place. However, a small hysteresis loop has been

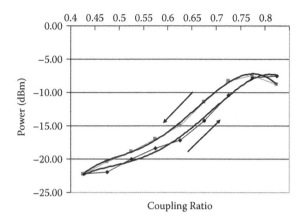

FIGURE 7.19
Hysteresis loop observed, laser output power versus the coupling ratio of one variable ratio coupler (VRC), while the other coupling remains unchanged operating under high pumping current.

observed in our laser setup even with a high pump current (i.e., near saturation region), when changing the coupling ratio of one VRC while that of the other one remains unchanged, as shown in Figure 7.19. Changing the coupling ratio of the VRC is directly altering the total power within the cavity, and hence modifying its gain and absorption behavior. As a result, a small hysteresis loop is observable even with a constant high pump power, which forms an additional member of the bistable state family. The power distribution of Output 1 and Output2 of the NOLM-NALM fiber laser obtained experimentally is depicted in Figure 7.20. We are able to obtain the switching between the outputs by tuning the coupling ratios of VRC1 and VRC2. The simulation results for transmittance of various coupling ratios under linear operation are shown in Figure 7.21, because the available pump power of our experimental setup is insufficient to create high power within the cavity. Both the experimental and numerical results have come to some agreements, but not all, because the model developed is simple and does not consider the polarization, dispersion characteristics, and so forth of the propagating lightwaves.

7.7.3 Amplitude Modulated NOLM-NALM Fiber Ring Laser

The schematic of AM modulated NOLM-NALM fiber ring laser is shown in Figure 7.22. A few new photonic components are added into the laser cavity. An asymmetric coplanar traveling wave 10Gb/s, Ti:LiNbO$_3$ Mach–Zehnder amplitude modulator is used in the inner loop of the cavity with half-wave voltage, V_π of 5.3 V. The modulator is DC biased near the quadrature point and not driven higher than V_π such that it operates on the linear region of its characteristic curve to ensure minimum chirp imposing on the modulated

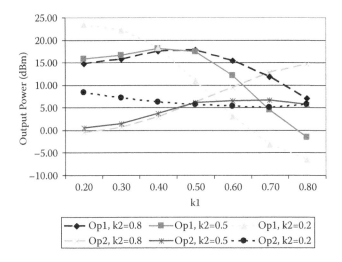

FIGURE 7.20
Laser output power as function of coupling coefficient, experimental results for Output 1 (Op1) and Output2 (Op2) with coupling ratios (k_1, coupling ratio of VRC1; k_2, coupling ratio of VRC2) as parameters.

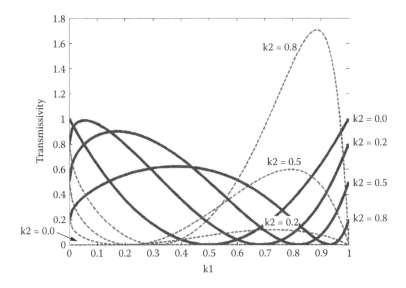

FIGURE 7.21
Simulated transmittances of Output 1 (solid line) and Output2 (dotted line) for various coupling ratios of VRC1 and VRC2.

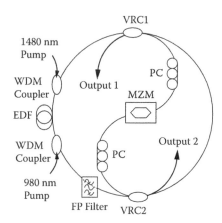

FIGURE 7.22
Experimental setup for amplitude modulation nonlinear optical loop mirror/nonlinear amplifying loop mirror (NOLM-NALM) fiber laser (VRC, variable ratio coupler; PC, polarization controller; MZM, Mach–Zehnder modulator).

lightwaves. It is driven by the sinusoidal signal derived from an Anritsu 68347C Synthesizer Signal Generator. The modulator has an insertion loss of £7 dB. Two polarization controllers are placed in front of the inputs of the modulator in both directions to ensure proper polarization alignment into the modulator. A wider bandwidth (i.e., 5 nm) tunable Fabry–Perot (FP) filter is used in this case to allow more longitudinal modes within the laser for possible mode-locking process.

With the insertion of AM modulator into the laser cavity, we are able to obtain the pulse operation from the laser, by means of active harmonic mode-locking technique. Both propagation lightwaves are observed. By proper adjustment of the modulation frequency, the polarization controllers and the variable ratio couplers, unidirectional pulse operation at modulating frequency is obtained. However, it is highly sensitive to the environmental change. The direction of the lightwave propagation of the laser can be controlled by the VRCs. The unidirectional pulse train propagation obtained experimentally and numerically is shown in Figure 7.23. However, with a slight deviation to the system parameters, either in modulation frequency detuning or polarization, period doubling and quasi-periodic operations in the laser are observed, as shown in Figure 7.24. We believe that the effect is due the interference between the bidirectional propagations lightwaves, which have suffered nonlinear phase shifts in each direction, because we are operating in the saturation region of the EDFA. Another factor for this formation is that the lightwave in one direction is the feedback signal for another one. Furthermore, the intensity modulation of the optical modulator is not identical for co- and counterinteractions between the lightwaves and the traveling microwaves on the surface of the optical waveguide. This would contribute to the mismatch of the locking condition of the laser.

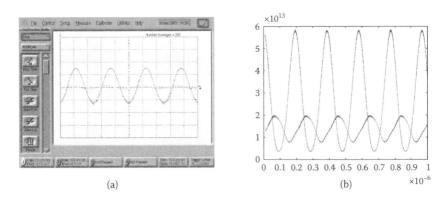

(a)   (b)

FIGURE 7.23
Unidirectional pulse operation in AM NOLM-NALM fiber laser: (a) Experimental results;
(b) simulation results.

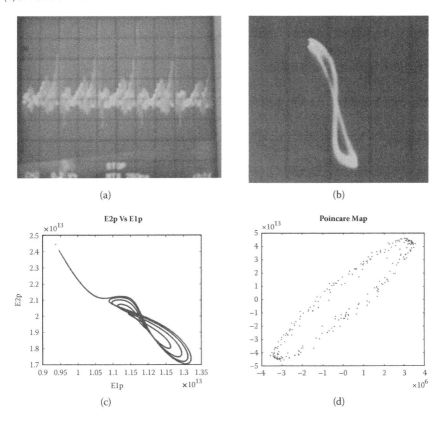

FIGURE 7.24
Quasi-periodic operation in amplitude modulation nonlinear optical loop mirror/nonlinear
amplifying loop mirror (NOLM-NALM) fiber laser: (a) photograph of the oscilloscope trace,
(b) XY plot of Output 1 and Output 2, (c) simulated XY plot, (d) simulated Poincaré map.

7.8 Remarks

Bidirectional optical bistability in a dual-pumped erbium-doped fiber ring laser without isolator has been studied. A ~70 mA 1480 nm pump current bistable region has also been obtained. With the bidirectional bistability characteristics of the NOLM-NALM fiber lasers, we experimentally demonstrate and numerically simulated the switching and bifurcation behaviors. From the simulated bifurcations maps, three basic operation regions can be identified, namely unidirectional, period-doubling, and chaos operations. The VRCs used in the setup not only control the lightwave directionality, but also its lasing wavelength. Unidirectional lightwave propagations, without isolator, were achieved in both continuous-wave and pulse operations by tuning the coupling ratios of the VRCs within the laser system. Bifurcation was also obtained from the AM NOLM-NALM fiber laser. However, chaotic operation was not observed experimentally due to the hardware limitation of our system, which required higher gain coefficient and input power as predicted in our simulation.

The configuration is somehow similar to the optical flip-flop concept, which can be useful in optical communication systems. The optical flip-flop concept has been used in optical packet switches and optical buffers [23–26]. NOLM-NALM fiber lasers possess several interesting optical behaviors, which may be deduced for photonic signal processing, such as photonic flip-flops, optical buffer loop, optical pulse sampling devices, and secured optical communications.

References

1. H. M. Gibbs, S. L. McCall, and T. N. C. Venkatesan, Differential Gain and Bistability Using a Sodium-Filled Fabry-Perot Interferometer, *Phys. Rev. Lett.*, 36(19), 1135–1138, 1976.
2. H. Nakatsuka, S. Asaka, H. Itoh, K. Ikeda, and M. Matsuoka, Observation of Bifurcation to Chaos in an All-Optical Bistable System, *Phys. Rev. Lett.*, 50(2), 109–112, 1983.
3. K. Tamura, E. P. Ippen, H. A. Haus, and L. E. Nelson, 77-fs Pulse Generation from a Stretched-Pulse Mode Locked All Fiber Ring Laser, *Opt. Lett.*, 18(13), 1080–1082, 1993.
4. G. Yandong, S. Ping, and T. Dingyuan, 298 fs Passively Mode Locked Ring Fiber Soliton Laser, *Microwave Opt. Technol. Lett.*, 32(5), 320–333, 2002.
5. K. K. Gupta, N. Onodera, K. S. Abedin, and M. Hyodo, Pulse Repetition Frequency Multiplication via Intracavity Optical Filtering in AM Mode-Locked Fiber Ring Lasers, *IEEE Photonics Technol. Lett.*, 14(3), 284–286, 2002.
6. A. E. Siegman, *Lasers*, University Science Books, Mill Valley, CA, 1986.

7. Y. Shi, M. Sejka, and O. Poulsen, A Unidirectional Er³⁺-Doped Fiber Ring Laser without Isolator, *IEEE Photonics Technol. Lett.*, 7(3), 290–292, 1995.
8. M. Oh and D. Lee, Strong Optical Bistability in a Simple L-Band Tunable Erbium-Doped Fiber Ring Laser, *J. Quantum Electron.*, 40(4), 374–377, 2004.
9. Q. Mao and J. W. Y. Lit, L-Band Fiber Laser with Wide Tuning Range Bases on Dual-Wavelength Optical Bistability in Linear Overlapping Grating Cavities, *IEEE J. Quantum Electron.*, 39(10), 1252–1259, 2003.
10. L. Luo, T. J. Tee, and P. L. Chu, Bistability of Erbium Doped Fiber Laser, *Opt. Commun.*, 146, 151–157, 1998.
11. Y. Shi, M. Sejka, and O. Poulsen, A Unidirectional Er³⁺-Doped Fiber Ring Laser without Isolator, *IEEE Photonics Technol. Lett.*, 7(3), 290–292, 1995.
12. P. P. Banerjee, *Nonlinear Optics—Theory, Numerical Modeling, and Applications*, Marcel Dekker, New York, 2004.
13. G. P. Agrawal, *Applications of Nonlinear Fiber Optics*, 2nd ed., Academic Press, New York, 2006.
14. P. Meystre, On the Use of the Mean-Field Theory in Optical Bistability, *Opt. Commun.*, 26(2), 277–280, 1978.
15. L. A. Orozco, H. J. Kimble, A. T. Rosenberger, L. A. Lugiato, M. L. Asquini, M. Brambilla, and L. M. Narducci, Single-Mode Instability in Optical Bistability, *Phys. Rev. A*, 39(3), 1235–1252, 1989.
16. K. Ikeda, Multiple-Valued Stationary State and Its Instability of the Transmitted Light by a Ring Cavity, *Opt. Commun.*, 30(2), 257–261, 1979.
17. M. J. Ogorzalek, Chaos and Complexity in Nonlinear Electronic Circuits, *World Scientific Series on Nonlinear Science, Series A*, no. 22, 1997.
18. M. P. Kennedy, Three Steps to Chaos—Part I: Evolution, *IEEE Trans. Circuits and Systems—I: Fundam. Theory Appl.*, 40(10), 640–656, 1993.
19. N. J. Doran and D. Wood, Nonlinear-Optical Loop Mirror, *Opt. Lett.*, 13(1), 56–58, 1988.
20. M. E. Fermann, F. Haberl, M. Hofer, and H. Hochreiter, Nonlinear Amplifying Loop Mirror, *Opt. Lett.*, 15(13), 752–754, 1990.
21. M. E. Fermann, A. Galvanauskas, and G. Sucha, *Ultrafast Lasers—Technology and Applications*, Marcel Dekker, New York, 2003.
22. M. Mohebi, J. G. Mejia, and N. Jamasbi, Bidirectional Action of a Titanium-Sapphire Ring Laser with Mode-Locking by a Kerr Lens, *J. Opt. Technol.*, 69(5), 312–316, 2002.
23. X. Zhang, M. Karlsson, and P. A. Andrekson, Design Guidelines of Actively Mode-Locked Fiber Ring Lasers, *IEEE Photonics Technol. Lett.*, 10(8), 1103–1105, 1998.
24. N. J. Doran and D. Wood, Nonlinear-Optical Loop Mirror, *Opt. Lett.*, 13(1), 56–58, 1988.
25. R. Langenhorst, M. Eiselt, W. Pieper, G. Grobkopf, R. Ludwig, L. Küller, E. Dietrich, and H. C. Weber, Fiber Loop Optical Buffer, *J. Lightwave Technol.*, 14(3), 324–335, 1996.
26. A. Liu, C. Wu, Y. Gong, and P. Shum, Dual-Loop Optical Buffer (DLOB) Based on a 3 × 3 Collinear Fiber Coupler, *IEEE Photonics Technol. Lett.*, 16(9), 2129–2131, 2004.

8

Nonlinear Fiber Ring Lasers

Le Nguyen Binh

Hua Wei Technologies, European Research Center, Munich, Germany

CONTENTS

8.1 Overview

This chapter presents the operational principles and implementation of mode-locked fiber lasers operating in the nonlinear region, whether via the optical saturated amplification or the photonic interactions of the pump sources and generated lightwaves. Harmonic and regenerative mode-locked types for 10 and 40 G-pulse/s employing the harmonic detuning technique for generating up to 200 G-pulses/s, and harmonic repetition multiplication using temporal diffraction are described. An ultrastable mode-locked laser operating at 10 GHz repetition rate has been designed, constructed, and tested. The laser generates an optical pulse train of 4.5 ps pulse width when the modulator is biased at the phase quadrature quiescent region. Long-term stability of amplitude and phase noise indicates that the optical pulse source can produce an error-free pattern in a self-locking mode for more than 20 hours, the most stable photonic fiber ring laser reported to date.

The repetition rate is demonstrated up to 200 G-pulse/s using the harmonic detuning mechanism in a nonlinear fiber ring laser. In this system, the system operation under the rational harmonic mode-locking is analyzed using the phase plane technique in control engineering. Furthermore, we examine the harmonic distortion contribution to this system performance. The multiplication factor of 660× and 1230× of the fundamental rate of a 100 MHz pulse train can be achieved, hence 66 GHz and 123 GHz pulse rate. The system behavior of group velocity dispersion repetition rate multiplication is proven as one of the principal mechanisms. Stability and the transient response of the multiplied pulses are studied using the phase plane technique.

Nonlinear fiber lasers can be used to generate bistable operations and multibound solitons based on the nonlinearity in the ring cavity. Experimental and theoretical generation of multisoliton bound states in an active frequency modulation (FM) mode-locked fiber laser will be described in Chapter 3 in which not only bound soliton pairs but also triple- and quadruple-soliton bound states can be generated.

8.2 Introduction

Generation of ultrashort optical pulses with a multiple gigabits repetition rate is critical for ultrahigh bit rate optical communications, particularly for the next generation of terabits/sec optical fibers systems. As the demand for the bandwidth of the optical communication systems increases, the generation of short pulses with an ultrahigh repetition rate becomes increasingly important in the coming decades.

The mode-locked fibers laser offers a potential source of such a pulse train. Although the generation of ultrashort pulses by mode locking of a multimodal ring laser is well known, the applications of such short pulse trains in multigigabits/sec optical communications challenges its designers on its stability and spectral properties. Recent reports on the generation of short pulse trains at repetition rates in order of 40 Gb/s, possibly higher in the near future [1], motivates us to design and experiment with these sources in order to evaluate whether they can be employed in practical optical communications systems.

Further, the interest of multiplexed transmission at 160 Gb/s and higher in the foreseeable future requires us to experiment with an optical pulse source having a short pulse duration and high repetition rates. This report describes laboratory experiments of a mode-locked fiber ring laser (MLFRL), initially with a repetition rate of 10 GHz and preliminary results of higher multiple repetition rates up to 40 GHz. The mode-locked ring lasers reported hereunder adopt an active mode-locking scheme whereby partial optical power of the output optical waves is detected and filtered and a clock signal is recovered at the desired repetition rate. It is then used as a radio-frequency (RF) drive signal to the intensity modulator incorporated in the ring laser. A brief description on the principle of operation of the MLFRL is given in the next section followed by a description of the mode-locked laser experimental setup and characterization.

Active mode-locked fiber lasers remain as a potential candidate for the generation of such pulse trains. However, the pulse repetition rate is often limited by the bandwidth of the modulator used or the RF oscillator that generates the modulation signal. Hence, some techniques have been proposed to increase the repetition frequency of the generated pulse trains. Rational harmonic mode locking is widely used to increase the system repetition frequency [1–3]. A 40 GHz repetition frequency has been obtained with fourth-order rational harmonic mode locking at 10 GHz base band modulation frequency [2]. Wu and Dutta [3] reported 22nd-order rational harmonic detuning in the active mode-locked fiber laser, with 1 GHz base frequency, leading to 22 GHz pulse operation. This technique is simple and achieved by applying a slight deviated frequency from the multiple of fundamental cavity frequency. Nevertheless, it is well known that it suffers from inherent pulse amplitude instability as well as poor long-term stability. Therefore, pulse amplitude equalization techniques are often applied to achieve better system performance [3–5].

Other than this rational harmonic detuning, other techniques have been reported and used to achieve the same objective. The fractional temporal Talbot-based repetition rate multiplication technique [6,7] uses the interference effect between the dispersed pulses to achieve the repetition rate multiplication. The essential element of this technique is the dispersive medium, such as linearly chirped fiber grating (LCFG) [6,8] and dispersive fiber [9–11]. Intracavity optical filtering [12,13] uses modulators and a high finesse Fabry–Perot filter (FFP) within the laser cavity to achieve a higher repetition rate by filtering out certain lasing modes in the mode-locked laser. Other techniques used in repetition rate multiplication include higher-order FM mode locking [14], optical time domain multiplexing [15], and so forth.

The stability of a high repetition rate pulse train generated is one of the main concerns for practical multi-Giga bits/sec optical communications system. Qualitatively, a laser pulse source is considered as stable if it is operating at a state where any perturbations or deviations from this operating point are not increased but suppressed. Conventionally the stability analyses of such laser systems are based on the linear behavior of the laser in which we can analyze the system behavior in both time and frequency domains. However, when the mode-locked fiber laser is operating under a nonlinear regime, none of these standard approaches can be used, because the direct solution of a nonlinear different equation is generally impossible, hence frequency domain transformation is not applicable. Some inherent nonlinearities in the fiber laser may affect its stability and performance, such as the saturation of the embedded gain medium, nonquadrature biasing of the modulator, nonlinearities in the fiber, and so forth, hence, a nonlinear stability approach should be used in any laser stability analysis.

In the next section, we focus on the stability and transient analyses of the rational harmonic mode locking in the fiber ring laser system using the phase plane method, which is commonly used in nonlinear control systems. This technique was previously used in [11] to study the system performance of the fractional temporal Talbot repetition rate multiplication systems. It has been shown that it is an attractive tool in system behavior analysis. However, it has not been used in the rational harmonic mode-locking fiber laser system. In the next section, the rational harmonic detuning technique is briefly discussed.

Rational harmonic detuning [3,16] is achieved by applying a slight deviated frequency from the multiple of fundamental cavity frequency. A 40 GHz repetition frequency has been obtained by [3] using a 10 GHz base band modulation frequency with fourth-order rational harmonic mode locking. This technique is simple in nature. However, this technique suffers from inherent pulse amplitude instability, which includes both amplitude noise and inequality in pulse amplitude; furthermore, it gives poor long-term stability. Hence, pulse amplitude equalization techniques are often applied to achieve better system performance [2,4,5]. The fractional temporal Talbot-based repetition rate multiplication technique [4–8] uses the interference effect between the dispersed pulses to achieve the repetition rate multiplication. The essential element of this technique is the dispersive medium, such as linearly chirped fiber grating (LCFG) [8,16] and single-mode fiber [8,9]. This technique will be discussed further in Section 8.2. Intracavity optical filtering [13,14] uses modulators and a high finesse Fabry–Perot filter (FFP) within the laser cavity to achieve a higher repetition rate by filtering out certain lasing modes in the mode-locked laser. Other techniques used in repetition rate multiplication include higher-order FM mode locking [13], optical time domain multiplexing, and so forth.

Although Talbot-based repetition rate multiplication systems are based on the linear behavior of the laser, there are still some inherent nonlinearities affecting its stability, such as the saturation of the embedded gain medium, nonquadrature biasing of the modulator, nonlinearities in the fiber, and so forth; hence, a nonlinear stability approach must be adopted. In Section 8.3, we focus on the stability and transient analyses of the group velocity dispersion (GVD) multiplied pulse train using the phase plane analysis of nonlinear control analytical technique [2]. This section uses the phase plane analysis described in Chapter 2 to study the stability and transient performances of the GVD repetition rate multiplication systems. In Section 8.2, the GVD repetition rate multiplication technique is briefly given. Section 8.3 describes the experimental setup for the repetition rate multiplication. This section also investigates the dynamic behavior of the phase plane of the GVD multiplication system, followed by some simulation results. Finally, some concluding remarks and possible future developments for this type of laser are given.

8.3 Active Mode-Locked Fiber Ring Laser by Rational Harmonic Detuning

In this section we investigate the system behavior of rational harmonic mode-locking in the fiber ring laser using the phase plane technique of the nonlinear control engineering. Furthermore, we examine the harmonic distortion contribution to this system performance. We also demonstrate 660× and 1230× repetition rate multiplications on 100 MHz pulse train generated from an active harmonically mode-locked fiber ring laser, hence achieving 66 GHz and 123 GHz pulse operations by using rational harmonic detuning, which is the highest rational harmonic order reported to date.

8.3.1 Rational Harmonic Mode Locking

In an active harmonically mode-lock fiber ring laser, the repetition frequency of the generated pulses is determined by the modulation frequency of the modulator, $f_m = qf_c$, where q is the qth harmonic of the fundamental cavity frequency, f_c, which is determined by the cavity length of the laser, $f_c = c/nL$, where c is the speed of light, n is the refractive index of the fiber, and L is the cavity length. Typically, f_c is in the range of kHz or MHz. Hence, in order to generate a GHz pulse train, mode locking is normally performed by modulation in the states of $q \gg 1$ (i.e., q pulses circulating within the cavity), which is known as harmonic mode locking. By applying a slight deviation or a fraction of the fundamental cavity frequency, $Df = f_c/m$, where m is the integer, the modulation frequency becomes

$$f_m = qf_c \pm \frac{f_c}{m} \tag{8.1}$$

This leads to an m-times increase in the system repetition rate, $f_r = mf_m$, where f_r is the repetition frequency of the system [2]. When the modulation frequency is detuned by an m fraction, the contributions of the detuned neighboring modes are weakened, only every mth lasing mode oscillates in phase and the oscillation waveform maximums accumulate, hence achieving *m* times higher repetition frequency. However, the small but not negligible detuned neighboring modes affect the resultant pulse train, which leads to uneven pulse amplitude distribution and poor long-term stability. This is considered as harmonic distortion in our modeling, and it depends on the laser linewidth and amount detuned (i.e., a fraction *m*). The amount of the allowable detunable range or rather the obtainable increase in the system repetition rate by this technique is very much limited by the amount of harmonic distortion. When the amount of frequency detuned is too small relative to the modulation frequency, that is very high *m*, contributions of the neighboring lasing modes become prominent, thus reducing the repetition rate multiplication capability significantly. In other words, no repetition frequency multiplication is achieved when the detuned frequency is unnoticeably small. Often the case, it is considered as the system noise due to improper modulation frequency tuning. In addition, the pulse amplitude fluctuation is determined by this harmonic distortion.

8.3.2 Experimental Setup

In general, the experimental setup of the active harmonically mode-locked fiber ring laser is similar to Figure 8.1. The principal element of the laser is an optical open loop with an optical gain medium, a Mach–Zehnder amplitude modulator (MZM), an optical amplifier to supply sufficient energy of photons, an optical polarization controller (PC), an optical bandpass filter (BPF), optical couplers, and other associated optics.

The gain medium used in our fiber laser system is an erbium-doped fiber amplifier (EDFA) with the saturation power of 16 dBm. A polarization-independent optical isolator is used to ensure unidirectional lightwave propagation as well as to eliminate back-reflections from the fiber splices and optical connectors. A free space filter with 3 dB bandwidth of 4 nm at 1555 nm is inserted into the cavity to select the operating wavelength

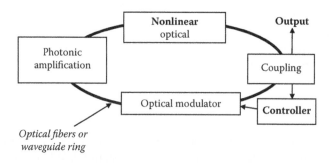

FIGURE 8.1
Setup of a THz regenerative mode-locked fiber ring laser using nonlinear effects such as parametric amplification sequence incorporating the proposed controller.

of the generated signal and to reduce the noise in the system. In addition, it is responsible for the longitudinal modes selection in the mode-locking process. The birefringence of the fiber is compensated by a polarization controller, which is also used for the polarization alignment of the linearly polarized lightwave before entering the planar structure modulator for better output efficiency. Pulse operation is achieved by introducing an asymmetric coplanar traveling wave 10 Gb/s lithium niobate, Ti:LiNbO$_3$ Mach–Zehnder amplitude modulator into the cavity with half-wave voltage, V_p of 5.8 V and insertion loss of £7 dB. The modulator is DC biased near the quadrature point and not more than the V_p such that it operates around the linear region of its characteristic curve. The modulator is driven by a 100 MHz, 100 ps step recovery diode (SRD), which is in turn driven by an RF amplifier (RFA), a RF signal generator. The modulating signal generated by the step recovery diode is a ~1% duty cycle Gaussian pulse train. The output coupling of the laser is optimized using a 10/90 coupler. Then 90% of the optical field power is coupled back into the cavity ring loop, while the remaining portion is taken out as the output of the laser and is analyzed.

8.3.3 Phase Plane Analysis

A nonlinear system frequently has more than one equilibrium point. It can also oscillate at a fixed amplitude and fixed period without external excitation. This oscillation is called the limit cycle. However, limit cycles in nonlinear systems are different from linear oscillations. First, the amplitude of self-sustained excitation is independent of the initial condition, while the oscillation of a marginally stable linear system has its amplitude determined by the initial conditions. Second, marginally stable linear systems are very sensitive to changes, while limit cycles are not easily affected by parameter changes [31].

As described in Chapter 2, phase plane analysis is a graphical method of studying second-order nonlinear systems. The result is a family of system motion of trajectories on a two-dimensional plane, which allows us to visually observe the motion patterns of the system. Nonlinear systems can display more complicated patterns in the phase plane, such as multiple equilibrium points and limit cycles. In the phase plane, a limit cycle is defined as an isolated closed curve. The trajectory has to be both closed, indicating the periodic nature of the motion, and isolated, indicating the limiting nature of the cycle [17–31].

The system modeling of the rational harmonic mode-locked fiber ring laser system is implemented on the following assumptions: (1) detuned frequency is perfectly adjusted according to the fraction number required, (2) there is small harmonic distortion, (3) no fiber nonlinearity is included in the analysis, (4) no other noise sources are involved in the system, and (5) there is Gaussian lasing mode amplitude distribution analysis.

The phase plane of a perfect 10 GHz mode-locked pulse train without any frequency detune is shown in Figure 8.2 and the corresponding pulse train is shown in Figure 8.3a. The shape of the phase plane exposes the phase

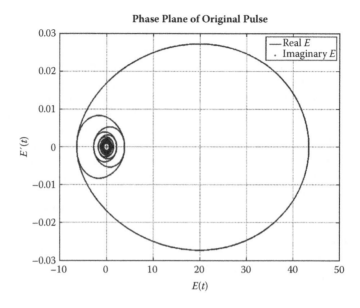

FIGURE 8.2
Phase plane of a 10 GHz mode-locked pulse train. (Solid line, real part of the energy; dotted line, imaginary part of the energy; x-axes, $E(t)$; and y-axes, $E'(t)$)

between the displacement and its derivative. From the phase plane obtained, one can easily observe that the origin is a stable node and the limit cycle around that vicinity is a stable limit cycle, hence leading to stable system trajectory. The 4× multiplication pulse trains (i.e., $m = 4$) without and with 5% harmonic distortion are shown in Figures 8.3b and 8.3c. Their corresponding phase planes are shown in Figures 8.4a and 8.4b. For the case of zero harmonic distortion, which is the ideal case, the generated pulse train is perfectly multiplied with equal amplitude and the phase plane has stable symmetry periodic trajectories around the origin. However, for the practical case (i.e., with 5% harmonic distortion), it is obvious that the pulse amplitude is unevenly distributed, which can be easily verified with the experimental results obtained in [3]. Its corresponding phase plane shows more complex asymmetry system trajectories.

One may naively think that the detuning fraction, m, could be increased to a very large number, so a very small frequency deviated, Δf, so as to obtain a very high repetition frequency. This is only true in the ideal world, if no harmonic distortion is present in the system. However, this is unreasonable for a practical mode-locked laser system.

We define the percentage fluctuation, %F as follows:

$$\%F = \frac{E_{max} - E_{min}}{E_{max}} \times 100\% \qquad (8.2)$$

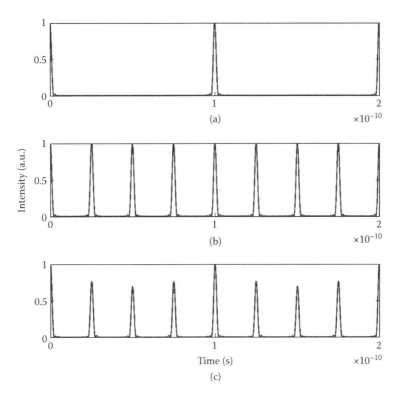

FIGURE 8.3
Normalized pulse propagation of original pulse (a) detuning fraction of 4, with 0%, (b) 5%, (c) harmonic distortion noise.

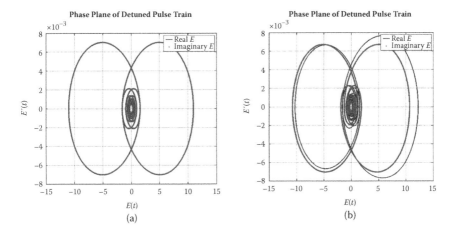

FIGURE 8.4
Phase plane of detuned pulse train, $m = 4$, 0% harmonic distortion (a), and 5% harmonic distortion (b) (solid line, real part of the energy; dotted line, imaginary part of the energy; x-axes, $E(t)$; and y-axes, $E'(t)$).

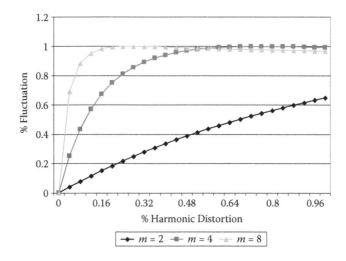

FIGURE 8.5
Relationship between the amplitude fluctuation and the percentage harmonic distortion (diamond, $m = 2$; square, $m = 4$; triangle, $m = 8$).

where E_{max} and E_{min} are the maximum and minimum peak amplitudes of the generated pulse train. For any practical mode-locked laser system, fluctuations above 50% should be considered as poor laser system design. Therefore, this is one of the limiting factors in a rational harmonic mode-locking fiber laser system. The relationships between the percentage fluctuation and harmonic distortion for three multipliers ($m = 2, 4$, and 8) are shown in Figure 8.4. Thus, the obtainable rational harmonic mode-locking is very much limited by the harmonic distortion of the system. For 100% fluctuation, it means no repetition rate multiplication, but with additional noise components (see Figure 8.5); a typical pulse train and its corresponding phase plane are shown in Figure 8.6 (lower plot) and Figure 8.7 with $m = 8$ and 20% harmonic distortion. The asymmetric trajectories of the phase graph explain the amplitude unevenness of the pulse train. Furthermore, it shows a more complex pulse formation system. Thus, it is clear that for any harmonic mode-locked laser system, the small side pulses generated are largely due to improper or not exact tuning of the modulation frequency of the system. An experimental result is depicted in Figure 8.10 for a comparison.

8.3.4 Demonstration

By careful adjustment of the modulation frequency, polarization, gain level, and other parameters of the fiber ring laser, we managed to obtain the 660th and 1230th order of rational harmonic detuning in the mode-locked fiber ring laser with a base frequency of 100 MHz, hence achieving 66 GHz and 123 GHz repetition frequency pulse operation. The autocorrelation traces

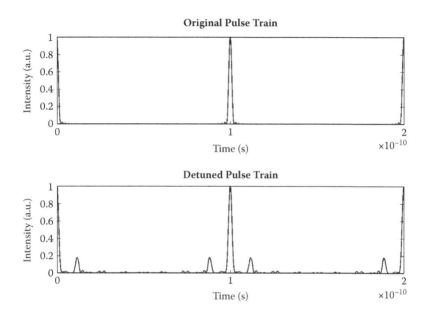

FIGURE 8.6
10 GHz pulse train (upper plot); pulse train with $m = 8$ and 20% harmonic distortion (lower plot).

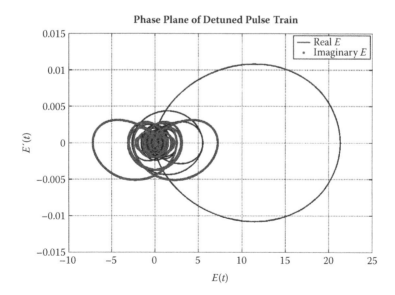

FIGURE 8.7
Phase plane of the pulse train with $m = 8$ and 20% harmonic distortion.

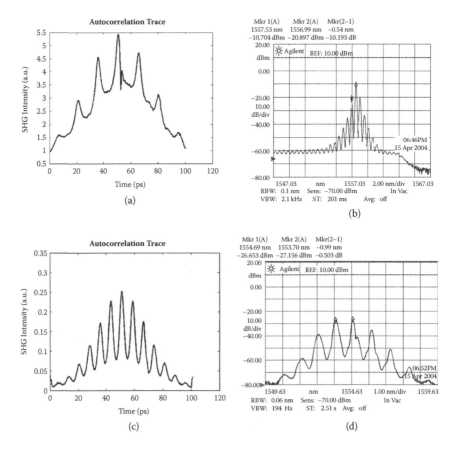

FIGURE 8.8
Autocorrelation traces of 66 GHz (a) and 123 GHz (c) pulse operation; optical spectrums of 66 GHz (b) and 123 GHz (d).

and optical spectrums of the pulse operations are shown in Figure 8.8. With Gaussian pulse assumption, the obtained pulse widths of the operations are 2.5456 ps and 2.2853 ps, respectively. For the 100 MHz pulse operation (i.e., without any frequency detune), the generated pulse width is about 91 ps. Thus, we achieved not only an increase in the pulse repetition frequency, but also a decrease in the generated pulse widths. This pulse narrowing effect is partly due to the self-phase modulation effect of the system, as observed in the optical spectrums. Another reason for this pulse shortening is stated by Haus in [18], where the pulse width is inversely proportional to the modulation frequency as follows:

$$\tau^4 = \frac{2g}{M\omega_m^2\omega_g^2} \tag{8.3}$$

where t is the pulse width of the mode-locked pulse, w_m is the modulation frequency, g is the gain coefficient, M is the modulation index, and w_g is the gain bandwidth of the system. In addition, the duty cycle of our Gaussian modulation signal is ~1%, which is very much less than 50%, this leads to a narrow pulse width, too. Besides the uneven pulse amplitude distribution, a high level of pedestal noise is also observed in the obtained results.

For 66 GHz pulse operation, a 4 nm bandwidth filter is used in the setup, but it is removed for the 123 GHz operation. It is done so to allow more modes to be locked during the operation and, thus, to achieve better pulse quality. In contrast, this increases the level of difficulty significantly in the system tuning and adjustment. As a result, the operation is very much determined by the gain bandwidth of the EDFA used in the laser setup.

The simulated phase planes for the above pulse operation are shown in Figure 8.9. They are simulated based on the 100 MHz base frequency, 10 round-trips condition, and 0.001% of harmonic distortion contribution. There is no stable limit cycle in the phase graphs obtained; hence, the system stability is hardly achievable, which is a known fact in the rational harmonic mode locking. Asymmetric system trajectories are observed in the phase planes of the pulse operations. This reflects the unevenness of the amplitude of the pulses generated. Furthermore, more complex pulse formation process is also revealed in the phase graphs obtained.

By a very small amount of frequency deviation, or improper modulation frequency tuning in the general context, we can generate a pulse train with ~100 MHz modulation frequency with very short adjacent side pulses as shown in Figure 8.10. It is rather similar to Figure 8.6 (lower plot) despite the level of

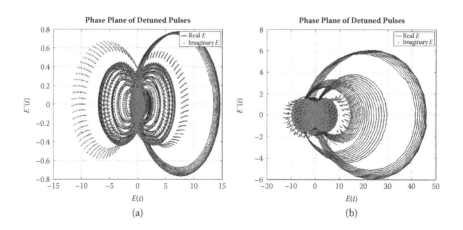

FIGURE 8.9
Phase plane of the 66 GHz (a) and 123 GHz (b) pulse train with 0.001% harmonic distortion noise.

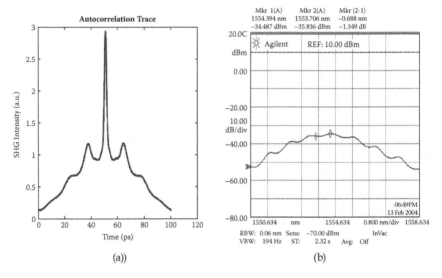

FIGURE 8.10
Autocorrelation trace (a) and optical spectrum (b) of slight frequency detune in the mode-locked fiber ring laser.

pedestal noise in the actual case. This is mainly because we do not consider other sources of noise in our modeling, except the harmonic distortion.

8.3.5 Remarks

We demonstrated a 660th and 1230th order of rational harmonic mode locking from a base modulation frequency of 100 MHz in the erbium-doped fiber ring laser, hence achieving 66 GHz and 123 GHz pulse repetition frequency. To the best of our knowledge, this is the highest rational harmonic order obtained to date. Besides the repetition rate multiplication, we also obtain a high pulse compression factor in the system, ~35× and 40× relative to the nonmultiplied laser system.

In addition, we use phase plane analysis to study the laser system behavior. From the analysis model, the amplitude stability of the detuned pulse train can only be achieved under a negligible or no harmonic distortion condition, which is the ideal situation. The phase plane analysis also reveals the pulse-forming complexity of the laser system.

8.4 Repetition-Rate Multiplication Ring Laser Using Temporal Diffraction Effects

The pulse repetition rate of a mode-locked ring laser is usually limited by the bandwidth of the intracavity modulator. Hence, a number of techniques have to be used to increase the repetition frequency of the generated pulse train. Rational harmonic detuning [3,16] is achieved by applying a slight deviated frequency from the multiple of a fundamental cavity frequency. A 40 GHz repetition frequency has been obtained by [3] using a 10 GHz base band modulation frequency with fourth-order rational harmonic mode locking. This technique is simple in nature. However, this technique suffers from inherent pulse amplitude instability, which includes both amplitude noise and inequality in pulse amplitude, furthermore, it gives poor long-term stability. Hence, pulse amplitude equalization techniques are often applied to achieve better system performance [2,4,5]. The fractional temporal Talbot-based repetition rate multiplication technique [4–8] uses the interference effect between the dispersed pulses to achieve the repetition rate multiplication. The essential element of this technique is the dispersive medium, such as linearly chirped fiber grating (LCFG) [8,16] and single-mode fiber [8,9]. This technique is discussed further in Section 8.2. Intracavity optical filtering [13,14] uses modulators and an FFP within the laser cavity to achieve a higher repetition rate by filtering out certain lasing modes in the mode-locked laser. Other techniques used in repetition rate multiplication include higher-order FM mode-locking [13], optical time domain multiplexing, and so forth.

The stability of a high repetition rate pulse train generated is one of the main concerns for practical multi-Giga bits/sec optical communications systems. Qualitatively, a laser pulse source is considered as stable if it is operating at a state where any perturbations or deviations from this operating point is not increased but suppressed. Conventionally the stability analyses of such laser systems are based on the linear behavior of the laser in which we can analytically analyze the system behavior in both time and frequency domains. However, when the mode-locked fiber laser is operating under a nonlinear regime, none of these standard approaches can be used, because the direct solution of a nonlinear different equation is generally impossible, hence frequency domain transformation is not applicable. Although Talbot-based repetition rate multiplication systems are based on the linear evolution of the laser, there are still some inherent nonlinearities affecting its stability, such as the saturation of the embedded gain medium, nonquadrature biasing of the modulator, nonlinearities in the fiber, and so forth; hence, a nonlinear stability approach must be adopted.

We investigate the stability and transient analyses of the GVD multiplied pulse train using the phase plane analysis of the nonlinear control analytical technique [2]. This is the first time, to the best of our knowledge, that the phase plane analysis in the field of digital control can be used to analyze the stability and transient performances of the GVD repetition rate multiplication systems.

The stability and the transient response of the multiplied pulses are studied using the phase plane technique of nonlinear control engineering. We also demonstrated a four times repetition rate multiplication on a 10 Gbits/s pulse train generated from the active harmonically mode-locked fiber ring laser, hence achieving a 40 Gbits/s pulse train by using the fiber GVD effect. It has been found that the stability of the GVD multiplied pulse train, based on the phase plane analysis, is hardly achievable even under perfect multiplication conditions. Furthermore, uneven pulse amplitude distribution is observed in the multiplied pulse train. In addition to that, the influences of the filter bandwidth in the laser cavity, nonlinear effect, and the noise performance are studied in our analyses.

When a pulse train is transmitted through an optical fiber, the phase shift of a kth individual lasing mode due to group velocity dispersion (GVD) is

$$\varphi_k = \frac{\pi \lambda^2 D z k^2 f_r^2}{c} \tag{8.4}$$

where λ is the center wavelength of the mode-locked pulses, D is the fiber's GVD factor, z is the fiber length, f_r is the repetition frequency, and c is the speed of light in a vacuum. This phase shift induces pulse broadening and distortion. At Talbot distance, $z_T = 2/\Delta\lambda f_r \frac{1}{2} D\frac{1}{2}$ [6] the initial pulse shape is restored, where $\Delta\lambda = f_r \lambda^2/c$ is the spacing between Fourier-transformed spectrum of the pulse train. When the fiber length is equal to $z_T/(2m)$, (where $m = 2,3,4,\ldots$), every mth lasing mode oscillates in phase and the oscillation waveform maximums accumulate. However, when the phases of other modes become mismatched, this weakens their contributions to pulse waveform formation. This leads to the generation of a pulse train with a multiplied repetition frequency with m-times. The pulse duration does not change that much even after the multiplication, because every mth lasing mode dominates in pulse waveform formation of m-times multiplied pulses. The pulse waveform therefore becomes identical to that generated from the mode-locked laser, with the same spectral property. Optical spectrum does not change after the multiplication process, because this technique utilizes only the change of the phase relationship between lasing modes and does not use the fiber's nonlinearity.

The effect of higher-order dispersion might degrade the quality of the multiplied pulses (i.e., pulse broadening, appearance of pulse wings, and pulse-to-pulse intensity fluctuation). In this case, any dispersive media to compensate the fiber's higher-order dispersion would be required in order to complete the multiplication process. To achieve higher multiplications the input pulses must have a broad spectrum, and the fractional Talbot length must be very precise in order to receive high-quality pulses. If the average power of the pulse train induces the nonlinear suppression and anomalous dispersion is experienced along the fiber, solitonic dynamics would occur and prevent the linear Talbot effect from occurring.

The highest repetition rate obtainable is limited by the duration of the individual pulses, as pulses start to overlap when the pulse duration becomes comparable to the pulse train period (i.e., $m_{max} = \Delta T/\Delta t$, where ΔT is the pulse train period and Δt is the pulse duration).

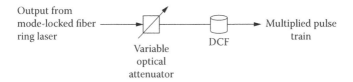

FIGURE 8.11
Experimental setup for group velocity dispersion repetition rate multiplication system.

GVD repetition rate multiplication is used to achieve a 40 Gbits/s operation. The input to the GVD multiplier is a 10.217993 Gbits/s laser pulse source, obtained from an active harmonically mode-locked fiber ring laser, operating at 1550.2 nm.

The principal element of the active harmonically mode-locked fiber ring laser is an optical closed loop with an optical gain medium, that is the erbium-doped fiber under a 980 nm pump source, an optical 10 GHz amplitude modulator, optical bandpass filter, optical fiber couplers, and other associated optics. The generic schematic construction of the active mode-locked fiber ring laser is shown in Figure 8.1 and now in Figure 8.11. In this case the active mode-locked fiber laser design is based on a fiber ring cavity where the 25 meter EDF with Er^{3+} ion concentration of 4.7×10^{24} ions/m^3 is pumped by two diode lasers at 980 nm: SDLO-27-8000-300 and CosetK1116 with maximum forward pump power of 280 mW and backward pump power of 120 mW. The pump lights are coupled into the cavity by the 980/1550 nm wavelength division multiplexing (WDM) couplers; with insertion loss for 980 nm and 1550 nm signals are about 0.48 dB and 0.35 dB, respectively. A polarization-independent optical isolator ensures the unidirectional lasing. The birefringence of the fiber is compensated by a polarization controller (PC). A tunable FP filter with 3 dB bandwidth of 1 nm and wavelength tuning range from 1530 nm to 1560 nm is inserted into the cavity to select the center wavelength of the generated signal as well as to reduce the noise in the system. In addition, it is used for the longitudinal mode selection in the mode-locking process. Pulse operation is achieved by introducing a JDS Uniphase 10Gb/s lithium niobate, Ti:LiNbO$_3$ Mach–Zehnder amplitude modulator into the cavity with half-wave voltage, V_p of 5.8 V. The modulator is DC biased near the quadrature point and not more than the V_p such that it operates on the linear region of its characteristic curve and is driven by the sinusoidal signal derived from an Anritsu 68347C Synthesizer Signal Generator. The modulating depth should be less than unity to avoid signal distortion. The modulator has an insertion loss of £7 dB. The output coupling of the laser is optimized using a 10/90 coupler. Then 90% of the optical field power is coupled back into the cavity ring loop, while the remaining portion is taken out as the output of the laser and is analyzed using a New Focus 1014B 40 GHz photodetector, Ando AQ6317B Optical Spectrum Analyzer, Textronix CSA 8000 80E01 50 GHz Communications Signal Analyzer, or Agilent E4407B RF Spectrum Analyzer.

One rim of about 3.042 km of a dispersion compensating fiber (DCF), with a dispersion value of –98 ps/nm/km was used in the experiment; the schematic of the experimental setup is shown in Figure 8.11. The variable optical attenuator used in the setup is to reduce the optical power of the pulse train generated by the mode-locked fiber ring laser, hence to remove the nonlinear effect of the pulse. A DCF (i.e., fiber of negative dispersion factor length for 4× multiplication factor on the ~10 GHz signal) is required and estimated to be 3.048173 km. The output of the multiplier (i.e., at the end of DCF) is then observed using the Textronix CSA 8000 80E01 50 GHz Communications Signal Analyzer.

8.4.1 Phase Plane Analysis

A nonlinear system frequently has more than one equilibrium point. It can also oscillate at a fixed amplitude and fixed period without external excitation. This oscillation is called the limit cycle. However, limit cycles in nonlinear systems are different from linear oscillations. First, the amplitude of self-sustained excitation is independent of the initial condition, while the oscillation of a marginally stable linear system has its amplitude determined by the initial conditions. Second, marginally stable linear systems are very sensitive to changes, while limit cycles are not easily affected by parameter changes [26–31].

Phase plane analysis is a graphical method of studying second-order nonlinear systems. The result is a family of system motion of trajectories on a two-dimensional plane, which allows us to visually observe the motion patterns of the system. Nonlinear systems can display more complicated patterns in the phase plane, such as multiple equilibrium points and limit cycles. In the phase plane, a limit cycle is defined as an isolated closed curve. The trajectory has to be both closed, indicating the periodic nature of the motion, and isolated, indicating the limiting nature of the cycle.

The system modeling for the GVD multiplier is done based on the following assumptions: (1) perfect output pulse from the mode-locked fiber ring laser without any timing jitter, (2) multiplication achieved under ideal conditions (i.e., exact fiber length for a certain dispersion value), (3) no fiber nonlinearity included in the analysis of the multiplied pulse, (4) no other noise sources involved in the system, and (5) uniform or Gaussian lasing mode amplitude distribution.

8.4.2 Uniform Lasing Mode Amplitude Distribution

Uniform lasing mode amplitude distribution is assumed at the first instance (i.e., ideal mode-locking condition). The simulation is done based on the 10 Gbits/s pulse train, centered at 1550 nm, with fiber dispersion value of –98 ps/km/nm, 1 nm flat-top passband filter is used in the cavity of mode-locked fiber laser. The estimated Talbot distance is 25.484 km.

The original pulse (direct from the mode-locked laser) propagation behavior and its phase plane are shown in Figure 8.12a and Figure 8.13a.

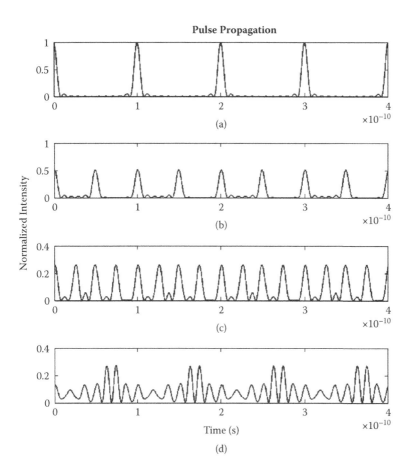

FIGURE 8.12
Pulse propagation of (a) original pulse, (b) 2× multiplication, (c) 4× multiplication, and (d) 8× multiplication with 1 nm filter bandwidth and equal lasing mode amplitude analysis.

From the phase plane obtained, one can observe that the origin is a stable node and the limit cycle around that vicinity is a stable limit cycle. This agrees with our first assumption of an ideal pulse train at the input of the multiplier. Also, we present the pulse propagation behavior and phase plane for 2×, 4×, and 8× GVD multiplication system in Figures 8.12 to 8.16. The shape of the phase graph exposes the phase between the displacement and its derivative.

As the multiplication factor increases, the system trajectories are moving away from the origin. As for the 4× and 8× multiplications (see Figures 8.18 to 8.21), there is neither a stable limit cycle nor a stable node on the phase planes even with the ideal multiplication parameters. Here we see the system trajectories spiral out to an outer radius and back to an inner radius again. The change in the radius of the spiral is the transient response of the

Phase Plane

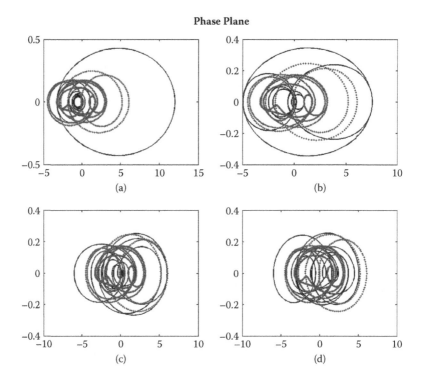

(a)   (b)

(c)   (d)

FIGURE 8.13
Phase plane of (a) original pulse, (b) 2x multiplication, (c) 4x multiplication, and (d) 8x multiplication with 1 nm filter bandwidth and equal lasing mode amplitude analysis (solid line, real part of the energy; dotted line, imaginary part of the energy; x-axes, $E(t)$; and y-axes, $E'(t)$).

system. Hence, with the increase in the multiplication factor, the system trajectories become more sophisticated. Although GVD repetition rate multiplication uses only the phase change effect in multiplication process, the inherent nonlinearities still affect its stability indirectly. Despite the reduction in the pulse amplitude, we observe uneven pulse amplitude distribution in the multiplied pulse train. The percentage of unevenness increases with the multiplication factor in the system.

8.4.3 Gaussian Lasing Mode Amplitude Distribution

This set of the simulation models the practical filter used in the system. It gives us a better insight on the GVD repetition rate multiplication system behavior. The parameters used in the simulation are the same except the filter of the laser has been changed to a 1 nm (125 GHz at 1550 nm) Gaussian-profile passband filter. The spirals of the system trajectories and uneven pulse amplitude distribution are more severe than those in the uniform lasing mode amplitude analysis.

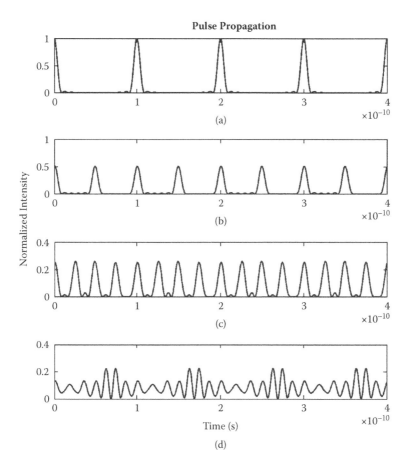

FIGURE 8.14
Pulse propagation of (a) original pulse, (b) 2x multiplication, (c) 4x multiplication, and (d) 8x multiplication with 1 nm filter bandwidth and Gaussian lasing mode amplitude analysis.

8.4.4 Effects of Filter Bandwidth

Filter bandwidth used in the mode-locked fiber ring laser will affect the system stability of the GVD repetition rate multiplication system as well. The analysis done above is based on 1 nm filter bandwidth. The number of modes locked in the laser system increases with the bandwidth of the filter used, which gives us a better quality of the mode-locked pulse train. The simulation results shown below are based on the Gaussian lasing mode amplitude distribution, 3 nm filter bandwidth used in the laser cavity, and other parameters that remain unchanged.

With wider filter bandwidth, the pulse width and the percentage pulse amplitude fluctuation decrease. This suggests a better stability condition. Instead of spiraling away from the origin, the system trajectories move

Phase Plane

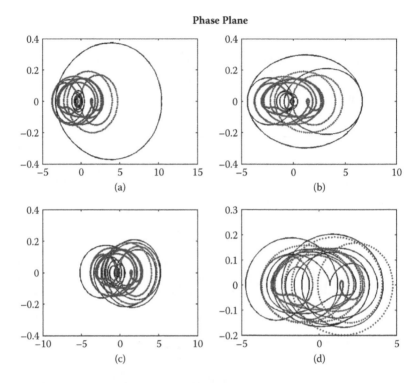

FIGURE 8.15
Phase plane of (a) original pulse, (b) 2x multiplication, (c) 4x multiplication, and (d) 8x multiplication with 1 nm filter bandwidth and Gaussian lasing mode amplitude analysis (solid line, real part of the energy; dotted line, imaginary part of the energy; x-axes, $E(t)$, and y-axes, $E'(t)$).

inward to the stable node. However, this leads to a more complex pulse formation system.

8.4.5 Nonlinear Effects

When the input power of the pulse train enters the nonlinear region, the GVD multiplier loses its multiplication capability as predicted. The additional nonlinear phase shift due to the high-input power is added to the total pulse phase shift and destroys the phase change condition of the lasing modes required by the multiplication condition. Furthermore, this additional nonlinear phase shift changes the pulse shape and the phase plane of the multiplied pulses.

8.4.6 Noise Effects

The above simulations are all based on the noiseless situation. However, in the practical optical communication systems, noises are always sources of

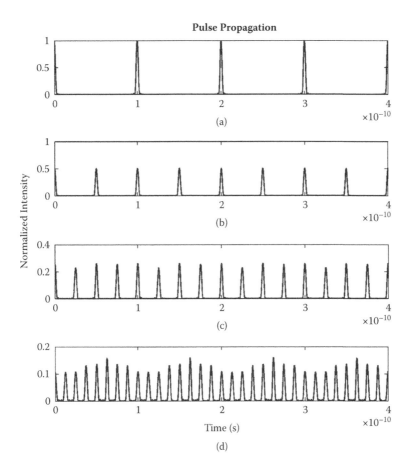

FIGURE 8.16
Pulse propagation of (a) original pulse, (b) 2x multiplication, (c) 4x multiplication, and (d) 8x multiplication with 3 nm filter bandwidth and Gaussian lasing mode amplitude analysis.

nuisance that can cause system instability; therefore, it must be taken into consideration for system stability studies.

Because the optical intensity of the m-times multiplied pulse is m-times less than the original pulse, it is more vulnerable to noise. The signal is difficult to differentiate from the noise within the system if the power of the multiplied pulse is too small. The phase plane of the multiplied pulse is distorted due to the presence of the noise, which leads to poor stability performance.

8.4.7 Demonstration

The obtained 10 GHz output pulse train from the mode-locked fiber ring laser is shown in Figure 8.22. Its spectrum is shown in Figure 8.23. This output was then used as the input to the dispersion compensating fiber,

Phase Plane

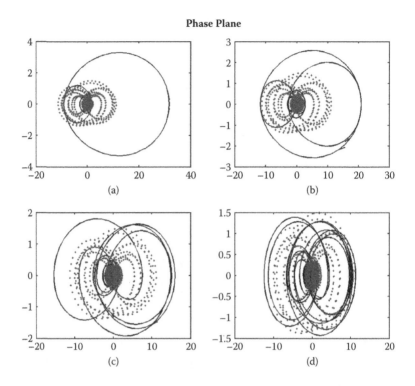

(a)

(b)

(c)

(d)

FIGURE 8.17
Phase plane of (a) original pulse, (b) 2x multiplication, (c) 4x multiplication, and (d) 8x multiplication with 3 nm filter bandwidth and Gaussian lasing mode amplitude analysis (solid line, real part of the energy; dotted line, imaginary part of the energy; x-axes, $E(t)$; and y-axes, $E'(t)$).

which acts as the GVD multiplier in our experiment. The obtained 4x multiplication by the GVD effect and its spectrum are shown in Figures 8.24 and Figure 8.25. The spectra for both cases (original and multiplied pulse) are the same because this repetition rate multiplication technique utilizes only the change of phase relationship between lasing modes and does not use the fiber's nonlinearity.

The multiplied pulse suffers an amplitude reduction in the output pulse train; however, the pulse characteristics should remain the same. The instability of the multiplied pulse train is mainly due to the slight deviation from the required DCF length (0.2% deviation). Another reason for the pulse instability, which derived from our analysis, is the divergence of the pulse energy variation in the vicinity around the origin, as the multiplication factor gets higher. The pulse amplitude decreases with the increase in, multiplication factor, as the fact of energy conservation; when it reaches a certain

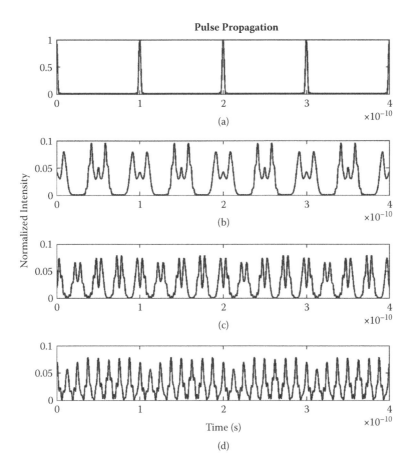

FIGURE 8.18
Pulse propagation of (a) original pulse, (b) 2x multiplication, (c) 4x multiplication, and (d) 8x multiplication with 3 nm filter bandwidth, Gaussian lasing mode amplitude analysis and input power = 1 W.

energy level, which is indistinguishable from the noise level in the system, the whole system will become unstable and noisy.

8.4.8 Remarks

In this section, 4× repetition rate multiplication by using the fiber GVD effect is demonstrated; hence, a 40 GHz pulse train can be obtained from 10 GHz mode-locked fiber laser source. However, its stability is of great concern for practical use in optical communications systems. Although

Phase Plane

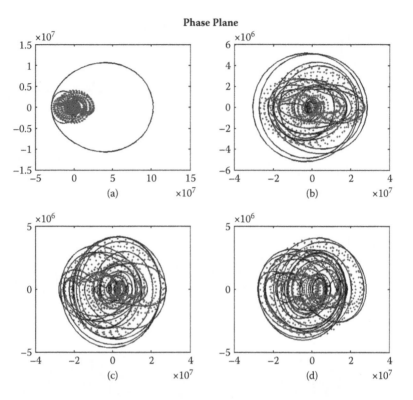

FIGURE 8.19

Phase plane of (a) original pulse, (b) 2x multiplication, (c) 4x multiplication, and (d) 8x multiplication with 3 nm filter bandwidth, Gaussian lasing mode amplitude analysis and input power = 1 W (solid line, real part of the energy; dotted line, imaginary part of the energy; x-axes, $E(t)$; and y-axes, $E'(t)$).

the GVD repetition rate multiplication technique is linear in nature, the inherent nonlinear effects in such a system may disturb the stability of the system. Hence, any linear approach may not be suitable in deriving the system stability. Stability analysis for this multiplied pulse train has been studied by using the nonlinear control stability theory, which is the first time, to the best of our knowledge, that phase plane analysis is being used to study the transient and stability performance of the GVD repetition rate multiplication system. Surprisingly, from the analysis model, the stability of the multiplied pulse train can hardly be achieved even under perfect multiplication conditions. Furthermore, we observed uneven pulse amplitude distribution in the GVD multiplied pulse train, which is due to the energy variations between the pulses that cause some energy beating between them. Another possibility is the divergence of the pulse energy variation in the vicinity around the equilibrium point that leads to instability.

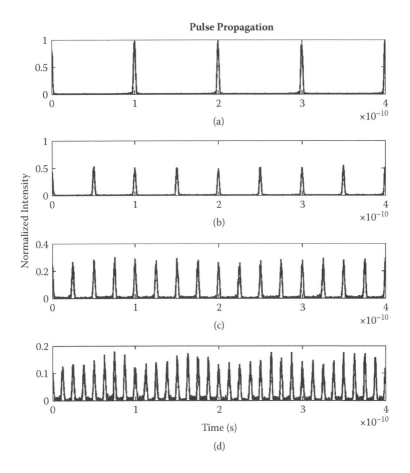

FIGURE 8.20
Pulse propagation of (a) original pulse, (b) 2x multiplication, (c) 4x multiplication, and (d) 8x multiplication with 3 nm filter bandwidth, Gaussian lasing mode amplitude analysis, and 0 dB signal-to-noise ratio.

The pulse amplitude fluctuation increases with the multiplication factor. Also, with the wider filter bandwidth used in the laser cavity, better stability conditions can be achieved. The nonlinear phase shift and noises in the system challenge the system stability of the multiplied pulses. They not only change the pulse shape of the multiplied pulses, they also distort the phase plane of the system. Hence, the system stability is greatly affected by the self-phase modulation as well as the system noises.

This stability analysis model can further be extended to include some system nonlinearities, such as the gain saturation effect, nonquadrature biasing of the modulator, fiber nonlinearities, and so forth. The chaotic behavior of the system may also be studied by applying different initial phase and injected energy conditions to the model.

Phase Plane

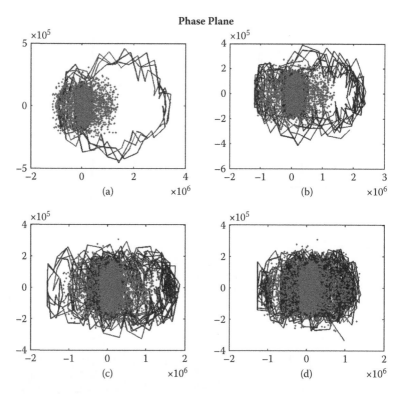

FIGURE 8.21

Phase plane of (a) original pulse, (b) 2x multiplication, (c) 4x multiplication, and (d) 8x multiplication with 3 nm filter bandwidth, Gaussian lasing mode amplitude analysis, and 0 dB signal-to-noise ratio (solid line, real part of the energy; dotted line, imaginary part of the energy; x-axes, $E(t)$; and y-axes, $E'(t)$).

8.5 Conclusions

We have successfully demonstrated a mode-locked laser operating under the open-loop condition and with optical to electrical (O/E) RF feedback providing regenerative mode locking. The O/E feedback can certainly provide a self-locking mechanism under the condition that the polarization characteristics of the ring laser are manageable. The regenerative MLFRL can self-lock even under the DC drifting effect of the modulator bias voltage (over 20 hours).* The generated pulse trains of 4.5 ps duration can be, with minimum difficulty, compressed further to less than 3 ps for 160 Gb/s optical communication systems.

---

* Typically the DC bias voltage of a LiNbO$_3$ intensity modulator is drifted by 1.5 volts after 15 hours of continuous operation.

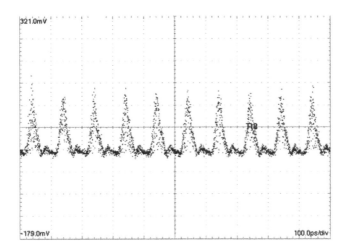

FIGURE 8.22
10 GHz pulse train from mode-locked fiber ring laser (100 ps/div, 50 mV/div).

We also demonstrated 660th and 1230th orders of rational harmonic mode locking from a base modulation frequency of 100 MHz in the optically amplified fiber ring laser, hence achieving 66 GHz and 123 GHz pulse repetition frequency. Besides the repetition rate multiplication, we obtained a high pulse compression factor in the system, ~35× and 40× relative to the nonmultiplied laser system. In addition, we use phase plane analysis to study the laser system behavior. From the analysis model, the amplitude

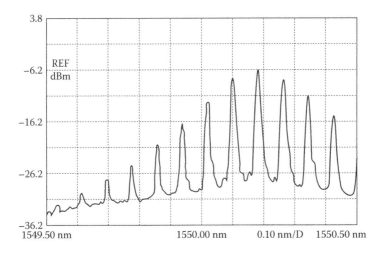

FIGURE 8.23
10 GHz pulse spectrum from mode-locked fiber ring laser.

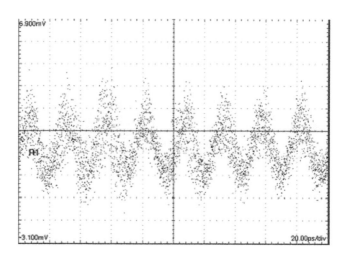

FIGURE 8.24
40 GHz multiplied pulse train (20 ps/div, 1 mV/div).

stability of the detuned pulse train can only be achieved under negligible or no harmonic distortion condition, which is the ideal situation. The phase plane analysis also reveals the pulse-forming complexity of the laser system.

Furthermore, the *N*-times, multiplication of the repetition rate can be achieved by using the fiber, GVD effect, which is a linear diffraction mechanism; hence, a 40 GHz pulse train is obtained from a 10 GHz mode-locked fiber laser source. Stability analysis for this multiplied pulse train has been studied by using the nonlinear control stability theory in which a phase plane

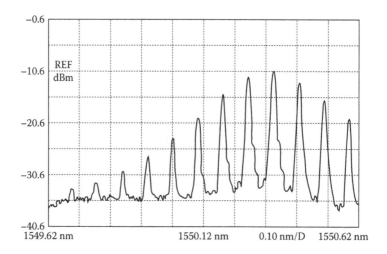

FIGURE 8.25
40 GHz pulse spectrum from group velocity dispersion multiplier.

analysis is being used to study the transient and stability performance of the GVD repetition rate multiplication system. Surprisingly, from the analysis model, the stability of the multiplied pulse train can hardly be achieved even under perfect multiplication conditions. Furthermore, we observed uneven pulse amplitude distribution in the GVD multiplied pulse train, which is due to the energy variations between the pulses that cause some energy beating between them. Another possibility is the divergence of the pulse energy variation in the vicinity around the equilibrium point that leads to instability.

The pulse amplitude fluctuation increases with the multiplication factor. Furthermore, with wider filter bandwidth used in the laser cavity, better stability condition can be achieved. The nonlinear phase shift and noises in the system challenge the system stability of the multiplied pulses. They not only change the pulse shape of the multiplied pulses, they also distort the phase plane of the system. Hence, the system stability is greatly affected by the self-phase modulation as well as the system noises. This stability analysis model can be extended to include some system nonlinearities, such as the gain saturation effect, nonquadrature biasing of the modulator, fiber nonlinearities, and so forth. The chaotic behavior of the system may also be studied by applying different initial phase and injected energy conditions to the model.

Currently the design and demonstration of multiwavelength mode-locked lasers to generate ultrashort and ultrahigh repetition-rate pulse sequences are under consideration by employing a multispectral filter demultiplexer and multiplexer incorporated within the fiber ring. Furthermore, the locking in the THz region will also be reported. The principal challenge is the conversion from the THz photonic to electronic domain for stabilization. This can be implemented in either electronic or photonic sampling.

References

1. K. Kuroda and H. Takakura, Mode-Locked Ring Laser with Output Pulse Width of 0.4 ps, *IEEE Trans. Inst. Meas.*, 48, 1018–1022, 1999.
2. G. Zhu, H. Chen, and N. Dutta, Time Domain Analysis of a Rational Harmonic Mode-Locked Ring Fiber Laser, *J. Appl. Phys.*, 90, 2143–2147, 2001.
3. C. Wu and N.K. Dutta, High Repetition Rate Optical Pulse Generation Using a Rational Harmonic Mode-Locked Fiber Laser, *IEEE J. Quantum Electron.*, 36, 145–150, 2000.
4. K. K. Gupta, N. Onodera, and M. Hyodo, Technique to Generate Equal Amplitude, Higher-Order Optical Pulses in Rational Harmonically Mode Locked Fiber Ring Lasers, *Electron. Lett.*, 37, 948–950, 2001.
5. Y. Shiquan, L. Zhaohui, Z. Chunliu, D. Xiaoyi, Y. Shuzhong, K. Guiyun, and Z. Qida, Pulse-Amplitude Equalization in a Rational Harmonic Mode-Locked Fiber Ring Laser by Using Modulator as Both Mode Locker and Equalizer, *IEEE Photonics Technol. Lett.*, 15, 389–391, 2003.

6. J. Azana and M.A. Muriel, Technique for Multiplying the Repetition Rates of Periodic Trains of Pulses by Means of a Temporal Self-Imaging Effect in Chirped Fiber Gratings, *Opt. Lett.*, 24, 1672–1674, 1999.

7. S. Atkins and B. Fischer, All Optical Pulse Rate Multiplication Using Fractional Talbot Effect and Field-to-Intensity Conversion with Cross Gain Modulation, *IEEE Photonics Technol. Lett.*, 15, 132–134, 2003.

8. J. Azana and M. A. Muriel, Temporal Self-Imaging Effects: Theory and Application for Multiplying Pulse Repetition Rates, *IEEE J. Quantum Electron.*, 7, 728–744, 2001.

9. D. A. Chestnut, C. J. S. de Matos, and J. R. Taylor, 4× Repetition Rate Multiplication and Raman Compression of Pulses in the Same Optical Fiber, *Opt. Lett.*, 27, 1262–1264, 2002.

10. S. Arahira, S. Kutsuzawa, Y. Matsui, D. Kunimatsu, and Y. Ogawa, Repetition Frequency Multiplication of Mode-Locked Using Fiber Dispersion, *J. Lightwave Technol.*, 16, 405–410, 1998.

11. W. J. Lai, P. Shum, and L. N. Binh, Stability and Transient Analyses of Temporal Talbot-Effect-Based Repetition-Rate Multiplication Mode-Locked Laser Systems, *IEEE Photonics Technol. Lett.*, 16, 437–439, 2004.

12. K. K. Gupta, N. Onodera, K. S. Abedin, and M. Hyodo, Pulse Repetition Frequency Multiplication via Intracavity Optical Filtering in AM Mode-Locked Fiber Ring Lasers, *IEEE Photonics Technol. Lett.*, 14, 284–286, 2002.

13. K. S. Abedin, N. Onodera, and M. Hyodo, Repetition-Rate Multiplication in Actively Mode-Locked Fiber Lasers by Higher-Order FM Mode-Locking Using a High Finesse Fabry Perot Filter, *Appl. Phys. Lett.*, 73, 1311–1313, 1998.

14. K. S. Abedin, N. Onodera, and M. Hyodo, Higher Order FM Mode-Locking for Pulse-Repetition-Rate Enhancement in Actively Mode-locked Lasers: Theory and Experiment, *IEEE J. Quantum Electron.*, 35, 875–890, 1999.

15. W. Daoping, Z. Yucheng, L. Tangjun, and J. Shuisheng, 20 Gb/s Optical Time Division Multiplexing Signal Generation by Fiber Coupler Loop-Connecting Configuration, presented at Fourth Optoelectronics and Communications Conference, 1999.

16. D. L. A. Seixasn and M. C. R. Carvalho, 50 GHz Fiber Ring Laser Using Rational Harmonic Mode-Locking, presented at Microwave and Optoelectronics Conference, Nis, Yugoslav, 2001.

17. R. Y. Kim, Fiber Lasers and Their Applications, presented at Laser and Electro-Optics, CLEO/Pacific Rim '95, Jeju, Korea, 1995.

18. H. Zmuda, R. A. Soref, P. Payson, S. Johns, and E. N. Toughlian, Photonic Beamformer for Phased Array Antennas Using a Fiber Grating Prism, *IEEE Photonics Technol. Lett.*, 9, 241–243, 1997.

19. G. A. Ball, W. W. Morey, and W. H. Glenn, Standing-Wave Monomode Erbium Fiber Laser, *IEEE Photonics Technol. Lett.*, 3, 613–615, 1991.

20. D. Wei, T. Li, Y. Zhao, et al., Multi-Wavelength Erbium-Doped Fiber Ring Laser with Overlap-Written Fiber Bragg Gratings, *Opt. Lett.*, 25, 1150–1152, 2000.

21. S. K. Kim, M. J. Chu, and J. H. Lee, Wideband Multi-Wavelength Erbium-Doped Fiber Ring Laser with Frequency Shifted Feedback, *Opt. Commun.*, 190, 291–302, 2001.

22. Z. Li, L. Caiyun, and G. Yizhi, A Polarization Controlled Multi-Wavelength Er-Doped Fiber Laser, presented at APCC/OECC99, Wuhan, China 1999.

23. R. M. Sova, C. S. Kim, and J. U. Kang, Tunable Dual-Wavelength All-PM Fiber Ring Laser, *IEEE Photonics Technol. Lett.*, 14, 287–289, 2002.

24. I. D. Miller, D. B. Mortimore, P. Urquhart, et al., A Nd_{3+}-Doped CW Fiber Laser Using All-Fiber Reflectors, *Appl. Opt.*, 26, 2197–2201, 1987.

25. X. Fang and R. O. Claus, Polarization-Independent All-Fiber Wavelength-Division Multiplexer Based on a Sagnac Interferometer, *Opt. Lett.*, 20, 2146–2148, 1995.

26. X. Fang, H. Ji, C. T. Aleen, et al., A Compound High-Order Polarization-Independent Birefringence Filter, *IEEE Photonics Technol. Lett.*, 19, 458–460, 1997.

27. X. P. Dong, Li, S., K. S. Chiang et al., Multi-Wavelength Erbium-Doped Fiber Laser Based on a High-Birefringence Fiber Loop, *Electron. Lett.*, 36, 1609–1610, 2000.

28. D. Jones, H. Haus, and E. Ippen, Subpicosecond Solitons in an Actively Mode Locked Fiber Laser, *Opt. Lett.*, 21(22), 1818–1820, 1996.

29. X. Zhang, M. Karlson, and P. Andrekson, Design Guideline for Actively Mode Locked Fiber Ring Lasers, *IEEE Photonics Tech. Lett.*, 1103–1105, 1998.

30. A. E. Siegman, *Laser*. University Press, Mill Valley, CA, 1986.

31. J. J. E. Slotine and W. Li, *Applied Nonlinear Control*. Prentice Hall, Englewood Cliffs, NJ, 1991.

9

Nonlinear Photonic Signal Processing Using Third-Order Nonlinearity

Le Nguyen Binh

Hua Wei Technologies, European Research Center, Munich, Germany

CONTENTS

9.1 Introduction

This chapter describes the applications of nonlinear effects in guided wave devices, especially the channel and rib waveguide structure, so that the converted lightwave beam can be coupled with optical transmission systems to generate phase conjugation for compensation of distorted data pulses or as a bispectrum photonic preprocessing or in a fiber ring laser to generate mode-locked pulse sequences. We also illustrate the bispectrum property of optical multibound solitons described in Chapter 4.

9.2 Nonlinear Effects on Optical Waveguides and Photonic Signal Processing

9.2.1 Introductory Remarks

With increasing demand for high capacity, communication networks are facing several challenges, especially in signal processing at the physical layer at ultrahigh speed. When the processing speed is over that of the electronic limit or requires massive parallel and high-speed operations, the processing in the optical domain offers significant advantages. Thus, all-optical signal processing is a promising technology for future optical communication networks. An advanced optical network requires a variety of signal processing functions including optical regeneration, wavelength conversion, optical switching, and signal monitoring. An attractive way to realize these processing functions in transparent and high-speed mode is to exploit the third-order nonlinearity in optical waveguides, particularly parametric processes.

Nonlinearity is a fundamental property of optical waveguides including channel, rib-integrated structures, or circular fibers. The origin of nonlinearity comes from the third-order nonlinear polarization in optical transmission media [1]. It is responsible for various phenomena such as self-phase modulation (SPM), cross-phase modulation (XPM), and four-wave mixing (FWM) effects. In these effects, the parametric FWM process is of special interest because it offers several possibilities for signal processing applications [2–16]. To implement all-optical signal processing functions, highly nonlinear optical waveguides are required where the field of the guided waves is concentrated in its core region, hence efficient nonlinear effects. Therefore, the highly nonlinear fibers (HNLFs) are commonly employed for this purpose because the nonlinear coefficient of HNLF is about 10-fold higher than that of standard transmission fibers. The third-order nonlinearity of conventional fibers is often very small to prevent degradation of the transmission signal from nonlinear distortion. Recently, nonlinear chalcogenite and tellurite glass waveguides have emerged as a promising device for ultra-high-speed photonic processing [2,17]. Because of their geometries, these waveguides are called planar waveguides. A planar waveguide can confine a high intensity of light within an area comparable to the wavelength of light, over a short distance of a few centimeters. Hence, they are very compact for signal processing.

In this chapter we demonstrate a number of important applications in optical signal processing using third-order nonlinearity, especially parametric FWM processes by simulation. This section is organized as follows: Section 9.2.2 gives us a mathematical review of third-order nonlinearity in optical waveguides. We particularly focus on the parametric process FWM. The basic propagation equations that describe propagation of optical signal as well as interactions between optical waves in optical waveguides, are also given in this section. Then a MATLAB® model of nonlinear waveguides is developed and integrated into the Simulink® platform as described in Section 9.2.3 enabling our investigation of the parametric processes in nonlinear waveguides. This model is used as a functional block in all-optical signal-processing applications. System applications and performance evaluations based on parametric processes are demonstrated in Section 9.2.4. Besides important applications such as optical amplification, ultra-high-speed switching, and distortion compensation, we also demonstrated the potential of a triple-correlation as high-order spectrum estimation for ultrasensitive optical receivers based on FWM as given in Sections 9.3.4 through 9.3.6.

9.2.2 Third-Order Nonlinearity and Parametric Four-Wave Mixing Process

9.2.2.1 Nonlinear Wave Equation

In optical waveguides including optical fibers, the third-order nonlinearity is of special importance because it is responsible for all nonlinear effects. The confinement of lightwaves and their propagation in optical waveguides are generally governed by the nonlinear wave equation (NLE) that can be derived from the Maxwell's equations under the coupling of the nonlinear polarization. The nonlinear wave propagation in nonlinear waveguide in the time-spatial domain in vector form can be expressed as [1] (see also Appendix A for further details):

$$\nabla^2 \vec{E} - \frac{1}{c^2}\frac{\partial^2 \vec{E}}{\partial t^2} = \mu_0 \left(\frac{\partial^2 \overrightarrow{P_L}}{\partial t^2} + \frac{\partial^2 \overrightarrow{P_{NL}}}{\partial t^2} \right) \tag{9.1}$$

where $\vec{E}$ is the electric field vector of the lightwave; μ_0 is the vacuum permeability assuming a nonmagnetic waveguiding medium; c is the speed of light in a vacuum; P_L, P_{NL} are, respectively, the linear and nonlinear polarization vectors that are formed as

$$\overrightarrow{P_L}(\vec{r},t) = \varepsilon_0 \chi^{(1)}.\vec{E}(\vec{r},t) \tag{9.2}$$

$$\overrightarrow{P_{NL}}(\vec{r},t) = \varepsilon_0 \chi^{(3)} : \vec{E}(\vec{r},t)\vec{E}(\vec{r},t)\vec{E}(\vec{r},t) \tag{9.3}$$

where $\chi^{(3)}$ is the third-order susceptibility. Thus, the linear and nonlinear coupling effects in optical waveguides can be described by (9.1). The second term on the right-hand side is responsible for nonlinear processes including interaction between optical waves through the third-order susceptibility.

In most telecommunication applications, only the complex envelope of an optical signal is considered in analysis because bandwidth of the optical signal is much smaller than the optical carrier frequency. To model the evolution of the light propagation in optical waveguides, it requires that Equation (9.1) be further modified and simplified by some assumptions that are valid in most telecommunication applications [1]. Hence, the electrical field $\vec{E}$ can be written as

$$\vec{E}(\vec{r},t) = \tfrac{1}{2}\hat{x}\{F(x,y)A(z,t)\exp[i(kz - \omega t)] + c.c.\} \tag{9.4}$$

where $A(z,t)$ is the slowly varying complex envelope propagating along z in the waveguide, and k is the wave number. After some algebra using a method of separating variables, the following equation for propagation in the optical waveguide is obtained as

$$\frac{\partial A}{\partial z} + \frac{\alpha}{2} A - i \sum_{n=1}^{\infty} \frac{i^n \beta_n}{n!} \frac{\partial^n A}{\partial t^n} = i\gamma \left(1 + \frac{i}{\omega_0} \frac{\partial}{\partial t}\right) \times A \int_{-\infty}^{\infty} g(t') |A(z, t - t')|^2 dt' \quad (9.5)$$

where the effect of propagation constant β around ω_0 is the Taylor-series expanded, and $g(t)$ is the nonlinear response function including the electronic and nuclear contributions. For the optical pulses wide enough to contain many optical cycles, Equation (9.5) can be simplified as

$$\frac{\partial A}{\partial z} + \frac{\alpha}{2} A + \frac{i\beta_2}{2} \frac{\partial^2 A}{\partial \tau^2} - \frac{\beta_3}{6} \frac{\partial^3 A}{\partial \tau^3} = i\gamma \left[|A|^2 A + \frac{i}{\omega_0} \frac{\partial(|A|^2 A)}{\partial \tau} - T_R A \frac{\partial(|A|^2)}{\partial \tau} \right] \quad (9.6)$$

where a frame of reference moving with the pulse at the group velocity v_g is used by making the transformation $\tau = t - z/v_g \equiv t - \beta_1 z$; and A is the total complex envelope of propagation waves; a, b_k are the linear loss and dispersion coefficients, respectively; $\gamma = \omega_0 n_2 / c A_{eff}$ is the nonlinear coefficient of the guided wave structure; and the first moment of the nonlinear response function is defined as

$$T_R \equiv \int_0^{\infty} t g(t') dt' \quad (9.7)$$

Equation (9.6) is the basic propagation equation, commonly known as the nonlinear Schrödinger equation (NLSE) that is very useful for investigating the evolution of the amplitude of the optical signal and the phase of the lightwave carrier under the effect of third-order nonlinearity in optical waveguides. The left-hand side (LHS) in (9.6) contains all linear terms, while all nonlinear terms are contained on the right-hand side (RHS). In this equation, the first term on the RHS is responsible for the intensity-dependent refractive index effects including FWM.

9.2.2.2 Four-Wave Mixing Coupled-Wave Equations

Four-wave mixing (FWM) is a parametric process through the third-order susceptibility $\chi^{(3)}$. In the FWM process, the superposition and generation of the propagating of the waves with different amplitudes A_k, frequencies ω_k, and wave numbers k_k through the waveguide can be represented as

$$A = \sum_n A_n e^{[j(k_n z - \omega_n \tau)]} \quad \text{with} \quad n = 1, \ldots, 4 \quad (9.8)$$

By ignoring the linear and scattering effects and with the introduction of (9.8) into Equation (9.6), the NLSE can be separated into coupled differential equations, each of which is responsible for one distinct wave in the waveguide:

$$\frac{\partial A_1}{\partial z} + \frac{\alpha}{2} A_1 = i\gamma A_1 \left[|A_1|^2 + 2\sum_{n\neq 1} |A_n|^2 \right] + i\gamma 2 A_3 A_4 A_2^* \exp(-i\Delta k_1 z)$$

$$\frac{\partial A_2}{\partial z} + \frac{\alpha}{2} A_2 = i\gamma A_2 \left[|A_2|^2 + 2\sum_{n\neq 2} |A_n|^2 \right] + i\gamma 2 A_3 A_4 A_1^* \exp(-i\Delta k_2 z) \qquad (9.9)$$

$$\frac{\partial A_3}{\partial z} + \frac{\alpha}{2} A_3 = i\gamma A_3 \left[|A_3|^2 + 2\sum_{n\neq 3} |A_n|^2 \right] + i\gamma 2 A_1 A_2 A_4^* \exp(-i\Delta k_3 z)$$

$$\frac{\partial A_4}{\partial z} + \frac{\alpha}{2} A_4 = i\gamma A_4 \left[|A_4|^2 + 2\sum_{n\neq 4} |A_n|^2 \right] + i\gamma 2 A_1 A_2 A_3^* \exp(-i\Delta k_4 z)$$

where $\Delta k = k_1 + k_2 - k_3 - k_4$ is the wave vector mismatch. The equation system (9.9) thus describes the interaction between different waves in nonlinear waveguides. The interaction represented by the last term in (9.9) can generate new waves. For three waves with different frequencies, a fourth wave can be generated at frequency $\omega_4 = \omega_1 + \omega_2 - \omega_3$. The waves at frequencies ω_1 and ω_2 are called pump waves, whereas the wave at frequency ω_3 is the signal and generated wave at ω_4 is called the idler wave as shown in Figure 9.1a. If all three waves have the same frequency $\omega_1 = \omega_2 = \omega_3$, the interaction is called a degenerate FWM with the new wave at the same frequency. If only two of the three waves are at the same frequency ($\omega_1 = \omega_2 \neq \omega_3$), the process is called partly degenerate FWM, which is important for some applications like the wavelength converter and parametric amplifier.

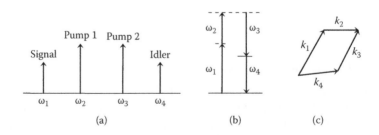

FIGURE 9.1
(a) Position and notation of the distinct waves, (b) diagram of energy conservation, and (c) diagram of momentum conservation in four-wave mixing.

9.2.2.3 Phase Matching

In parametric nonlinear processes such as FWM, the energy conservation and momentum conservation must be satisfied to obtain a high efficiency of the energy transfer as shown in Figure 9.1a. The phase matching condition for the new wave requires

$$\Delta k = k_1 + k_2 - k_3 - k_4 = \frac{1}{c}(n_1\omega_1 + n_2\omega_2 - n_3\omega_3 - n_4\omega_4) = 2\pi\left(\frac{n_1}{\lambda_1} + \frac{n_2}{\lambda_2} - \frac{n_3}{\lambda_3} - \frac{n_4}{\lambda_4}\right)$$

(9.10)

During propagation in optical waveguides, the relative phase difference $\theta(z)$ between four involved waves is determined by [3,5]

$$\theta(z) = \Delta kz + \phi_1(z) + \phi_2(z) - \phi_3(z) - \phi_4(z)$$

(9.11)

where $\phi_k(z)$ relates to the initial phase and the nonlinear phase shift during propagation. An approximation of the phase-matching condition can be given as follows [4]:

$$\frac{\partial \theta}{\partial z} \approx \Delta k + \gamma(P_1 + P_2 - P_3 - P_4) = \kappa$$

(9.12)

where P_k is the power of the waves, and k is the phase mismatch parameter. Thus, the FWM process has maximum efficiency for $k = 0$. The mismatch comes from the frequency dependence of the refractive index and the dispersion of optical waveguides. Depending on the dispersion profile of the nonlinear waveguides, it is very important in selection of pump wavelengths to ensure that the phase mismatch parameter is minimized.

9.2.3 Transmission Models and Nonlinear Guided Wave Devices

To model the parametric FWM process between multiwaves, the basic propagation equations described in Section 9.2 are used. There are two approaches to simulate the interaction between waves. The first approach, named the separating channel technique, is to use the coupled equations system (9.9) in which the interactions between different waves are obviously modeled by certain coupling terms in each coupled equation. Thus, each optical wave considered as one separated channel is represented by a phasor. The coupled equations system is then solved to obtain the solutions of the FWM process. The outputs of the nonlinear waveguide are also represented by separated phasors; hence, the desired signal can be extracted without using a filter.

The second or alternating approach is to use the propagation equation, Equation (9.6) that allows us to simulate all evolutionary effects of the optical waves in the nonlinear waveguides. In this technique a total field is used instead of individual waves. The superimposed complex envelope A is represented by only one phasor, which is the summation of individual complex amplitudes of different waves given as

$$A = \sum_k A_k e^{\left[j(\omega_k - \omega_0)\tau \right]}$$ (9.13)

where ω_0 is the defined angular central frequency, A_n, ω_n are the complex envelope and the carrier frequency of individual waves, respectively. Hence, various waves at different frequencies are combined into only a total signal vector that facilitates integration of the nonlinear waveguide model into the Simulink platform. Equation (9.6) can be numerically solved by the split-step Fourier method (SSFM). The Simulink block representing the nonlinear waveguide is implemented with an embedded MATLAB program. Because only complex envelopes of the guided waves are considered in the simulation, each different optical wave is shifted by a frequency difference between the central frequency and the frequency of the wave to allocate the wave in the frequency band of the total field. Then the summation of individual waves, which is equivalent to the combination process at an optical coupler, is performed prior to entering the block of nonlinear waveguide as depicted in Figure 9.3. The output of a nonlinear waveguide will be selected by an optical bandpass filter (BPF). In this way, the model of nonlinear waveguide can be easily connected to other Simulink blocks available in the platform for simulation of optical fiber communication systems [18].

9.2.4 System Applications of Third-Order Parametric Nonlinearity in Optical Signal Processing

In this section, a range of signal processing applications are demonstrated through simulations that use the model of a nonlinear waveguide to model the wave-mixing process.

9.2.4.1 Parametric Amplifiers

One of the important applications of the $\chi^{(3)}$ nonlinearity is parametric amplification. The optical parametric amplifiers (OPAs) offer a wide gain bandwidth, high differential gain and optional wavelength conversion, and operation at any wavelength [4–7]. These important features of OPAs are obtained because the parametric gain process does not rely on energy transitions between energy states, but it is based on highly efficient FWM in which

TABLE 9.1

Critical Parameters of the Parametric Amplifier in 40 Gb/s System

RZ 40 Gb/s Transmitter
$\lambda_s = 1520$ nm – 1600 nm, $\lambda_0 = 1559$ nm
Modulation: RZ-OOK, $P_s = 0.01$ mW (peak), $B_r = 40$ Gb/s
Parametric Amplifier
Pump source: $P_p = 1$ W (after EDFA), $\lambda_p = 1560.07$ nm
HNLF: $L_f = 500$ m, $D = 0.02$ ps/km/nm, $S = 0.09$ ps/nm²/km, $\alpha = 0.5$ dB/km, $A_{eff} = 12$ µm², $\gamma = 13\ 1/W/km$
BPF: $\Delta\lambda_{BPF} = 0.64$ nm
Receiver
Bandwidth $B_e = 28$ GHz, $i_{eq} = 20$ pA/Hz$^{1/2}$, $i_d = 10$ nA

Note: RZ-OOK (return to zero-on-off keying), EDFA (erbium-doped fiber amplifier), HNLF (high nonlinear fiber), BPF (band pass filter).

two photons at one or two pump wavelengths interact with a signal photon. The fourth photon, the idler, is formed with a phase such that the phase difference between the pump photons and the signal and idler photons satisfies the phase matching condition (9.10). The schematic of the fiber-based parametric amplifier (Table 9.1) is shown in Figures 9.2 and 9.3.

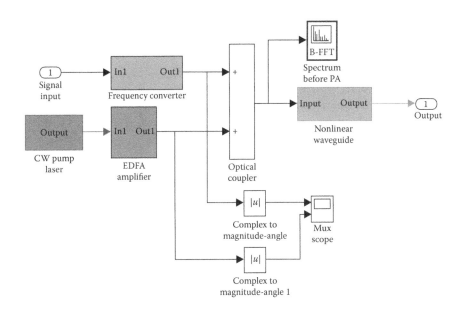

FIGURE 9.2
Typical Simulink setup of the parametric amplifier using the model of nonlinear waveguide.

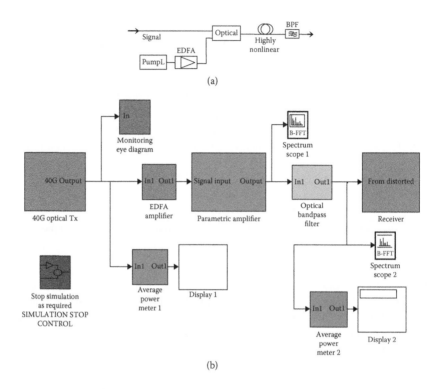

FIGURE 9.3
(a) A typical setup of an optical parametric amplifier; (b) Simulink model of optical parametric amplifier.

For a parametric amplifier using one pump source, from the coupled equations (9.9) with $A_1 = A_2 = A_p$, $A_3 = A_s$, and $A_4 = A_i$, it is possible to derive three coupled equations for complex field amplitude of the three waves $A_{p,s,i}$:

$$\frac{\partial A_p}{\partial z} = -\frac{\alpha}{2} A_p + i\gamma A_p \left[|A_p|^2 + 2\left(|A_s|^2 + |A_i|^2 \right) \right] + i2\gamma A_s A_i A_p^* \exp(-i\Delta kz)$$

$$\frac{\partial A_s}{\partial z} = -\frac{\alpha}{2} A_s + i\gamma A_s \left[|A_s|^2 + 2\left(|A_p|^2 + |A_i|^2 \right) \right] + i\gamma A_p^2 A_i^* \exp(-i\Delta kz) \quad (9.14)$$

$$\frac{\partial A_i}{\partial z} = -\frac{\alpha}{2} A_i + i\gamma A_i \left[|A_i|^2 + 2\left(|A_s|^2 + |A_p|^2 \right) \right] + i\gamma A_p^2 A_s^* \exp(-i\Delta kz)$$

The analytical solution of these coupled equations determines the gain of the amplifier [1,3]:

$$G_s(L) = \frac{|A_s(L)|^2}{|A_s(0)|^2} = 1 + \left[\frac{\gamma P_p}{g} \sinh(gL) \right]^2 \quad (9.15)$$

with L the length of the highly nonlinear fiber/waveguide, P_p the pump power, and g the parametric gain coefficient:

$$g^2 = -\Delta k \left(\frac{\Delta k}{4} + \gamma P_p \right) \tag{9.16}$$

where the phase mismatch Δk can be approximated by extending the propagation constant in a Taylor series around ω_0:

$$\Delta k = -\frac{2\pi c}{\lambda_0^2} \frac{dD}{d\lambda} (\lambda_p - \lambda_0)(\lambda_p - \lambda_s)^2 \tag{9.17}$$

Here, $dD/d\lambda$ is the slope of the dispersion factor $D(\lambda)$ evaluated at the zero-dispersion of the guided wave component (i.e., at the optical wavelength, $\lambda_k = 2\pi c/\omega_k$).

Figure 9.4b shows the Simulink® setup of the 40 Gb/s return to zero (RZ) transmission system using the parametric amplifier. The setup contains a 40 Gb/s optical RZ transmitter, an optical receiver for monitoring, a parametric amplifier block, and a bandpass filter that filters the desired signal from the total field output of the amplifier. Details of the parametric amplifier block can be seen in Figure 9.3. The block setup of a parametric amplifier consists of a continuous-wave (CW) pump laser source, an optical coupler to combine the signal and the pump, and a highly nonlinear fiber block that contains the embedded MATLAB® model for nonlinear propagation. The important simulation parameters of the system are listed in Table 9.2.

Figure 9.5 shows the signals before and after the amplifier in the time domain. The time trace indicates the amplitude fluctuation of the amplified signal as a noisy source from the wave-mixing process. Their corresponding spectra are shown in Figure 9.6. The noise floor of the output spectrum of amplifier shows the gain profile of OPA. Simulated dependence of OPA gain on the wavelength difference between the signal and the pump is shown in Figure 9.7 together with theoretical gain using (9.15). The plot shows agreement between theoretical and simulated results. The peak gain is achieved at the phase-matched condition where the linear phase mismatch is compensated for by the nonlinear phase shift.

9.2.4.2 Wavelength Conversion and Nonlinear Phase Conjugation

In addition to the signal amplification in a parametric amplifier, the idler is generated after the wave-mixing process. Therefore, this process can also be applied to wavelength conversion. Due to very fast response of the third-order nonlinearity in optical waveguides, the wavelength conversion based on this effect is transparent to the modulation format and the bit rate of signals. For a flat wideband converter, which is a key device in wavelength-division multiplexing (WDM) networks, a short-length HNLF with a low dispersion slope is required in design. By a suitable selection of the pump wavelength, the

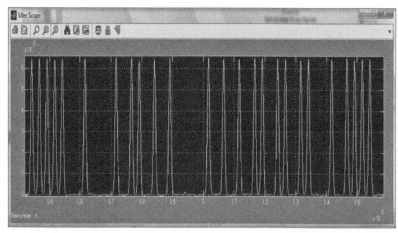

(a)

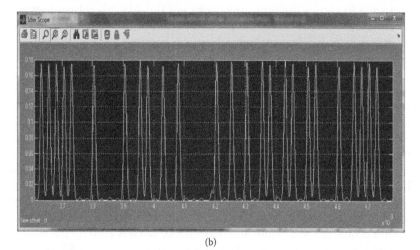

(b)

FIGURE 9.4

Time traces of the 40 Gb/s signal before (a) and after (b) the parametric amplifier.

TABLE 9.2

Critical Parameters of the Parametric Amplifier in 40 Gb/s Transmission System

RZ 40 Gb/s Signal
$\lambda_0 = 1559$ nm, $\lambda_s = \{1531.12, 1537.4, 1543.73, 1550.12\}$ nm
$P_s = 1$ mW (peak), $B_r = 40$ Gb/s
Parametric Amplifier
L pump source: $P_p = 100$ mW (after EDFA), $\lambda_p = 1560.07$ nm
HNLF: $L_f = 200$ m, $D = 0.02$ ps/km/nm, $S = 0.03$ ps/nm²/km, $\alpha = 0.5$ dB/km, $A_{eff} = 12$ μm², $\gamma = 13$ 1/W/km
BPF: $\Delta\lambda_{BPF} = 0.64$ nm, $\lambda_i = \{1587.91, 1581.21, 1574.58, 1567.98\}$ nm

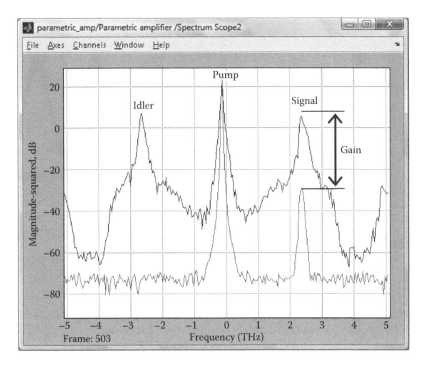

FIGURE 9.5
Optical spectra at the input (red, lower) and the output (black, higher) of the optical parametric amplifiers.

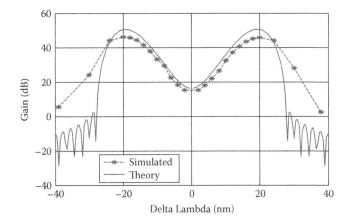

FIGURE 9.6
Calculated and simulated gain of the optical parametric amplifier at $P_p = 30$ dBm.

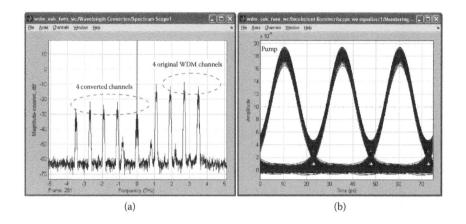

FIGURE 9.7
(a) The wavelength conversion of four wavelength-division multiplexing channels. (b) Eye diagram of the converted 40 Gb/s signal after bandpass filter.

wavelength converter can be optimized to obtain a bandwidth of 200 nm [7]. Therefore, the wavelength conversion between bands such as C and L bands can be performed in WDM networks. Figure 9.7 shows an example of the wavelength conversion for four WDM channels at the C-band. The important parameters of the wavelength converter are shown in Table 9.2. The WDM signals are converted into the L-band with the conversion efficiency of –12 dB.

Another important application with the same setup is the nonlinear phase conjugation (NPC). A phase conjugated replica of the signal wave can be generated by the FWM process. From Equation (9.8), the idler wave is approximately given in case of degenerate FWM for simplification: $E_i \sim A_p^2 A_s^* e^{-j\Delta kz}$ or $E_i \sim r A_s^* e^{\left[j(-kz-\omega\tau) \right]}$ with the signal wave $E_s \sim A_s e^{\left[j(kz-\omega\tau) \right]}$. Thus, the idler field is a complex conjugate of the signal field. In appropriate conditions, optical distortions can be compensated by using NPC, and optical pulses propagating in the fiber link can be recovered. The basic principle of distortion compensation with NPC refers to spectral inversion. When an optical pulse propagates in an optical fiber, its shape will be spread in time and distorted by the group velocity dispersion. The phase-conjugated replica of the pulse is generated in the middle point of the transmission link by the nonlinear effect. On the other hand, the pulse is spectrally inverted where spectral components in the lower-frequency range are shifted to the higher-frequency range and vice versa. If the pulse propagates in the second part of the link with the same manner in the first part, it is inversely distorted again and can cancel the distortion in the first part to recover the pulse shape at the end of the transmission link. With using NPC for distortion compensation, a 40% to 50% increase in transmission distance compared to a conventional transmission link can be obtained [8–11]. Figure 9.8 shows the setup of a long-haul 40 Gb/s transmission system demonstrating the distortion compensation using NPC.

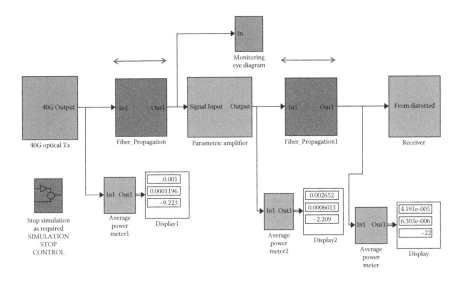

FIGURE 9.8
Simulink setup of a long-haul 40 Gb/s transmission system using nonlinear phase conjugation for distortion compensation.

The fiber transmission link of the system is divided into two sections by an NPC based on the parametric amplifier. Each section consists of five spans with 100 km standard single-mode fiber (SSMF) in each span. Figure 9.9a shows the eye diagram of the signal after propagating through the first fiber section. After the parametric amplifier at the midpoint of the link, the idler signal, a phase-conjugated replica of the original signal, is filtered for transmission in the next section. The signal in the second section suffers the same

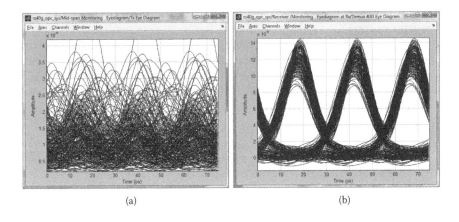

(a)   (b)

FIGURE 9.9
Eye diagrams of the 40 Gb/s signal at the end (a) of the first section and (b) of the transmission link.

TABLE 9.3

Critical Parameters of the Long-Haul Transmission System Using Nonlinear Phase Conjugation (NPC) for Distortion Compensation

RZ 40 Gb/s Transmitter
$\lambda_s = 1547$ nm, $\lambda_0 = 1559$ nm
Modulation: RZ-OOK, $P_s = 1$ mW (peak), $B_r = 40$ Gb/s
Fiber transmission link
SMF: $L_{SMF} = 100$ km, $D_{SMF} = 17$ ps/nm/km, $\alpha = 0.2$ dB/km
EDFA: Gain = 20 dB, NF = 5 dB
Number of spans: 10 (5 in each section), $L_{link} = 1000$ km
NPC based on OPA
Pump source: $P_p = 1$ W (after EDFA), $\lambda_p = 1560.07$ nm
HNLF: $L_f = 500$ m, $D = 0.02$ ps/km/nm, $S = 0.09$ ps/nm^2/km, $\alpha = 0.5$ dB/km, $A_{eff} = 12$ μm^2, $\gamma = 13$ 1/W/km
BPF: $\Delta\lambda_{BPF} = 0.64$ nm
Receiver
Bandwidth: $B_e = 28$GHz, $i_{eq} = 20$ pA/Hz$^{1/2}$, $i_d = 10$ nA.

Note: SMF (single mode fiber), OPA (optical parametric amplifier).

dispersion as in the first section. At the output of the transmission system the optical signal is regenerated as shown in Figure 9.9b. Due to the change in the wavelength of the signal in NPC, a tunable dispersion compensator can be required to compensate the residual dispersion after transmission in real system. The parameters of the transmission system with NPC are given in Tables 9.3 and 9.4.

9.2.4.3 High-Speed Optical Switching

When the pump is an intensity-modulated signal instead of the CW signal, the gain of the OPA is also modulated due to its exponential dependence on the pump power in a phase-matched condition. The width of gain profile in the time domain is inversely proportional to the product of the gain slope (S_p) or the nonlinear coefficient and the length of the nonlinear waveguide (L) [3]. Therefore, an OPA with high gain or large S_pL operates as an optical switch

TABLE 9.4

Parameters of the 40 GHz Short-Pulse Generator

Short-Pulse Generator
Signal: $P_s = 0.7$ mW, $\lambda_s = 1535$ nm, $\lambda_0 = 1559$ nm
Pump source: $P_p = 1$ W (peak), $\lambda_p = 1560.07$ nm, $f_m = 40$ GHz
HNLF: $L_f = 500$ m, $D = 0.02$ ps/km/nm, $S = 0.03$ ps/nm^2/km
$\alpha = 0.5$ dB/km, $A_{eff} = 12$ μm^2, $\gamma = 13$ 1/W/km
BPF: $\Delta\lambda_{BPF} = 3.2$ nm

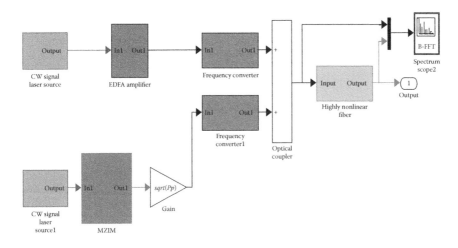

FIGURE 9.10
Simulink setup of the 40 GHz short-pulse generator.

with an ultrahigh bandwidth, which is very important in some signal-processing applications such as pulse compression or short-pulse generation [12,13]. A Simulink setup for a 40 GHz short-pulse generator is built with the configuration as shown in Figure 9.10. In this setup, the input signal is a CW source with low power and the pump is amplitude modulated by a Mach–Zehnder intensity modulator (MZIM), which is driven by a radio-frequency (RF) sinusoidal wave at 40 GHz. The waveform of the modulated pump is shown in Figure 9.11a. Important parameters of the FWM-based short-pulse generator are shown in Table 9.5. Figure 9.11b shows the generated short-pulse sequence with the pulse width of 2.6 ps at the signal wavelength after the optical BPF.

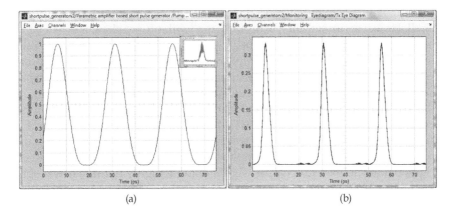

FIGURE 9.11
Time traces of (a) the sinusoidal amplitude modulated pump, and (b) the generated short-pulse sequence (Inset: the pulse spectrum).

TABLE 9.5

Important Parameters of the Four-Wave Mixing (FWM) Based Optical Time-Division Multiplexing (OTDM) Demultiplexer Using a Nonlinear Waveguide

OTDM Transmitter
MLL: $P_0 = 1$ mW, $T_p = 2.5$ ps, $f_m = 40$ GHz
Modulation formats: OOK and DQPSK; OTDM multiplexer: 4×40 GSymbols/s
FWM-Based Demultiplexer
Pumped control: $P_p = 500$ mW, $T_p = 2.5$ ps, $f_m = 40$ GHz, $\lambda_p = 1556.55$ nm
Input signal: $P_s = 10$ mW (after EDFA), $\lambda_s = 1548.51$ nm
Waveguide: $L_w = 7$ cm, $D_w = 28$ ps/km/nm, $S_w = 0.003$ ps/nm²/km
$\alpha = 0.5$ dB/cm, $\gamma = 10^4$ 1/W/km
BPF: $\Delta\lambda_{BPF} = 0.64$ nm

Note: MLL (mode locked laser), QPSK (quadrature phase shift keying).

Another important application of the optical switch based on the FWM process is the demultiplexer, a key component in ultra-high-speed optical time-division multiplexing (OTDM) systems. OTDM is a key technology for Tb/s Ethernet transmission, which can meet the increasing demand of traffic in future optical networks. A typical scheme of the OTDM demultiplexer in which the pump is a mode-locked laser (MLL) to generate short pulses for control is shown in Figure 9.12a. The working principle of the FWM-based demultiplexing is described as follows: The control pulses generated from a MLL at the tributary rate are pumped and copropagated with the OTDM signal through the nonlinear waveguide. Mixing processes between the control pulses and the OTDM signal during propagation through the nonlinear waveguide converts the desired tributary channel to a new idler wavelength. Then the demultiplexed signal at the idler wavelength is extracted by a bandpass filter before going to a receiver as shown in Figure 9.12a.

Using HNLF is relatively popular in structures of the OTDM demultiplexer [6,14]. However, its stability, especially the walk-off problem, is still a serious obstacle. Recently, planar nonlinear waveguides have emerged as promising devices for ultra-high-speed photonic processing [15,16]. These nonlinear waveguides offer a lot of advantages such as no free-carrier absorption, stable at room temperature, no requirement of quasi-phase matching, and the possibility of dispersion engineering. With the same operational principle, planar waveguide-based OTDM demultiplexers are very compact and suitable for photonic integrated solutions. Figure 9.12b shows the Simulink setup of the FWM-based demultiplexer of the on-off keying (OOK) 40 Gb/s signal from the 160 Gb/s OTDM signal using a highly nonlinear waveguide instead of HNLF. Important parameters of the OTDM system in Table 9.4 are used in the simulation. Figure 9.13a shows the spectrum at the output of the nonlinear waveguide. Then the demultiplexed signal is extracted by the bandpass filter as shown in Figure 9.13b. Figure 9.14 shows the time traces of the 160 Gb/s OTDM signal, the control signal, and the 40 Gb/s demultiplexed

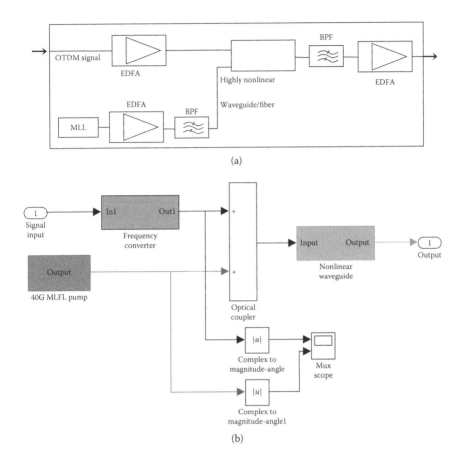

FIGURE 9.12
(a) A typical setup of the four-wave mixing–based optical time-division multiplexing (OTDM) demultiplexer. (b) Simulink model of the OTDM demultiplexer.

signal, respectively. The dots in Figure 9.14a indicate the time slots of the desired tributary signal in the OTDM signal. The developed model of the OTDM demultiplexer can be applied not only to the conventional OOK format but also to advanced modulation formats such as the differential quadrature phase shift keying (DQPSK), which increases the data load of the OTDM system without increase in bandwidth of the signal. By using available blocks developed for the DQPSK system [18], a Simulink model of the DQPSK-OTDM system is also set up for demonstration. The bit-rate of the OTDM system is doubled to 320 Gb/s with the same pulse repetition rate. Figure 9.15 shows the simulated performance of the demultiplexer in both 160 Gb/s OOK- and 320 Gb/s DQPSK-OTDM systems. The bit error rate (BER) curve in the case of the DQPSK-OTDM signal shows a low error floor that may result from the influence of nonlinear effects on phase-modulated signals in the waveguide.

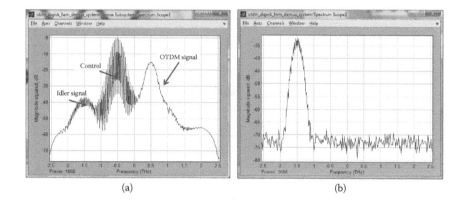

(a)   (b)

FIGURE 9.13
Spectra at the outputs (a) of nonlinear waveguide and (b) of bandpass filter.

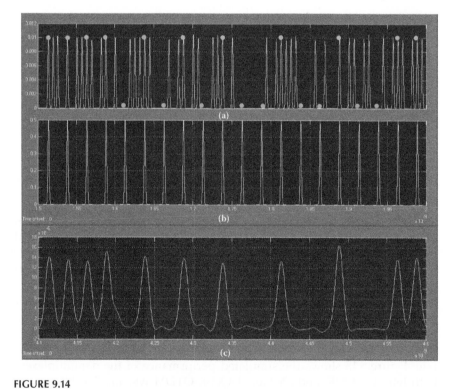

FIGURE 9.14
Time traces of (a) the 160 Gb/s optical time-division multiplexing signal, (b) the control signal, and (c) the 40 Gb/s demultiplexed signal.

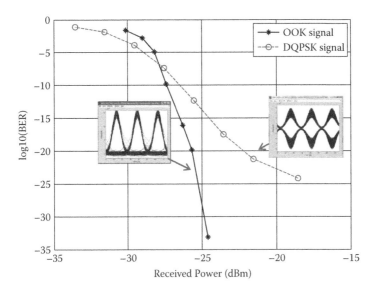

FIGURE 9.15
Simulated performance of the demultiplexed signals for 160 Gb/s on-off keying and 320 Gb/s differential quadrature phase shift keying/optical time-division multiplexing (DQPSK-OTDM) systems (Insets: eye diagrams at the receiver).

9.2.4.4 Triple Correlation

One promising application exploiting the $\chi^{(3)}$ nonlinearity is implementation of triple correlation in the optical domain. Triple correlation is a higher-order correlation technique, and its Fourier transform called bispectrum is very important in signal processing, especially in signal recovery [19,20]. The triple correlation of a signal $s(t)$ can be defined as

$$C^3(\tau_1, \tau_2) = \int s(t)s(t - \tau_1)s(t - \tau_2)\,dt \tag{9.18}$$

where τ_1, τ_2 are time-delay variables. To implement the triple correlation in the optical domain, the product of three signals including different delayed versions of the original signal need to be generated and then detected by an optical photodiode to perform the integral operation. From the representation of the nonlinear polarization vector (see Equation 9.3), this triple product can be generated by the $\chi^{(3)}$ nonlinearity. One way to generate the triple correlation is based on third-harmonic generation (THG) where the generated new wave containing the triple product is at the frequency of three times the original carrier frequency. Thus, if the signal wavelength is in the 1550 nm band, the new wave needs to be detected at around 517 nm. The triple-optical autocorrelation based on single-stage THG has been demonstrated in direct optical pulse shape measurement [21]. However, this way is hard to obtain high efficiency in the wave-mixing process due to the difficulty of phase matching between three signals. Moreover, the triple-product wave is in 517 nm where wideband photodetectors are not available for high-speed communication applications. Therefore, a possible alternative to generate the triple product is based on other nonlinear interactions such as FWM. From (9.9), the fourth wave is proportional to the product of three waves $A_4 \sim A_1 A_2 A_3^* e^{-j\Delta kz}$. If A_1 and A_2 are the delayed versions of the signal A_3, the mixing of three waves results in the fourth wave A_4, which is obviously the triple product of three signals. Dk is the phase mismatching and z is the propagation direction. As mentioned in Section 9.3, all three waves can take the same frequency; however, these waves should propagate into different directions to possibly distinguish the new generated wave in a diverse propagation direction that requires a strict arrangement of the signals in spatial domain. An alternative way we propose is to convert the three signals into different frequencies (ω_1, ω_2, and ω_3). Then the triple-product wave can be extracted at the frequency $\omega_4 = \omega_1 + \omega_2, -\omega_3$, which is still in the 1550 nm band.

Figure 9.15a shows the Simulink model for the triple correlation based on FWM in nonlinear waveguide (see Simulink model in Figure 9.17). The structural block consists of two variable delay lines to generate delayed versions of the original signal as shown in Figure 9.16 and frequency

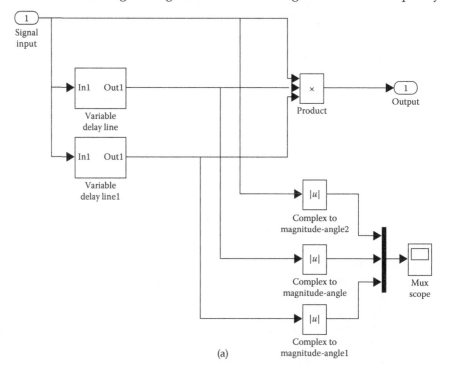

(a)

FIGURE 9.16

(a) Simulink setup of the four-wave mixing-based triple-product generation. (b) Simulink setup of the theory-based triple-product generation.

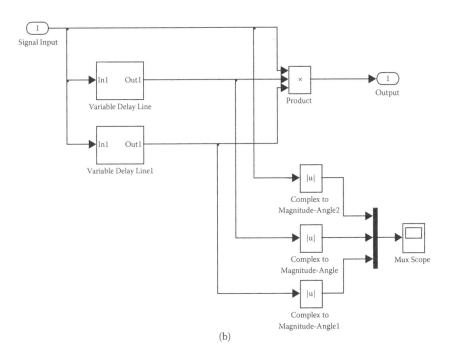

(b)

FIGURE 9.16
(Continued)

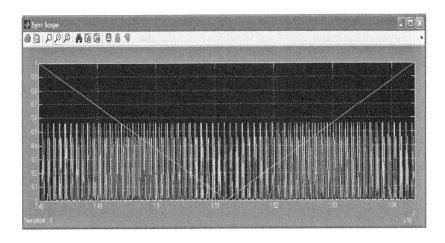

FIGURE 9.17
The variation in time domain of the time delay (slightly darker), the original signal (dark), and
the delayed signal (light).

TABLE 9.6

Important Parameters of the Four-Wave Mixing (FWM) Based Optical Time-Division Multiplexing (OTDM) Demultiplexer Using a Nonlinear Waveguide

Signal Generator
Single-pulse: $P_0 = 100$ mW, $T_p = 2.5$ ps, $f_m = 10$ GHz
Dual-pulse: $P_1 = 100$ mW, $P_2 = 2/3P_1$, $T_p = 2.5$ ps, $f_m = 10$ GHz
FWM based triple-product generator
Original signal: $\lambda_{s1} = 1550$ nm, $\lambda_{s1} = 1552.52$ nm
Delayed t_1 signal: $\lambda_{s2} = 1552.52$ nm
Delayed t_2 signal: $\lambda_{s3} = 1554.13$ nm
Waveguide: $L_w = 7$ cm, $D_w = 28$ ps/km/nm, $S_w = 0.003$ ps/nm²/km
$\alpha = 0.5$ dB/cm, g $= 10^4$ 1/W/km
BPF: $\Delta\lambda_{BPF} = 0.64$ nm

converters to convert the signal into three different waves before combining at the optical coupler to launch into the nonlinear waveguide (see Simulink model in Figure 9.17). Then the fourth wave signal generated by FWM is extracted by the passband filter. To verify the triple-product based on FWM, another model shown in Figure 9.16b to estimate the triple product by using (9.18) is also implemented for comparison. The integration of the generated triple-product signal is then performed at the photodetector in the optical receiver to estimate the triple correlation of the signal. A repetitive signal, which is a dual-pulse sequence with unequal amplitude, is generated for investigation. Important parameters of the setup are shown in Table 9.6. Table 9.6 shows the waveform of the dual-pulse signal and the spectrum at the output of the nonlinear waveguide. The wavelength spacing between three waves is unequal to reduce the noise from other mixing processes. The triple-product waveforms estimated by theory and FWM process are shown in Table 9.6.

In case of the estimation based on FWM, the triple-product signal is contaminated by the noise generated from other mixing processes as indicated in Table 9.6. The pulse sequence and its spectrum at the output of the NPC are shown in Figures 9.18a and b, respectively. Figure 9.19 shows the triple correlations of the signal after processing at the receiver in both cases based on theory and FWM. The triple correlation is represented by the three-dimensional plot that is displayed by the image. The x and y axes of the image represent the time-delay variables (τ_1 and τ_2) in terms of samples with the step-size of $T_m/32$, where T_m is the pulse period. The intensity of the triple correlation is represented by colors with scale specified by the color bar. Although the FWM-based triple correlation result is noisy, the triple-correlation pattern is still distinguishable as compared to the theory. Another signal pattern of the single pulse that is simpler has been also investigated as shown in Figures 19.20 and 19.21.

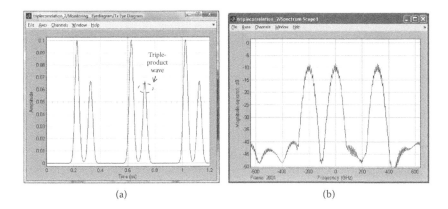

(a)   (b)

FIGURE 9.18
(a) Time trace of the dual-pulse sequence for investigation. (b) Spectrum at the output of the nonlinear waveguide.

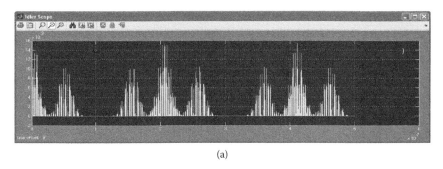

(a)

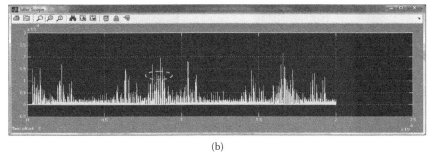

(b)

FIGURE 9.19
Generated triple-product waves in time domain of the dual-pulse signal based on (a) theory, and (b) four-wave mixing in nonlinear waveguide.

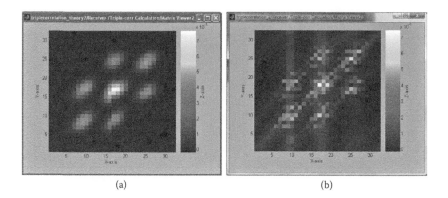

(a)   (b)

FIGURE 9.20
Triple correlation of the dual-pulse signal based on (a) theoretical estimation, and (b) four-wave mixing in nonlinear waveguide.

9.2.5 Application of Nonlinear Photonics in Advanced Telecommunications

This section looks at the uses of nonlinear effects and applications in modern optical communications networks in which 100 Gb/s optical Ethernet is expected to be deployed.

Typical performance of a photonic signal preprocessor employing no linear four-wave mixing is given, and that of an advanced processing of such received signals in the electronic domain processed by a digital triple correlation system. At least 10 dB improvement is achieved on the receiver sensitivity.

Regarding the nonlinear effects, the nonlinearity of the optical fibers hinders and limits the maximum level of the total average power of all the multiplexed channels for maximizing the transmission distance. These are due to the change of the refractive index of the guided medium as a function of

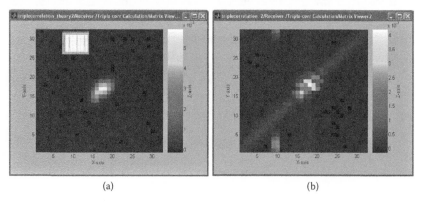

(a)   (b)

FIGURE 9.21
Triple correlation of the single-pulse signal based on (a) theoretical estimation, (b) four-wave mixing in nonlinear waveguide (Inset: the single-pulse pattern).

the intensity of the guided waves. This in turn creates the phase changes and hence different group delays, and then distortion. Furthermore, other associate nonlinear effects such as the four-wave mixing, Raman scattering, Brillouin scattering, intermodulation have also created jittering and distortion of the received pulse sequences after a long transmission distance.

However, recently we have been able to use to our advantage these nonlinear optical effects as a preprocessing element before the optical receiver to improve its sensitivity. A higher-order spectrum technique is employed with the triple correlation implemented in the optical domain via the use of the degenerate four-wave mixing effects in a high nonlinear optical waveguide. This may add additional optical elements and filtering in the processor and hence complicate the receiver structure. We can overcome this by bringing this nonlinear higher-order spectrum processing (see schematic diagram in Figure 9.22) to after the opto-electronic conversion and in the digital processing domain after a coherent receiving and electronic amplification subsystem is used.

In this section, we illustrate some uses of nonlinear effects and nonlinear processing algorithms for improving the sensitivity of optical receivers employing nonlinear processing at the front end of the photodetector and nonlinear processing algorithm in the electronic domain.

The spectral distribution of the FWM and the simulated spectral conversion can be achieved [22–26]. There is degeneracy of the frequencies of the waves so that efficient conversion can be achieved by satisfying the conservation of momentum. This can be detailed in another paper in the special session of this workshop [2]. The detected phase states and bispectral properties are depicted in Figure 9.23 in which the phases can be distinguished based on the diagonal spectral lines. Under noisy conditions these spectral distributions can be observed in Figure 9.24.

Alternating to the optical processing described above, a nonlinear processing technique using a high-order spectrum technique can be implemented in the electronic domain. This is implemented after the ADC, which samples the incoming electronic signals produced by the coherent optical receiver as shown in Figure 9.25 [27]. The operation of a third-order spectrum analysis is based on the combined interference of three signals (in this case the complex

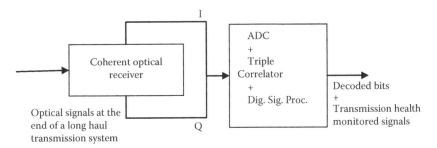

FIGURE 9.22
A high-order spectral optical receiver and electronic processing.

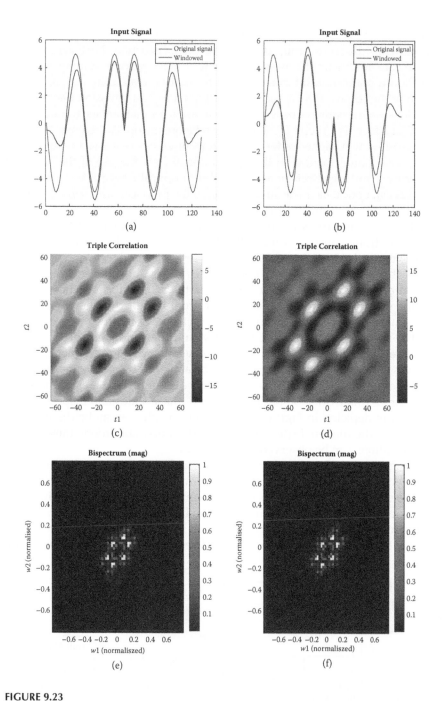

FIGURE 9.23
Input waveform with phase changes at the transitions (a,b), triple correlation and bispectrum (c–h) of both phase and amplitude.

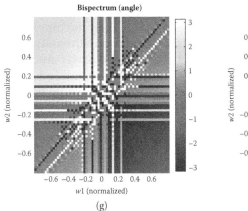
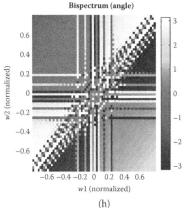

(g)

(h)

FIGURE 9.23
(Continued)

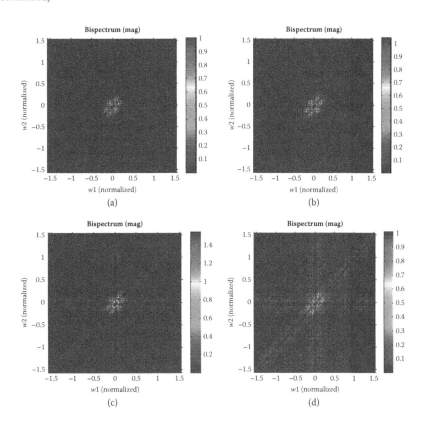

(a)

(b)

(c)

(d)

FIGURE 9.24
Effect of Gaussian noise on the bispectrum (a,c) amplitude distribution in two dim (b,d) phase spectral distribution.

signals produced at the output of the analog-to-digital converter, ADC), two of which are the delayed version of the original. Then the amplitude and phase distribution of the complex signals are obtained in three-dimensional graphs that allow us to determine the signal and noise power and the phase distribution. These distributions allow us to perform several functions necessary for evaluation of the performance of optical transmission systems. Simultaneously, the processed signals allow us to monitor the health of the transmission systems such as the effects due to nonlinear effects, the distortion due to chromatic dispersion of the fiber transmission lines, the noises contributed by inline amplifiers, and so forth. A typical curve that compares the performance of this innovative processing with convention detection techniques is shown in Figure 9.25. If we project the error rate of the receivers employing conventional techniques and our high-order spectral receiver employing digital signal processing techniques at a bit error rate of 10^{-9}, then at least 1000 times lower than digital receiver without using this type of processor. This is equivalent to at least one unit improvement on the quality factor of the eye opening. In turn, this is equivalent to about 10 dB in the signal-to-noise ratio.

These results are very exciting for network and system operators as significant improvement of the receiver sensitivity can be achieved, and this allows significant flexibility in the operation and management of the transmission systems and networks. Simultaneously, the monitored signals produced by the high-order spectral techniques can be used to determine the distortion

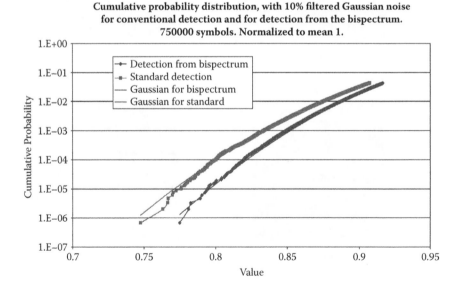

FIGURE 9.25
Error estimation version detection level of the high-order spectrum processor.

and noises of the transmission line and thus the management of the tuning of the operating parameters of the transmitter, the number of wavelength channels, and the receiver or inline optical amplifiers.

The effect of additive white Gaussian noise on the bispectrum magnitude is shown in Figure 9.24, the sequence of figures are as indicated in Figure 9.23. The uncorrupted bispectrum magnitude is shown in Figure 9.23a, while Figures 9.23b, 9.23c, and 9.23d were generated using signal-to-noise ratios of 10 dB, 3 dB, and 0 dB, respectively. This provides a method of monitoring the integrity of a channel and illustrates another attractive attribute of the bispectrum. It is noted that the bispectrum phase is more sensitive to Gaussian white noise than is the magnitude and quickly becomes indistinguishable below 6 dB.

The algorithm employing the nonlinear processing (e.g., optical preprocessing or electronic processing of the triple correlation and bispectrum) will involve hardware and soft implementation. From the point of view of industry, it needs to deliver to the market at the right time for systems and networks operating in the Tera-bits/s speed. One can thus be facing the following dilemma:

- The optical preprocessing requires an efficient nonlinear optical waveguide that must be in an integrated structure whereby efficient coupling and interaction can be achieved. If not then the gain of about 3 dB in signal-to-noise ratio would be defeated by this loss. Furthermore, the integration of the linear optical waveguiding section and a nonlinear optical waveguide is not matched due to differences in the waveguide structures of both regions. For a linear waveguide structure to be efficient for coupling with circular optical fibers, silica on silicon would be best suited due to the small refractive index difference and the technology of burying such waveguides to form an embedded structure whose optical spot size would match that of a single guided mode fiber. This silica on silicon would not match an efficient nonlinear waveguide made by As_2S_3 on silicon.

- On the other hand, if electronic processing is employed then it requires an ultrafast analog to digital converter (ADC) and then fast electronic signal processors. Currently a 56 GSamples/s ADC is available from Fujitsu as shown in Figure 9.26. It is noted that the data output samples of the ADC are structured in parallel forms with the referenced clock rate of 1.75 GHz. Thus, all processing of the digital samples must be in parallel form, and parallel processing algorithms must be structured in parallel. This is the most challenging problem we must overcome in the near future.

- An application-specific integrated circuit (ASIC) must also be designed for this processor.

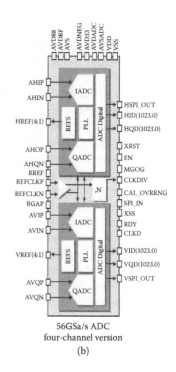

56GSa/s ADC two-channel version
using CHAIS architecture
(a)

56GSa/s ADC
four-channel version
(b)

FIGURE 9.26
Plane view of the Fujitsu analog-to-digital converter operating at 56 GSamples/s: (a) integrated view and (b) operation schematic.

- Hard decisions must be made as is fitting of such ASIC and associated optical and opto-electronic components into international standard compatible size. Thus, all designs and components must meet this requirement.

- Finally the laboratory and field testing must be demonstrated for market delivery.

These challenges will be met and we are currently progressing toward the final target for the delivery of such sensitive receivers for 100 Gb/s optical Internet.

9.2.6 Remarks

In this section we demonstrated a range of signal processing applications exploiting the parametric process in nonlinear waveguides. A brief mathematical description of the parametric process through third-order nonlinearity has been reviewed. A Simulink model of nonlinear waveguide has been developed to simulate interaction of multiwaves in optical waveguides including optical fibers. Based on the developed Simulink modeling platform, a range of signal processing applications exploiting parametric FWM processes has been investigated through simulation. With a CW pump source, the applications such as parametric amplifier, wavelength converter, and optical phase conjugator have been implemented for demonstration. The ultra-high-speed optical switching can be implemented by using an intensity-modulated pump to apply in the short-pulse generator and the OTDM demultiplexer. Moreover, the FWM process has been proposed to estimate the triple correlation that is very important in signal processing. The simulation results showed the possibility of the FWM-based triple correlation using the nonlinear waveguide with different pulse patterns. Although the triple-correlation is contaminated by noise from other FWM processes, it is possible to distinguish it. The wavelength positions as well as the power of three delayed signals need to be optimized to obtain the best results.

Furthermore, we also addressed the important issues of nonlinearity and its uses in optical transmission systems, the management of networks if the signals that indicate the health of the transmission system are available. It is no doubt that the nonlinear phenomena play several important roles in the distortion effects of signals transmitted but also allow us to improve the transmission quality of the signals. This has been briefly described in this paper on the optical processing using four-wave mixing effects and nonlinear signal processing using high-order spectral analysis and processing in the electronic domain. This ultra-high-speed optical preprocessing and electronic triple correlation and bispectrum receivers are the first system using nonlinear processing for 100 Gb/s optical Internet.

9.3 Nonlinear Photonic Preprocessing and Bispectrum Optical Receivers

In this section, we present the processing of optical signals before the optoelectronic detection in the optical domain in a nonlinear optical waveguide as an NL (NL) signal processing technique for the digital optical receiving system for long-haul optically amplified fiber transmission systems. The algorithm implemented is a high-order spectrum (HOS) technique in which the original signals and two delayed versions are correlated via the four-wave mixing or third harmonic conversion process. The optical receivers employing a higher-order spectral photonic preprocessor and very large scale integration (VLSI) electronic system for the electronic decoding and evaluation of the bit error rate of the transmission system are presented. A photonic signal preprocessing system is developed to generate the triple correlation via the third harmonic conversion in an NL optical waveguide. It is employed as the photonic preprocessor to generate the essential part of a triple correlator. The performance of an optical receiver incorporating the HOS processor is given for the long-haul phase-modulated fiber transmission.

9.3.1 Introductory Remarks

Recently, tremendous efforts have been pushing for reaching higher transmission bit rates and longer haul for optical fiber communication systems [29–32]. The bit rate can reach several hundreds of Gb/s to the Tb/s. In this extremely high-speed operational region, the limits of electronic speed processors have been surpassed, and optical processing is assumed to play an important part of the optical receiving circuitry. Novel processing techniques are required in order to minimize the bottlenecks of electronic processing and noises and distortion due to the impairment of the transmission medium and the linear and NL distortion effects.

This paper deals with the photonic processing of optically modulated signals prior to the electronic receiver for long-haul optically amplified transmission systems. NL optical waveguides in planar or channel structures are studied and employed as a third harmonic converter so as to generate a triple product of the original optical waves and its two delayed copies. The triple product is then detected by an opto-electronic receiver. Then the detected current would be electronically sampled and digitally processed to obtain the bispectrum of the data sequence, and a recovery algorithm is used to recover the data sequence. The generic structures of such high-order spectral optical receivers are shown in Figure 9.27. For the nonlinear photonic processor (Figure 9.27a), the optical signals at the input are delayed and then coupled to a nonlinear photonic device in which the nonlinear conversion process is implemented via the uses of third harmonic conversion or degenerate four-wave mixing. In the nonlinear digital

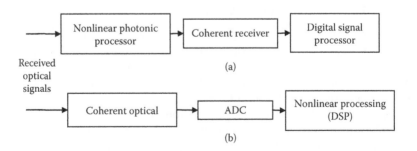

FIGURE 9.27
Generic structure of high-order spectrum optical receiver, (a) photonic preprocessor, (b) nonlinear digital signal processing in electronic domain.

processor (see Figure 9.27b) the optical signals are detected coherently and then sampled and processed using the nonlinear triple correlation and decoding algorithm.

We propose and simulate this NL optical preprocessor optical receiver under the MATLAB/Simulink platform for differentially coded phase shift keying, the DQPSK modulation scheme.

9.3.2 Bispectrum Optical Receiver

Figure 9.27 shows the structure of a bispectrum optical receiver in which there are three main sections: an all-optical preprocessor, an optoelectronic detection, and amplification including an ADC to generate sampled values of the triple correlated product.

9.3.3 Triple Correlation and Bispectra

9.3.3.1 Definition

The power spectrum is the Fourier transform of the autocorrelation of a signal. The bispectrum is the Fourier transform of the triple correlation of a signal. Thus, both the phase and amplitude information of the signals is embedded in the triple-correlated product.

While autocorrelation and its frequency domain power spectrum does not contain the phase information of a signal, the triple correlation contains both, due to the definition of the triple correlation:

$$c(\tau_1, \tau_2) = \int S(t)S(t + \tau_1)S(t + \tau_2)dt \qquad (9.19)$$

where $S(t)$ is the continuous time domain signals to be recovered. $\tau_1; \tau_2$ are the delay time intervals. For the special case where $\tau_1 = 0$ or $\tau_2 = 0$, the triple correlation is proportional with the autocorrelation. It means that the amplitude information is also contained in the triple correlation. The benefit of holding phase and amplitude information is that it gives a potential to recover the signal from its triple correlation. In practice, the delays τ_1 and τ_2 indicate the path difference between the three optical waveguides. These delay times are corresponding to the frequency regions in the spectral domain. Thus, a different time interval would determine the frequency lines in the bispectrum.

9.3.3.2 Gaussian Noise Rejection

Given a deterministic sampled signal $S(n)$, the sampled version of the continuous signals $S(t)$, are corrupted by Gaussian noise $w(n)$, with n the sampled time index. The observed signal takes the form $Y(n) = S(n) + w(n)$. The polyspectra of any Gaussian process is zero for any order greater than two [34]. The bispectrum is the third-order polyspectrum and offers a significant advantage for signal processing over the second-order polyspectrum, commonly known as the power spectrum, which is corrupted by Gaussian noise. Theoretically speaking, the bispectral analysis allows us to extract a non-Gaussian signal from the corrupting affects of Gaussian noise.

Thus, for a signal arriving at the optical receiver, the steps to recover the amplitude and phase of the lightwave modulated signals are [27] as follows:

- Estimate the bispectrum of $S(n)$ based on observations of $Y(n)$.
- From the amplitude and phase bispectra form an estimate of the amplitude and phase distribution in a one-dimensional frequency of the Fourier transform of $S(n)$. This forms the constituents of the signal $S(n)$ in the frequency domain.
- Recover the original signal $S(n)$ by taking the inverse Fourier transform.

This type of receiver is termed the *bispectral optical receiver*.

9.3.3.3 Encoding of Phase Information

The bispectra contain almost complete information about the original signal (magnitude and phase). If the original signal $x(n)$ is real and finite, it can be reconstructed except for a shift a. Equivalently, the Fourier transform can be determined except for a linear shift factor of $e^{-j2\pi\omega a}$. By determining two adjacent pulses any differential phase information will be readily available [28]. In other words, the bispectra, hence the triple correlation, contain the phase information of the original signal allowing it to pass through the square law photodiode that would otherwise destroy this information. The encoded phase information can then be recovered up to a linear phase term, thus necessitating a differential coding scheme.

9.3.3.4 Eliminating Gaussian Noise

For any processes that have zero mean and the symmetrical probability density function (pdf), their third-order cummulant are equaled to zero. Therefore, in a triple correlation, those symmetrical processes are eliminated. Gaussian noise is assumed to affect to signal quality. Mathematically, the third cummulant is defined as

$$c_3(\tau_1, \tau_2) = m_3(\tau_1, \tau_2) - m_1 \begin{bmatrix} m_2(\tau_1) + m_2(\tau_2) \\ + m_2(\tau_1 - \tau_2)] + 2(m_1)^3 \end{bmatrix} \tag{9.20}$$

where m_k is the kth order moment of the signal. Thus, for the zero mean and symmetrical pdf, its third-order cummulant becomes zero [29]. Theoretically,

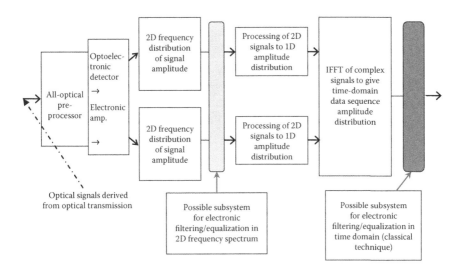

FIGURE 9.28
Generic structure of an optical preprocessing receiver employing bispectrum processing technique.

considering the signal as $u(t) = s(t) + n(t)$, where $n(t)$ is an additive Gaussian noise, the triple correlation of $u(t)$ will reject Gaussian noise affecting the $s(t)$ [30,31].

9.3.4 Bispectral Optical Structures

Figure 9.28 shows the generic and detailed structure of the bispectral optical receiver, respectively, which consists of (1) an all-optical preprocessor front end followed by (2) a photodetector and electronic amplifier to transfer the detected electronic current to voltage level appropriate for sampling by an analog to digital converter, and thus the signals at this stage are in sampled form; (3) the sampled triple correlation product is then transformed to the Fourier domain using the fast Fourier transform (FFT). The product at this stage is the row of the matrix of the bispectral amplitude and phase plane (see Figure 9.29). A number of parallel structures may be required if passive delay paths are used. (4) A recovery algorithm is used to derive the one-dimensional distribution of the amplitude and phase as a function of the frequency that are the essential parameters required for taking the inverse Fourier transform to recover the time domain signals.

9.3.4.1 Principles

The physical process of mixing the three waves to generate the fourth wave whose amplitude and phase are proportional to the product of the three input waves is well known in literatures of NL optics.

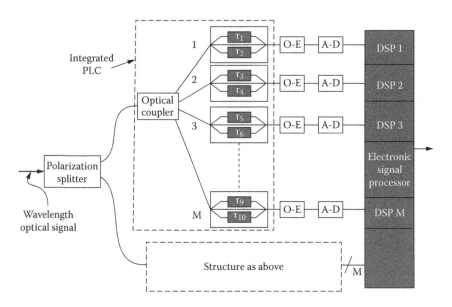

FIGURE 9.29
Parallel structures of photonic preprocessing to generate triple correlation product in the optical domain.

This process requires

- Highly NL medium so as to efficiently convert the energy of the three waves to that of the fourth wave
- Phase-matching conditions of the three input waves of the same frequency (wavelength) to satisfy the conservation of momentum

9.3.4.2 Technological Implementation

9.3.4.2.1 Nonlinear (NL) Optical Waveguides

In order to satisfy the condition of four-wave mixing we will be using

- A rib-waveguide for guiding whose NL refractive index coefficient is about 100,000 times greater than that of silica. The material used in this waveguide is a chalcogenide glass type (e.g., AS_2S_3) or TeO_2. The three waves are guided in this waveguide structure. Their optical fields are overlapped. The cross section of the waveguide is in the order of 4 mm × 0.4 mm.
- The waveguide cross section is designed such that the dispersion is "flat" over the spectral range of the input waves, ideally from 1520 nm to 1565 nm. This can be done by adjusting the thickness of the rib structure.

9.3.4.2.2 Mixing and Integrating

The fourth wave generated from the FWM waveguide is then detected by the photodetector that acts as an integrating device. Thus, the output of this detector is the triple correlation product in the electronic domain that we are looking for.

9.3.4.2.3 Equalization and Filtering

If equalization or filtering is required, then these functional blocks can be implemented in the bispectral domain as shown in Figure 9.28. Figure 9.29 shows the parallel structures of the bispectral receiver so as to obtain all rows of the bispectral matrix. The components of the structure are almost similar except the delay time of the optical preprocessor.

9.3.4.2.4 Four-Wave Mixing (FWM) in Highly Nonlinear Media

In the NL channel waveguide fabricated using TeO_2 (tellurium oxide) on silica the interaction of the three waves, one original and two delayed beams, happens via the electronic processes with highly NL coefficient [χ^3] will convert to the fourth wave.

When the three waves are copropagating, the conservation of the momentum of the three waves and the fourth wave is satisfied to produce efficient FWM. Phase matching can also be satisfied by one forward wave and two backward propagating waves (delayed version of the first wave), leading to almost 100% conversion efficiency to generate the fourth wave.

The interaction of the three waves via electronic process and the $\chi^{(3)}$ gives rise to the polarization vector P that couples to the electric field density of the lightwaves and then to the NL Schrödinger wave equation. By solving and modeling this wave equation with the FWM term on the right-hand side of the equation one can obtain the wave output (the fourth wave) at the output of the NL waveguide section.

9.3.4.2.5 Third Harmonic Conversion

Third harmonic conversion may happen but at extremely low efficiency, at least one thousand times less than that of FWM due to the nonmatching of the effective refractive indices of the guided modes at 1550 nm (fundamental wave) and 517 nm (third harmonic wave).

NOTE: The common term for this process is the matching of the dispersion characteristics (i.e., k/omega with omega the radial frequency of the waves at 1550 nm and 517 nm) versus the thickness of the waveguide.

9.3.4.2.6 Conservation of Momentum

The conservation of momentum and thus the phase matching condition for the FWM is satisfied without much difficulty as the wavelengths of the three input waves are the same. The optical NL channel waveguide is to be designed such that there is mismatching of the third harmonic conversion and is most efficient for FWM. It is considered that a single polarized mode, either traverse electric (TE) or traverse magnetic (TM) will be used to achieve efficient FWM. Thus, the dimension of the channel waveguide would be estimated at about 0.4 mm (height) × 4 mm (width).

9.3.4.2.7 Estimate of Optical Power Required for Four-Wave Mixing

In order to achieve the most efficient FWM process the NL coefficient n_2, which is proportional to $\chi^{(3)}$ by a constant ($8n/3$, with n the refractive index of the medium or approximately the effective refractive index of the guided mode). This NL coefficient is then multiplied by the intensity of the guided waves to give an estimate of the phase change and estimation of the efficiency of the FWM. With the cross section estimated in (9.3) and the well confinement of the guided mode, the effective area of the guided waves is very close to the cross-sectional area. Thus, an average power of the guide waves would be about 3 to 5 mW or about 6 dBm.

With the practical data of the loss of the linear section (section of multimode interference and delay split, similar to array waveguide grating technology) estimated at 3 dB, the input power of the three waves required for efficient FWM is about 10 dBm (maximum).

9.3.5 Mathematical Principles of Four-Wave Mixing and the Wave Equations

9.3.5.1 The Phenomena of Four-Wave Mixing

The origin of the FWM comes from the parametric processes that lie in the NL responses of bound electrons of a material to applied optical fields. More specifically the polarization induced in the medium is not linear in the applied field but contains NL terms whose magnitude is governed by the NL susceptibilities [32–34]. The first-, second-, and third-order parametric processes can occur due to these NL susceptibilities [$\chi^1 \chi^2 \chi^3$]. The coefficient χ^3 is responsible for the FWM that is exploited in this work. Simultaneously with this FWM, there is also a possibility of generating third harmonic waves with mixing the three waves and parametric amplification. The third harmonic generation is normally very small due to the phase mismatching of the guided wave number (the momentum vector) between the fundamental waves and the third harmonic waves. FWM in a guided wave medium such as single-mode optical fibers have been extensively studied due to its efficient mixing to give the fourth wave [35,36]. The exploitation of the FWM process has not been extensively exploited yet in channel optical waveguides. In this work, we demonstrate this theoretically and experimentally (and investigated for optical signal processing for the bispectrum analyzer).

The three lightwaves are mixed to generate the polarization vector $\vec{P}$ due to the NL third-order susceptibility given as

$$\vec{P}_{NL} = \varepsilon_0 \chi^{(3)} \vec{E}.\vec{E} \ .\vec{E} \tag{9.21}$$

where ε_0 is the permittivity in the vacuum; $\vec{E}_1, \vec{E}_2, \vec{E}_3$ are the electric field components of the lightwaves; $\vec{E} = \vec{E}_1 + \vec{E}_2 + \vec{E}_3$; is the total field entering the NL waveguide, and $\chi^{(3)}$ is the third-order susceptibility of the NL medium. P_{NL} is the product of the three total optical fields of the three optical waves that give the triple product of the waves required for the bispectrum receiver in which the NL waveguide acts as a multiplier of the three waves considered as the pump waves in this section. The mathematical analysis of the coupling equations via the wave equation is complicated but straightforward. Let ω_1, ω_2, ω_3, and ω_4 be the angular frequencies of the four waves of the FWM process and linearly polarized along the horizontal direction y of the channel waveguides and propagating along the z-direction. The total electric field vector of the four waves is given by

$$\vec{E} = \frac{1}{2}\vec{a}_y \sum_{i=1}^{4} \vec{E}_j e^{j(k_i z - \omega_i t)} + c.c. \tag{9.22}$$

with..$\vec{a}_y$ = unit..vector..along..y..axis;..and _ c.c. = complex..conjugate

The propagation constant can be obtained by $k_i = \frac{n_{eff,i}\omega_i}{c}$ with $n_{eff,i}$ the effective index of the ith guided waves E_i ($i = 1,...4$) a which can be either, TE or TM polarized guided mode propagating along the channel NL optical waveguide and all four waves are assumed propagating along the same direction. Substituting (9.22) into (9.21), we have

$$\vec{P}_{NL} = \frac{1}{2}\vec{a}_y \sum_{i=1}^{4} P_i e^{j(k_i z - \omega_i t)} + c.c. \tag{9.23}$$

where P_i ($i = 1,2..4$) consists of a large number of terms involving the product of three electric fields of the optical guided waves, for example, the term P_4 can be expressed as

$$P_4 = \frac{3\varepsilon_0}{4}\chi^{(3)}_{xxxx} \left\{ \begin{array}{l} \left|E_4\right|^2 E_4 + 2\left(\left|E_1\right|^2 + \left|E_2\right|^2 + \left|E_3\right|^2\right)E_4 \\ +2E_1 E_2 E_3 e^{j\varphi^+} + 2E_1 E_2 E_3^* e^{j\varphi^-} + c.c. \end{array} \right\}$$

with $\tag{9.24}$

$$\varphi^+ = (k_1 + k_2 + k_3 + k_4)z - (\omega_1 + \omega_2 + \omega_3 + \omega_4)t$$

$$\varphi^- = (k_1 + k_2 - k_3 - k_4)z - (\omega_1 + \omega_2 - \omega_3 - \omega_4)t$$

The first four terms of Equation (9.24) represents the self-phase modulation (SPM) and cross-phase modulation effects (XPM) that are dependent on the intensity of the waves. The remaining terms results into FWM. Thus, the question is which terms are the most effective components resulting from the parametric mixing process? The effectiveness of the parametric coupling depends on the phase-matching terms governed by φ^+ *and* φ^- or a similar quantity.

It is obvious that significant FWM would occur if the phase matching is satisfied. This requires the matching of both the frequency as well as the wave vectors as given in (9.24). From (9.24) we can see that the term φ^+ corresponds to the case in which three waves are mixed to give the fourth wave whose frequency is three times that of the original wave. This is the third harmonic generation. However, the matching of the wave vector would not normally be satisfied due to the dispersion effect. Furthermore, the propagation constants of the guided modes at different wavelength of the four wave mixing process would not be matched due to the dispersion characteristics of the optical waveguide. This would only allow a minute conversion to the third harmonic waves.

The conservation of momentum derived from the wave vectors of the four waves requires that

$$\Delta k = k_1 + k_2 - k_3 - k_4 = \frac{n_{eff,1}\omega_1 + n_{eff,2}\omega_2 - n_{eff,3}\omega_3 - n_{eff,4}\omega_4}{c} = 0 \qquad (9.25)$$

The effective refractive indices of the guided modes of the three waves E_1, E_2, and E_3 must be the same at their frequencies so as to achieve the most efficient conversion. This condition is automatically satisfied provided that the NL waveguide is designed such that it supports only a single polarized mode TE or TM and with minimum dispersion difference within the band of the signals.

9.3.5.2 Coupled Equations and Conversion Efficiency

To derive the wave equations to represent the propagation of the three waves to generate the fourth wave, we can resort to the Maxwell equations. It is lengthy to write down all the steps involved in this derivation so we summarize the standard steps usually employed to derive the wave equations as follows: First add the NL polarization vector given in (9.21) into the electric field density vector **D**. Then taking the curl of the Maxwell first equation and use the second equation of the Maxwell four equations and substituting the electric field density vector and using the fourth equation, one would then come up with the vectorial wave equation.

For the FWM process occurring during the interaction of the three waves along the propagation direction of the NL optical channel waveguide, the evolution of the amplitudes, A_1 through A_4, of the four waves, E_1 to E_4, is given by (only A_1 term is given)

$$\frac{dA_1}{dz} = \frac{jn_2\omega_1}{c}\left[\begin{pmatrix}\Gamma_{11}\left|A_1\right|^2 \\ +2\sum_{k\neq1}\Gamma_{1k}\left|A_k\right|^2\end{pmatrix}A_1 + 2\Gamma_{1234}A_2^*A_3A_4e^{j\Delta kz}\right] \tag{9.26}$$

where the wave vector mismatch Δk is given in (9.25), and the * denotes the complex conjugation. Note that the coefficient n_2 in Equation 9.26 is the NL coefficient related to the nonlinear susceptibility coefficient, defined as

$$n_2 = \frac{3}{8n\,\mathrm{Re}(\chi_{xxxx}^3)} \tag{9.27}$$

9.3.5.3 Evolution of FWM along the NL Waveguide Section

Once the fourth wave is generated, the interaction of the four waves along the section of the waveguide continues happening, thus the NL Schrödinger equation must be used to investigate the evolution of the waves. The NLSE is well known and presented in Bartelt et al. [34] and given for the temporal amplitude of the waves as

$$\frac{dA_j}{dz} \rightarrow \frac{dA_j}{dz} + \beta_1\frac{dA_j}{dz} + \frac{j}{2}\beta_2\frac{d^2A_j}{dz^2} + \frac{1}{2}\alpha_jA_j \tag{9.28}$$

This makes the four equations complicated and only numerical simulations can offer the evolution of the complex amplitude and the power of the fourth wave at the output of the NL waveguide. This takes into account the dispersion of the waveguide and material of the waveguide under chromatic dispersion.

9.3.6 Transmission and Detection

9.3.6.1 Optical Transmission Route and Simulation Platform

Shown in Figure 9.30 is the schematic of the transmission link over a total length of 700 km with sections from Melbourne City (of Victoria, Australia) to Gippsland, the inland section in Victoria of Australia, an undersea section of more than 300 km crossing the Bass Strait to George Town of Tasmania and then inland transmission to Hobart of Tasmania—the Gippsland/George Town Link (see Figure 9.30). Other inland sections in Victoria and Tasmania of Australia are

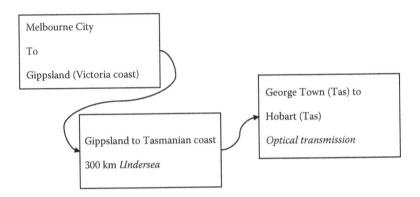

FIGURE 9.30
The transmission link including inland and undersea sections between Melbourne (Victoria) and Hobart (Tasmania) of Australia.

structured with optical fibers and lumped optical amplifiers (Er:doped fiber amplifiers, EDFA). Raman-distributed optical amplification (ROA) is used by pump sources located at both ends of the Melbourne, Victoria, to Hobart of Tasmania link including the 300 km undersea section. The undersea section of nearly 300 km consists of only the transmission and dispersion compensating fibers, no active subsystems are included. Only Raman amplification is used with pump sources located at both sides of the section and installed inland. This 300 km distance is fairly long, and only the Raman distributed gain is used. Simulink models of the transmission system include the optical transmitter, the transmission line, and the bispectrum optical receiver.

9.3.6.2 Four-Wave Mixing and Bispectrum Receiving

We also integrated into the MATLAB Simulink of the transmission system so that would enable us to investigate the NL parametric conversion system very close to practice. The spectra of the optical signals before and after this amplification are shown in Figures 9.31 and 9.32. The temporal distribution of the pulses at the input and output are also shown in Figures 9.33(a) and (b), respectively, indicating the conversion efficiency. This indicates the performance of the bispectrum optical receiver.

9.3.6.3 Performance

We implement the models for both techniques for binary phase shift keying modulation format for serving as a guideline for phase modulation optical transmission systems using nonlinear preprocessing. We note the following:

1. The arbitrary white Gaussian noise (AWGN) block in the Simulink platform can be set in different operating modes. This block then accepts the signal input and assumes the sampling rate of the input signal, then estimates the noise variance based on Gaussian

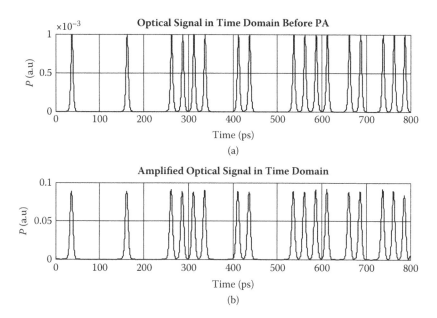

FIGURE 9.31
Time traces of the optical signal (a) before and (b) after the parametric amplifier.

distribution and the specified signal-to-noise ratio (SNR). This is then superimposed on the amplitude of the sampled value. Thus, we believe at that stage, the noise is contributed evenly across the entire band of the sampled time (converted to spectral band).

2. The ideal curve SNR versus BER plotted in the graph provided is calculated using the commonly used formula in several textbooks on communication theory. This is evaluated based on the geometrical distribution of the phase states and then the noise distribution over those states. That means that all the modulation and demodulation are assumed to be perfect. However, in the digital system simulation, the signals must be sampled. This is even more complicated when a carrier is embedded in the signal, especially when the phase shift keying modulation format is used.

3. We thus reset the models of (1) AWGN in a complete binary phase shift keying (BPSK) modulation format with both the ideal coherent modulator and demodulator and any necessary filtering required and (2) AWGN blocks with the coherent modulator and demodulator incorporating the triple correlator and necessary signal processing block. This is done in order to make a fair comparison between the two pressing systems.

4. In our former model, the AWGN block was being used incorrectly in that it was being used in the SNR mode that applies the noise power over the entire bandwidth of the channel, which of course is larger

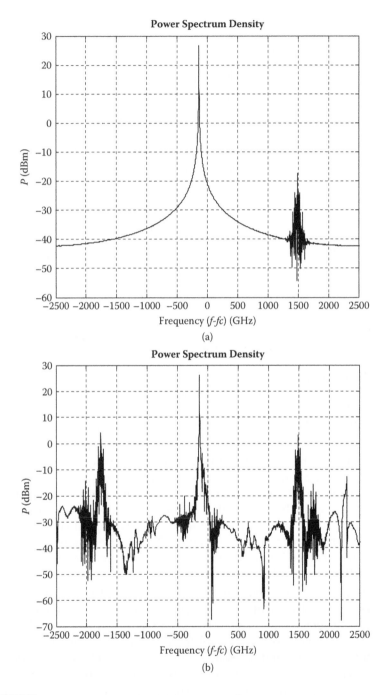

FIGURE 9.32
Corresponding spectra of the optical signal (a) before and (b) after the parametric amplifier.

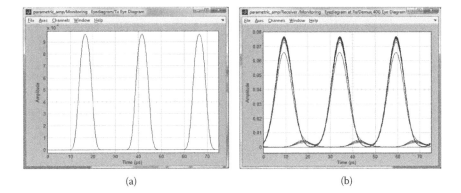

FIGURE 9.33
(a) Input data sequence. (b) Detected sequence processed using triple-correlation nonlinear photonic processing and recovery scheme bispectrum receiver.

than the data bandwidth, meaning that the amount of noise in the data band was a fraction of the total noise applied. We accept that this was an unfair comparison to the theoretical curve that is given against E_b/N_0 as defined in Mendel [37].

5. In the current model, we provide a fair comparison noise added to the modulated signal using the AWGN block in the E_b/N_0 mode (E_0 is the energy per bit and N_0 is the noise contained within the bit period) with the symbol period set to the carrier period, in effect this set the carrier-to-noise ratio (CNR). Also the triple correlation receiver was modified a little from the original—namely, the addition on the bispectrum product (BP) filter and some tweaking of the triple correlation delays this result in the BER curve shown in Figure 9.34. Also an ideal homodyne receiver model was constructed with noise added and measured in the same method as the triple correlation model. This provided a benchmark against which to compare the triple correlation receiver.

6. We can compare the simulated BER values with the theoretical limit set by

$$P_b = \frac{1}{2} erfc \sqrt{\frac{E_b}{N_0}} \tag{9.29}$$

By relating the CNR to E_b/N_0 so

$$\frac{E_b}{N_0} = CNR \frac{B_W}{f_s} \tag{9.30}$$

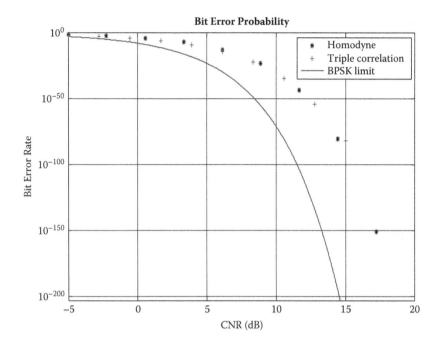

FIGURE 9.34
BER versus carrier—noise ratio for nonlinear triple correlation, ideal binary phase shift keying (BPSK) under coherent detection and ideal BPSK limit.

where channel bandwidth B_W is 1600 Hz set by the sampling rate; f_s is the symbol rate, in our case a symbol is one carrier period (100 Hz) as we are adding noise to the carrier. These frequencies are set at the normalized level so as to scale to wherever the spectral regions would be of interest. As can be seen in Figure 9.34 the triple correlation receiver matches the performance of the ideal homodyne case and closely approaches the theoretical limit of BPSK (approximately 3 dB at BER of 10^{-10}). As discussed the principal benefit from the triple correlation over the ideal homodyne case will be the characterization of the noise of the channel that is achieved by analysis of the regions of symmetry in the two-dimensional bispectrum. Finally, we still expect possible performance improvement when symbol identification is performed directly from the triple correlation matrix as opposed to the traditional method that involves recovering the pulse shape first. It is not possible at this stage to model the effect of the direct method.

As a demonstration of the triple correlation we would like to insert here the spectra of the bound solitons described in Chapter 4. Figure 3.35 shows a description of power spectrum regions (a) and its corresponding bispectrum regions (b). Figure 9.36 shows the triple correlations in

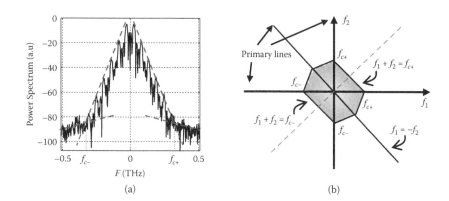

FIGURE 9.35
A description of (a) power spectrum regions and (b) bispectrum regions for explanation.

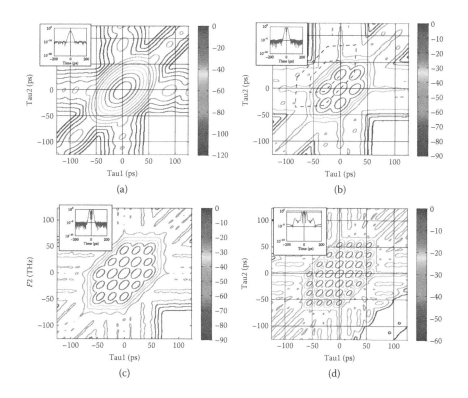

FIGURE 9.36
Triple correlations in contour plot view of various bound soliton states: (a) single soliton, (b) dual-bound solitons, (c) triple-bound solitons, and (d) quadruple-bound solitons, respectively (Insets: the temporal waveforms in logarithm scale showing the enhancement of the pedestal in higher-order multibound solitons).

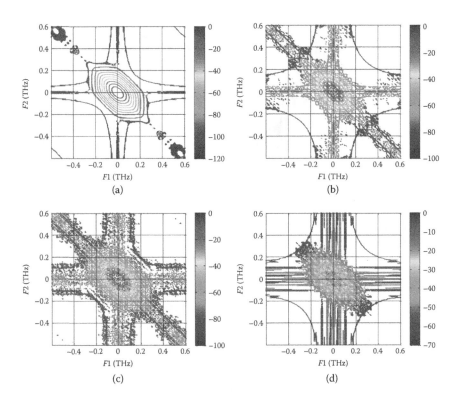

FIGURE 9.37
The magnitude bispectra in logarithm scale of different multibound soliton states: (a) single soliton, (b) dual-bound solitons, (c) triple-bound solitons, and (d) quadruple-bound solitons, respectively.

contour plot view of various bound soliton states: (a) single soliton, (b) dual-bound solitons, (c) triple-bound solitons, and (d) quadruple-bound solitons, respectively. The insets show the temporal waveforms in logarithm scale showing the enhancement of the pedestal in higher-order multibound solitons. Furthermore Figures 9.37 9a), (b), (c) and (d) show the magnitude bispectra in logarithm scale of different multibound soliton states: single soliton, dual-bound solitons, triple-bound solitons, and quadruple-bound solitons, respectively.

9.3.6.4 Remarks

This section demonstrates the employment of an NL optical waveguide and associated NL effects such as parametric amplification, four-wave mixing, and third harmonic generation for the implementation of the triple correlation, and the bispectrum creation and signal recovery techniques to reconstruct the data sequence transmitted over a long haul optically amplified fiber transmission link.

9.4 Bispectrum of Multibound Solitons

9.4.1 Bispectrum

In signal processing, the power spectrum estimation showing the distribution of power in the frequency domain is a useful and popular tool to analyze or characterize a signal or process, however the phase information between frequency components is suppressed in the power spectrum. Therefore, it is necessarily useful to exploit higher-order spectra known as multidimensional spectra instead of the power spectrum in some cases, especially in nonlinear processes or systems [1,2]. Different from the power spectrum, the Fourier transform of the autocorrelation, multidimensional spectra are known as Fourier transforms of high-order correlation functions, hence they provide us not only the magnitude information but also the phase information.

In particular, the two-dimensional spectrum also called the bispectrum is by definition the Fourier transform of the triple correlation or the third-order statistics [2]. For a signal $x(t)$ its triple-correlation function C_3 is defined as

$$C_3(\tau_1, \tau_2) = \int x(t)x(t - \tau_1)x(t - \tau_2)\,dt \tag{9.31}$$

where τ_1, τ_2 are the time-delay variables. Thus, the bispectrum can be estimated through the Fourier transform of C_3 as follows:

$$B_i(f_1, f_2) \equiv F\{C_3\} = \iint C_3(\tau_1, \tau_2)\exp(-2\pi j(f_1\tau_1 + f_2\tau_2))\,d\tau_1\,d\tau_2 \tag{9.32}$$

where $F\{\}$ is the Fourier transform, and f_1, f_2 are the frequency variables. From the definitions (9.31) and (9.32), both the triple correlation and the bispectrum are represented in a three-dimensional graph with two variables of time and frequency, respectively. Figure 9.35 shows the regions of the power spectrum and bispectrum, respectively, and their relationship. The cut-off frequencies are determined by the intersection between the noise and spectral lines of the signal. These frequencies also determine the distinct areas basically bounded by a hexagon in the bispectrum. The area inside the hexagon shows the relationship between frequency components of the signal only, otherwise the area outside shows the relationship between the signal components and noise. Due to a two-dimensional representation in the bispectrum, the variation of the signal and the interaction between signal components can be easily identified.

Because of unique features of the bispectrum, it is really useful in characterizing the non-Gaussian or nonlinear processes and is applicable in many various fields such as signal processing [37], biomedicine [37], and image reconstruction [37]. Extension of a number of representation dimensions makes the bispectrum become more easily and significantly a representation of different types of signals and differentiation of various processes, especially nonlinear processes such as doubling and chaos. Multidimensional spectra are proposed as a useful tool to analyze the behaviors of signals generated from these systems such as multibound solitons, especially transition states in the formation process of multibound solitons.

9.4.2 Various States of Bound Solitons

For multibound solitons, the optical power spectrum is modulated with the appearance of main lobes and sublobes due to the phase relationship between bound pulses as demonstrated in Chapter 3. Therefore, various multibound states generated from the FM mode-locked fiber laser can be distinguished by the optical power spectrum analysis. However, as a result the bispectrum of each multibound soliton state obviously exhibits a distinct structure for characterization. In this section, various multibound solitons in steady state obtained from simulation are used to estimate their bispectra. Moreover, some interesting information of the multibound solitons (MBS) can be obtained from the bispectrum.

First, the triple correlations of various multibound solitons are estimated as shown in Figure 9.36. The triple correlation of a pulse can be represented by the elliptical contour lines corresponding to its magnitude as shown in Figure 9.36a. Depending on the number of solitons in the bound state, the triple correlation of the multibound soliton is represented by the layers of elliptical contour lines bounded in a hexagon as shown in Figures 9.36b through 9.2d. The quality and symmetry of pulses can be reflected by the uniformity of the ellipsis in the same layer. In the triple correlation, the presence of pedestals, which is commonly characterized by the high dynamic range autocorrelation measurement [6], can be easily observed by the contour lines outside the hexagon. When the gain is enhanced in the cavity for the higher multibound state, the pulses are not only shortened but also possibly degraded by the increase in pedestal energy, which is proportional to the level of the lines outside the hexagon area.

The Fourier transform of the triple correlations results in the bispectra of multibound solitons, which consist of the magnitude and the phase representations. Figure 9.37 shows the magnitude in contour plot view of different multibound soliton states with a π phase difference that circulates stably in the FM mode-locked fiber laser cavity. In the single soliton state only, the magnitude bispectrum contour is only closed contour lines inside the area bounded by the hexagon as shown in Figure 9.37a. While the bispectra of higher-order bound solitons from dual- to quadruple-bound states exhibit a periodic structure inside this area as shown in Figures 9.37b through 9.37d. Depending on the number of pulses in a bound states, the periodic structures of both phase and magnitude spectra can vary correspondingly. When the number of pulses in a bound state increases, the periodic structure of its bispectra is more sophisticated because of the interactive relationship

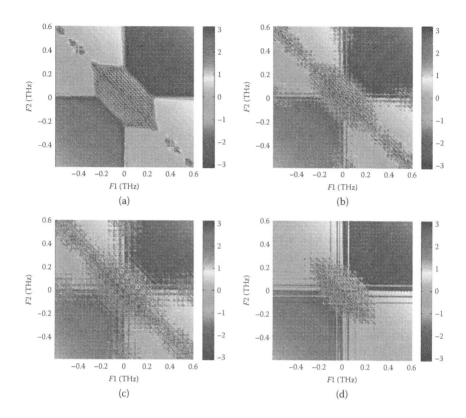

FIGURE 9.38

The phase bispectra of different multibound soliton states: (a) single soliton, (b) dual-bound solitons, (c) triple-bound solitons, and (d) quadruple-bound solitons, respectively.

between sidelobes or various frequency components. Moreover, the hexagonal area in the bispectrum that is broader at a higher-order bound state reflects the bandwidth of the signal that is broadened by the pulse compression due to the enhancement of the gain or power in the fiber cavity. Similar to the observation in the triple correlation, the stability of the bound states also exhibits obviously in the bispectrum. A lower stability of higher-order bound states is shown by the blurred contour lines inside the hexagon and the enhancement of the contour ridges outside the hexagon.

The phase bispectra provide additional information about the status and quality of the bound state. Figure 9.38 shows the phase bispectra of various soliton states from the single pulse to quadruple pulse. For the phase bispectrum, periodic phase variation of a stable state in the frequency domain is easily exhibited in two dimensions. The sharply periodic structure of the phase bispectrum inside the hexagon with two distinct regions that are complex conjugates of each other and separated by the line $f_1 = -f_2$ as shown in Figure 9.38a indicates high stability of the single soliton state. On the other hand, the periodic structure of the stable multibound state is corrupted by the change in operating conditions of the multibound state such as gain, the noise level in the fiber cavity. At a higher-order bound soliton state, the phase variation in the bispectrum is more blurred in the hexagon, however two conjugate regions can still be obviously identified.

9.4.3 Transitions in Multibound Soliton Formation

One of the important advantages of the bispectrum representation is to differentiate the linear and nonlinear responses or processes; therefore, it is especially useful in analyzing the transition processes from the noise to the steady state or from the unstable state to the stable. In transition processes of multibound solitons, the signal experiences strong fluctuations in phase and magnitude like a chaotic state before reaching a new steady state. Therefore, the average bispectrum of these processes also exhibits clearly the variation in terms of both phase and magnitude, which can be used to analyze multibound solitons at this stage. One of the most important transition processes is the formation of multibound solitons from the noise in the fiber cavity that has been numerically investigated in Chapter 3. The evolution of the signal from the noise into the multibound solitons can be split into three stages in which the first stage is the process that the noise builds up into the pulses, yet the amplitude and phase of the signal fluctuate strongly during circulating inside the cavity like a chaotic state. In order to analyze this stage, the bispectrum is averaged over the first 1000 round-trips in the evolution of the multibound soliton formation that was simulated in Chapter 3. Figure 9.39 shows the magnitude and phase bispectra of the dual-bound soliton and the triple-bound soliton evolutions, respectively. For the case of the dual-bound soliton, the bispectra in Figures 9.39a and 9.39b show nonperiodic structure in a hexagon with a small area that corresponds to the rapid variation of magnitude and phase of the signal from round-trip to round-trip. In particular, the phase structure shows a uniform distribution indicating an independence of the frequency components in the signal envelope because the signal behavior in this stage is similar to a chaotic state. While the magnitude bispectrum for the case of the triple-bound soliton shows the closed contours in a low- frequency region with the phase bispectrum having a periodic structure due to the existence of two dominant short pulses in this first stage. However, the state of these two pulses is unstable, as expressed by the ridges in the hexagon of the magnitude bispectrum and the unclarity of two distinct regions in phase structure of the bispectrum.

The second stage in the next 1000 round-trips is the process that the built-up pulses interact and bind together to form the multibound solitons. For the evolution of the dual-bound soliton, therefore the elliptical contours in magnitude bispectra and the periodic structure in phase bispectra in this stage appear obviously because of the presence of periodic components as

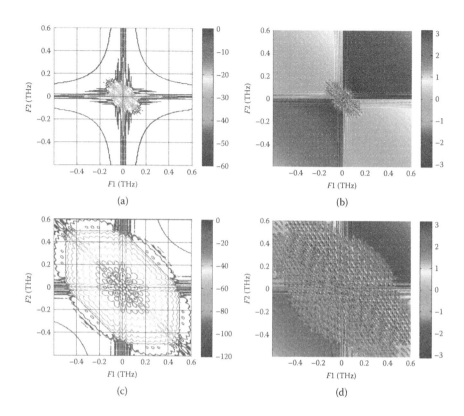

FIGURE 9.39
Magnitude and phase bispectra averaged over 1000 round-trips of evolution of the signal at the first stage in formation process from the noise of various multibound solitons: (a,b) dual-bound soliton and (c,d) triple-bound soliton.

shown in Figures 9.40a and 9.40b. The structure of closed contours in magnitude bispectra becomes finer and extended due to the appearance of three solitons in the evolution of the triple-bound soliton as shown in Figure 9.40c. However, the bound state in this stage is unstable during circulating in the cavity because the operating conditions such as gain factor are not optimized. Hence, the contamination in phase bispecta still exists, and the structure of ridges is dominant in the hexagon of the magnitude spectra. The transition process between different bound states due to the change in operating conditions behaves similar to the evolution in this stage, which is easily identified by the bispectral representation. When the fluctuations of multibound solitons occurs, the periodic lines of bispectrum at that process are commonly varied and smeared, before another periodic structure of the bispectrum is formed when a new stable state is established.

The last stage in the evolution of multibound solitons is the process in which the pulses do self-adjustments until multibound solitons reach a steady state after the gain of the cavity is optimally adjusted. Although there

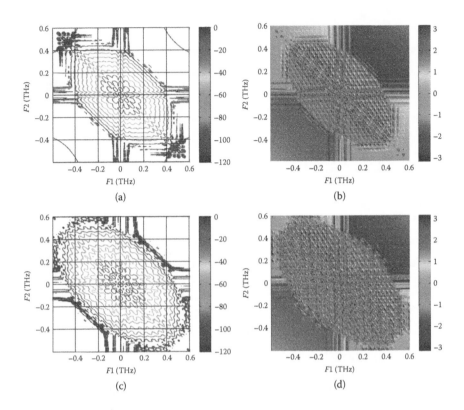

FIGURE 9.40
Magnitude and phase bispectra averaged over 1000 round-trips of evolution of the signal at the second stage in formation process of various multibound solitons: (a,b) dual-bound soliton and (c,d) triple-bound soliton.

is still the existence of small fluctuations of the phase and amplitude in this process, the pulses in bound states are well defined in a periodic evolution. Therefore, the magnitude bispectra in both the dual- and the triple-bound solitons show the periodic structure with the closed elliptical contours dominant in the hexagon as shown in Figures 9.41a and 9.41c. Figures 9.41b and 9.41d show the phase bispectra with a periodic relationship between frequency components of pulses and two clearly distinct conjugate regions separated by the diagonal $f_1 = -f_2$. On the other hand, the periodic structure in the bispectra is progressively widened until a stable final state during transition from the unstable to the stable states.

Another state of multipulse in the fiber cavity can be obtained when the gain factor is not optimized for a higher multibound soliton state. Therefore, the magnitude and the phase fluctuate strongly during circulation in the cavity. Nonuniformity and deviation of the phase difference between pulses from the value of the π result in collision of pulses in this state and formation of new pulses. Figure 9.42 shows both the magnitude and the

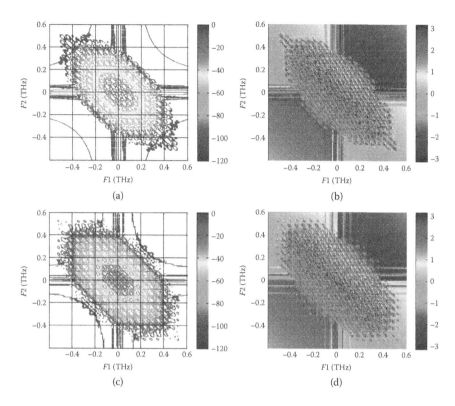

FIGURE 9.41
Magnitude and phase bispectra averaged over the last 1000 round-trips of evolution of the signal at the third stage in formation process of various multibound solitons: (a,b) dual-bound soliton and (c,d) triple-bound soliton.

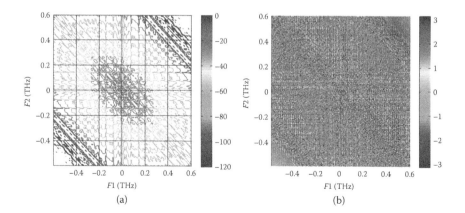

FIGURE 9.42
(a) Magnitude spectrum and (b) phase bispectra averaged over 2000 round-trips of the unstable multipulse state in the frequency modulation fiber ring cavity.

phase bispectra of this state averaged over 2000 round-trips. The presence of both the closed contours and the ridges in a jumble indicate the presence of a multipulsing operation and its dynamic state with the rapid variation in time due to the collision and the interaction of pulses in the cavity. The chaotic states as indicated by the uniform distribution of the phase bispectrum observed in Figure 9.42b.

9.5 Summary

In this chapter, a number of nonlinear wave propagation equations that describe the evolution as well as the parametric interaction of multioptical waves in nonlinear waveguides are developed for photonic signal processing. A range of signal processing applications based on the parametric conversion process of third-order nonlinearity are demonstrated by simulation, such as high-order spectrum analysis for ultrasensitive receivers, time division demultiplexing, amplification and wavelength conversion, and phase conjugation. Models of nonlinear waveguides are integrated into the MATLAB Simulink platform for processing performance simulations. The FWM is the central nonlinear physical process.

Furthermore, nonlinear effects in active mode-locked fiber lasers are described. A laser model including a gain media, a bandpass filter, optical fiber, and intensity modulator are introduced to investigate the performance of the laser under nonlinear effects. Lasers in both working conditions, nondetuning and detuning, are studied. We find that the nonlinear effects compress the pulse through the self-phase modulation (SPM) effect. However, there is a trade-off between pulse shortening and a large detuning range. A laser with a shorter pulse has a narrower detuning range and vice versa.

References

1. G. P. Agrawal, *Nonlinear Fiber Optics*, Academic Press, New York, 2004.
2. F. Luan et al., Dispersion Engineered As2S3 Planar Waveguides for Broadband Four-Wave Mixing Based Wavelength Conversion of 40 Gb/s Signals, *Opt. Express*, 17, 3514–3520, 2009.
3. M. E. Marhic et al., Broadband Fiber Optical Parametric Amplifiers, *Opt. Lett.*, 21, 573–575, 1996.
4. R. H. Stolen et al., Parametric Amplification and Frequency Conversion in Optical Fibers, *IEEE J. Selected Topics in Quantum Electron.*, 18, 1062–1072, 1982.

5. J. Hansryd et al., Fiber-Based Optical Parametric Amplifiers and Their Applications, *IEEE J. Selected Topics in Quantum Electron.*, 8(3), 506–520, 2002.
6. P. O. Hedekvist et al., Fiber Four-Wave Mixing Demultiplexing with Inherent Parametric Amplification, *J. Lightwave Technol.*, 19, 977–981, July 2001.
7. M. C. Ho et al., 200-nm-Bandwidth Fiber Optical Amplifier Combining Parametric and Raman Gain, *IEEE J. Lightwave Technol.*, 19, 977–981, July 2001.
8. R. M. Jopson et al., Compensation of Fiber Chromatic Dispersion by Spectral Inversion, *Electron. Lett.*, 29(7), 576–578,
9. S. L. Jansen et al., Long-Haul DWDM Transmission Systems Employing Optical Phase Conjugation, *IEEE J. Selected Topics Quantum Electron.*, 12, 505–520, 2006.
10. S. Watanabe et al., Simultaneous Wavelength Conversion and Optical Phase Conjugation of 200 Gb/s (5 × 40 Gb/s) WDM Signal Using a Highly Nonlinear Fiber Four-Wave Mixer, in *Proceedings of ECOC '97*, 1997, pp. 1–4.
11. S. Radic et al., Wavelength Division Multiplexed Transmission over Standard Single Mode Fiber Using Polarization Insensitive Signal Conjugation in Highly Nonlinear Optical Fiber, in *Proceedings of OFC '03*, Atlanta, Georgia, 2003, Paper PD12.
12. T. Yamamoto et al., Active Optical Pulse Compression with a Gain of 29 dB by Using Four-Wave Mixing in an Optical Fiber, *IEEE Photonics Technol. Lett.*, 9, 1595–1597, December 1997.
13. J. Hansryd et al., Wavelength Tunable 40 GHz Pulse Source Based on Fiber Optical Parametric Amplifier, *Electron. Lett.*, 37, 584–585, April 2001.
14. H. C. H. Mulvad et al., 1.28 Tbit/s Single-Polarization Serial OOK Optical Data Generation and Demultiplexing, *Electron. Lett.*, 45(5), 280–281, February 2009.
15. M. D. Pelusi et al., Applications of Highly-Nonlinear Chalcogenide Glass Devices Tailored for High-Speed All-Optical Signal Processing, *IEEE J. Selected Topics in Quantum Electron.*, 14, 529–539, 2008.
16. T. D. Vo et al., Photonic Chip Based 1.28 Tbaud Transmitter Optimization and Receiver OTDM Demultiplexing, *Proceedings of OFC 2010*, San Diego, 2010, Paper PDPC5.
17. M. Steve et al., Very Low Reactively Ion Etched Tellurium Dioxide Planar Rib Waveguides for Linear and Nonlinear Optics, *Opt. Express*, 17, 17645–17651, 2009.
18. L. N. Binh, *Optical Fiber Communications Systems: Theory and Practice with MATLAB and Simulink Models*, CRC Press, Boca Raton, FL, 2010.
19. C. L. Nikias and J. M. Mendel, Signal Processing with Higher-Order Spectra, *IEEE Sig. Proc. Mag.*, 10(3), 10–37, 1993.
20. G. Sundaramoorthy et al., Bispectral Reconstruction of Signals in Noise: Amplitude Reconstruction Issues, *IEEE Trans. Acoustics, Speech and Signal Proc.*, 38(7), 1297–1306, July 1990.
21. T. M. Liu et al., Triple-Optical Autocorrelation for Direct Optical Pulse-Shape Measurement, *App. Phys. Lett.*, 81(8), 1402–1404, 2002.
22. J. K. Yang and D. J. Kaup, Stability and Evolution of Solitary Waves in Perturbed Generalized Nonlinear Schrödinger Equations, *Siam J. Appl. Math.*, 60, 967–989, 2000.
23. J. C. Bronski, Nonlinear Scattering and Analyticity Properties of Solitons, *J. Nonlinear Sci.*, 8, 161–182, 1998.
24. G. P. Agrawal, *Applications of Nonlinear Fiber Optics*, Academic Press, Arlington, MA, 2001.

25. G. P. Agrawal, *Fiber-Optic Communication Systems*, 3rd ed., Wiley-Interscience, New York, 2002.

26. G. P. Agrawal, *Nonlinear Fiber Optics*, Springer, Berlin, 2001.

27. A. Hasegawa and F. Tappert, Transmission of Stationary Nonlinear Optical Pulses in Dispersive Dielectric Fibers. I. Anomalous Dispersion, *Appl. Phys. Lett.*, 23, 142–144, 1973.

28. C. T. Seaton, Xu Mai, G. I. Stegeman, and H. G. Winful, Nonlinear Guided Wave Applications, *Opt. Eng.*, 24, 593–599, 1985.

29. G. I. Stegeman, E. M. Wright, N. Finlayson, R. Zanoni, and C. T. Seaton, Third Order NL Integrated Optics, *IEEE J. Lightwave Technol.*, 6, 953–970, 1988.

30. G. I. Stegeman and R. H. Stolen, Waveguides and Fibers for Nonlinear Optics, *J. Opt. Soc. Am. B*, 6, 652–662, 1989.

31. K. Hayata and M. Koshiba, Full Vectorial Analysis of Nonlinear Optical Waveguides, *J. Opt. Soc. Am. B*, 5, 2494–2501, 1988.

32. D. R. Brillinger, Introduction to Polyspectra, *Ann. Math. Stat.*, 36, 1351–1374, 1965.

33. G. Sundaramoorthy, M. R. Raghuveer, and S. A. Dianat, Bispectral Reconstruction of Signals in Noise: Amplitude Reconstruction Issues, *IEEE Trans. Acoustics, Speech and Sig. Proc.*, 38(7), 1297–1306, July 1990.

34. H. Bartelt, A. W. Lohmann, and B. Wirnitzer, Phase and Amplitude Recovery from Bispectra, *Appl. Opt.*, 23, 3121–3129, 1984.

35. Chrysostomos L. Nikias et al., Signal Processing with Higher-Order Spectra, *IEEE Signal Proc. Mag.*, 10(3), July 1993.

36. B. M. Sadler et al., Acousto-optic Estimation of Correlations and Spectra Using Triple Correlations and Bispectra, *Opt. Eng.*, 31(10), 1992.

37. J. M. Mendel, Tutorial on High-Order Statistic (Spectra) in Signal Processing and System Theory: Theoretical Results and Some Applications, *Proc. IEEE*, 79(3), 1991.

10

Volterra Series Transfer Function in Optical
Transmission and Nonlinear Compensation

Le Nguyen Binh

Hua Wei Technologies, European Research Center, Munich, Germany

Liu Ling and Li Liangchuan

Hua Wei Technologies, Optical Networking Research Department, Shen Zhen, China

CONTENTS

10.1 Overview

Optical transmission over long-reach optical fibers and ultrashort pulse sequences has attracted the search of a number of efficient techniques for the modeling of such wideband signals propagating over very long distance with or without dispersion compensation, especially when multiwavelength channels are employed in the nonlinear operating region of the guide medium. The Volterra series transfer function is a technique that offers significant advantages over the most common split-step Fourier method (SSFM). This chapter thus presents the modeling of the transmission of multiplexed wavelength channels using the segmented Volterra series transfer function (S-VSTF) technique in which the divergence of the transfer series is ensured by the segmentation of the propagation length of the fiber link shorter than the divergence length of the Volterra series. The single-channel nonlinear Schrödinger wave equation is modified for the S-VSTF representation. Various nonlinear effects are incorporated in the nonlinear Schrödinger equation (NLSE) and the interchannel and cross-phase channel effects are represented in the frequency domain. The effectiveness of the S-VSTF over the SSFM is studied.

Equalization of nonlinear effects in fibers is very important in the ultrahigh bit rate optical transmission system. As an example of compensation of nonlinear effects in transmission fiber, these nonlinear impairments can be equalized in the electronic domain implemented in the digital signal processor as an algorithm specified by the inverse Volterra series or transfer function and its implementation in the electronic processors. This technique is presented in the last section of this chapter.

10.2 Introduction

A single-channel optical transmission system is capable of transmitting data of bit rate 40 Gb/s, 100 Gb/s, and Tb/s [1–6] for advanced optical communication systems [7–9] depending on the electronic parts used. However, the bandwidth of the optical fiber has far more capacity than can be utilized by one single channel. Under the dense wavelength division multiplexed (DWDM) transmission systems that may contain hundreds of channels multiplexed in a single fiber and under various forms of optical amplification, Volterra series transfer functions have attracted attention in recent years for the modeling of long-length fiber [10–18]. This kind of transfer function represents the dynamic behavior of the transmission medium in the frequency domain, thus both the phase and amplitude variations with respect to the frequency can be obtained to give insight into the transmission media. However, the convergence of such series transfer functions is not straightforward, and misleading results may be obtained if the convergence is not arrived at the end of the transmission system. Thus, a divergence criterion must be established and the length of the fiber to be presented by the VSTF must be restricted to a shorter length than this divergence limit. To the best of the author's knowledge, there has been no report of the segmentation of the fiber length into shorter sections that satisfy the limit distance of the divergence of the Volterra series. Hence, segmentation of the very long-reach fiber transmission system should be used to enhance the accuracy and avoid any numerical errors due to the series divergence. In this chapter, the theoretical model and the mathematical expression for simulating the transmission performance of multiwavelength channel systems are derived employing the Volterra series transfer functions (VSTFs) with segmentation. A multiwavelength channel NLSE is developed for both the split-step Fourier method (SSFM) and the VSTF models. Note that in the NLSE the carrier frequency cancels out in the process of derivation. This is because in single-channel optical transmission systems, there is no interchannel interference and hence the carrier frequency information is not critical. In the case of dense wavelength division multiplexed (DWDM) optical systems, the carrier frequencies are certainly important in the calculation of interchannel nonlinear interactions such as the cross-phase modulation and four-wave mixing (FWM).

In Section 10.3, the simultaneous NLSEs are used for the split step Fourier (SSF) method to account for the interaction between different channels. This is the method that is generally used for the SSF method-based simulations [19]. In Section 10.4, the total field formulation is used for the VSTF models because VSTF is a frequency domain approach in solving the NLSE. In Section 10.5, the mathematical models for both the SSF method and the VSTF method are derived for DWDM systems. Both the SSF and VSTF methods are used for simulating dual-channel wavelength division multiplexing (WDM) transmission systems. Two factors that affect the cross phase modulation (XPM) effects are also investigated. Nonlinear impairments affect the transmission performance and can be equalized by using the inverse Volterra series or transfer function and its implementation in the electronic processors—that is, in the electrical domain after the opto-electronic receiver and an analog-to-digital converter (ADC). This technique is presented in the last section of this chapter.

10.3 Nonlinear Wave Equation and Volterra Series Transfer Function Model

10.3.1 Nonlinear Wave Equation

In wavelength-division multiplexed transmission systems, more than one optical field is superimposed on each other with different wavelength propagating simultaneously through the single-mode fiber. This leads to a remarkable increase in the total optical power/intensity imposed on the fiber, and thus nonlinear effects occur. There are interactions between different optical fields leading to a number of scattering phenomena and distortion of the signal envelope [3]. Nonlinear effects may include, at new frequency, stimulated Brillouin scattering (SBS), and at high and shifted spectral regions, the stimulated Raman scattering (SRS), the harmonic generation, and four-wave mixing (FWM). The nonlinear effect responsible for coupling between two wavelength channels is the cross-phase modulation (XPM) that accompanies the self-phase modulation (SPM). Both XPM and SPM are dependent on the intensity of the optical fields.

In this section the mathematical expressions for XPM and SPM effects are derived because these two effects are the most dominant nonlinear effects in DWDM systems. Readers are referred to more detailed derivations given in Appendix A and Chapter 3. Assuming that two channels are copropagating along the optical fiber link, the total field of the two channels can be expressed by separating the carrier and the slowly varying envelope as

$$E_z(\mathbf{r},t) = \frac{1}{2}[E_1\, e^{j\omega_1 t} + E_2\, e^{j\omega_2 t}] + c.c. \tag{10.1}$$

where ω_1 and ω_2 are the carrier frequencies of the two different channels, and the corresponding amplitudes E_1 and E_2 are the slowly varying functions of time compared with the optical carriers. This quasi-monochromatic assumption is equivalent to assuming that the spectral width of each pulse satisfies the condition $\Delta\omega \ll \omega$. This is generally the case for pulse widths greater than 0.1 ps. In this work, we set the pulse width of 1.2 ps with a repetition period of 25 ps or equivalently to about 40 Gb/s to 640 Gb/s if the repetition period is reduced to about one pulse width. The pulse shape is assuming the Gaussian profile.

The XPM effects originated from the nonlinear polarization induced by the optical fields. By using (10.1), the nonlinear polarization can be written as

$$P_{NL}(\mathbf{r},t) = \frac{1}{2}\Big[P_{NL}(\omega_1)e^{j\omega_1 t} + P_{NL}(\omega_2)e^{j\omega_2 t} + P_{NL}(2\omega_1 - \omega_2)e^{j(2\omega_1 - \omega_2)t}$$
$$+ P_{NL}(2\omega_2 - \omega_1)e^{j(2\omega_2 - \omega_1)t} \Big] + c.c. \tag{10.2}$$

where the four terms of P_{NL} at different frequencies depend on E_1 and E_2 as

$$P_{NL}(\omega_1) = \chi_{eff}\left(|E_1|^2 + 2|E_2|^2\right)E_1 \tag{10.3}$$

$$P_{NL}(\omega_2) = \chi_{eff}\left(|E_2|^2 + 2|E_1|^2\right)E_2 \tag{10.4}$$

$$P_{NL}(2\omega_1 - \omega_2) = \chi_{eff} E_1^2 E_2^* \tag{10.5}$$

$$P_{NL}(2\omega_2 - \omega_1) = \chi_{eff} E_2^2 E_1^* \tag{10.6}$$

with $\chi_{eff} = (3\varepsilon_0/4)\chi_{xxxx}^{(3)}$ acting as an effective nonlinear parameter. The non-linear induced polarization generated in new frequencies at $2\omega_1 - \omega_2$ and $2\omega_2 - \omega_1$ in (10.2) can be given the term FWM. For FWM to be significant, phase-matching conditions between the two channels have to be satisfied, and generally that is not the case. The other two induced polarization terms in (10.5) and (10.6) represent the SPM and XPM nonlinear effects. They can be regarded as a nonlinear contribution to the refractive index [3]. The factor of two on the right-hand side shows that XPM is twice as effective as SPM for the same intensity.

The pulse-propagation equations for the two optical fields/carriers can be obtained by the following modified NLSE [3]:

$$\frac{\partial A_j}{\partial z} + \beta_{1j}\frac{\partial A_j}{\partial t} + \frac{j\beta_{2j}}{2}\frac{\partial^2 A_j}{\partial t^2} + \frac{\alpha_j}{2}A_j = jn_2\frac{\omega_j}{c}\left(f_{jj}\,|A_j|^2 + 2f_{jk}\,|A_k|^2\right) \tag{10.7}$$

where $k \ne j$; $b_{1j} = 1/v_{gj}$; v_{gj} is the group velocity; b_{2j} is the group velocity dispersion (GVD) coefficient; a_j is the loss coefficient; and n_2 is the nonlinear refractive index coefficient. The overlap integral f_{jk} is given as

$$f_{jk} = \frac{\int_{-\infty}^{\infty}\int_{-\infty}^{\infty}|F_j(x,y)|^2\,|F_k(x,y)|^2\,dx\,dy}{\left(\int_{-\infty}^{\infty}\int_{-\infty}^{\infty}|F_j(x,y)|^2\,dx\,dy\right)\left(\int_{-\infty}^{\infty}\int_{-\infty}^{\infty}|F_k(x,y)|^2\,dx\,dy\right)} \tag{10.8}$$

in which $F_i(x,y)$ and $F_j(x,y)$ indicate the field distribution of the guided and converted modes, respectively, or the modes of different wavelength channels, in the transverse plane (x,y). It is assumed that the phase matching of these modes is satisfied. In the case of a dual-channel DWDM transmission system, (10.7) can be obtained as a set of two coupled NLSEs as

$$\frac{\partial A_1}{\partial z} + \frac{1}{v_{g1}}\frac{\partial A_1}{\partial t} + \frac{j\beta_{21}}{2}\frac{\partial^2 A_1}{\partial t^2} + \frac{\alpha_1}{2}A_1 = j\gamma_1\left(|A_1|^2 + 2|A_2|^2\right)A_1 \tag{10.9}$$

$$\frac{\partial A_2}{\partial z} + \frac{1}{v_{g2}}\frac{\partial A_2}{\partial t} + \frac{j\beta_{22}}{2}\frac{\partial^2 A_2}{\partial t^2} + \frac{\alpha_2}{2}A_{2p} = j\gamma_2\left(|A_2|^2 + 2|A_1|^2\right)A_2 \tag{10.10}$$

where the nonlinear parameter g_j is defined as

$$\gamma_j = \frac{n_2 \omega_j}{c A_{eff}} \quad (j = 1, 2),$$

(10.11)

and A_{eff} is the effective core area [3]. Generally, the two optical pulses would be under different GVD coefficients and group velocities. The mismatching of the group velocity plays an important role as it limits the XPM interaction as pulses walk off from each other. The simultaneous set of equations (10.9) and (10.10) can be solved using the common method [5], the SSFM. With this approach, it is necessary to separate the NLSE for each channel carrying different carrier frequencies. As the number of channels increases, the number of simultaneous NLSEs increases, and hence the complexity of the model quickly grows accordingly. An obvious disadvantage of the SSFM approach is that the NLSE does not contain information about the carrier frequency; therefore, it is difficult to model other types of modulation schemes such as the carrier-suppressed RZ coding effectively.

On the other hand, the VSTF model for the DWDM transmission system takes the total field frequency approach. Instead of having separate NLSE for each channel, the VSTF model treats all channels in the fiber as one single wideband channel. Therefore, all existing channels are represented by one single wideband NLSE. Both the SSF and VSTF approaches are illustrated in Figures 10.1a and 10.1b, respectively.

SSF Approach for Multichannel Systems

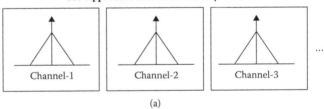

(a)

VSTF Approach for Multichannel Systems

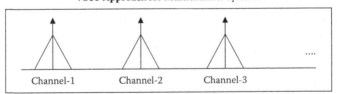

As a single wide-band channel

(b)

FIGURE 10.1
Spectrum representation (a) using split-step Fourier method approach (b) using Volterra series transfer function model approach.

The immediate advantage of the total field approach is that the mathematical model does not change when the number of channels change. Due to the wideband nature of the total field representation, it is logical to write the NLSE in the frequency domain. The total-field NLSE can be written as

$$\frac{\partial A(\omega, z)}{\partial z} = -j[\beta(\omega) - \beta_0]A(\omega, z) - \alpha(\omega)A(\omega, z)$$

$$+j\left(1 + \frac{\omega}{\omega_0}\right)\frac{\omega_0 n'}{cA_{eff}}[A(\omega, z) * A(\omega, z) * A^*(-\omega, z)] \qquad (10.12)$$

$$+j\left(1 + \frac{\omega}{\omega_0}\right)\frac{\omega_0}{cA_{eff}}G_R A(\omega, z) * [S_R(\omega)A(\omega, z) * A^*(-\omega, z)]$$

where $A(w, z)$ is the sum of all the optical fields; $\beta(\omega)$ and $\alpha(\omega)$ are the wavelength-dependent propagation constant and fiber loss; ω_0 is the central frequency of the optical fields; β_0 is the propagation constant at $\omega = \omega_0$; n' is the nonlinear refractive coefficient; A_{eff} is the effective area; G_R is the Raman gain coefficient factor; S_R is the Raman gain spectrum; and the * denotes the convolution operation.

Note that there are two distinct differences between the wideband NLSE shown in Equation (10.12) and the ordinary NLSE in Equation (10.7). These differences are (1) the slowly varying envelope approximation is not applicable under a wideband NLSE, and hence $A(\omega, z)$ contains all the information in the frequency domain including the optical carrier distributed along the propagation direction Z. (2) The frequency span of the wideband channel is large and hence the Taylor series representation of $\beta(\omega)$ around the central frequency is no longer valid.

10.3.2 Segmented Volterra Series Transfer Function Model

Conventionally the analysis of nonlinear systems often involves numerical methods. A number of analytic methods are available but only for weakly nonlinear systems. These methods include the describing function, phase plane analysis, and VSTF model. As demonstrated in Suzuki et al. [5], the convergence of the VSTF model for optical transmission systems is limited by two major factors, namely the input power level and the optical transmission distance. The convergence for the VSTF model decreases when the input power level or the transmission distance increases. These two factors are interrelated such that the longer the transmission distance is, the less convergent the VSTF model becomes for a specified input power level, and vice versa. The S-VSTF model is proposed based on the inverse-proportional relation between the input power level and the transmission distance. Theoretically, in order to improve the convergence for the VSTF model and hence improve the accuracy of its solution under the fixed input power level, the transmission distance represented by a VSTF model needs to be shortened. This can be done by dividing the total length of the fiber link into several segments and representing each segment with its own VSTF model. In doing so, the transmission distance each VSTF model represents is reduced, and hence it provides a better tolerance for the input power level for each of the VSTF segments.

10.3.2.1 Fiber Transmission Model by Volterra Series Transfer Function

The wave propagation inside a single-mode fiber can be governed by a simplified version of the NLSE [3] given in (10.7). The weakness of most of the recursive methods in solving the NLSE is that they do not provide much useful information to help the characterization of nonlinear effects. The Volterra series model provides an elegant way of describing a system's nonlinearities and enables the designers to see clearly where and how the nonlinearity affects the system performance. Although Ho and Grigoryan and Richter [8,9] have given an outline of the kernels of the transfer function using the Volterra series, it is necessary for clarity and physical representation of these functions, and brief derivations are given here on the nonlinear transfer functions of an optical fiber operating under nonlinear conditions. The Volterra series transfer function of a particular optical channel can be obtained in the frequency domain as a relationship between the input spectrum $X(\omega)$ and the output spectrum $Y(\omega)$, as

$$Y(\omega) = \sum_{n=1}^{\infty} \int_{-\infty}^{\infty} \cdots \int_{-\infty}^{\infty} H_n(\omega_1, \cdots, \omega_{n-1}, \omega - \omega_1 - \cdots - \omega_{n-1}) \times X(\omega_1) \cdots X(\omega_{n-1})$$

$$X(\omega - \omega_1 - \cdots - \omega_{n-1}) d\omega_1 \cdots d\omega_{n-1}$$

$$(10.13)$$

where $H_n(\omega_1, \cdots, \omega_n)$ is the nth-order frequency domain Volterra kernel including all signal frequencies of orders 1 to n. The proposed solution of the NLSE can be written with respect to the VSTF model of up to the fifth order as

$$A(\omega, z) = H_1(\omega, z)A(\omega) + \int_{-\infty}^{\infty} \int_{-\infty}^{\infty} H_3(\omega_1, \omega_2, \omega - \omega_1 + \omega_2, z)$$

$$\times A(\omega_1)A^*(\omega_2)A(\omega - \omega_1 + \omega_2)d\omega_1 d\omega_2$$

$$+ \int_{-\infty}^{\infty} \int_{-\infty}^{\infty} \int_{-\infty}^{\infty} \int_{-\infty}^{\infty} H_5(\omega_1, \omega_2, \omega_3, \omega_4, \omega - \omega_1 + \omega_2 - \omega_3 + \omega_4, z)$$

$$\times A(\omega_1)A^*(\omega_2)A(\omega_3)A^*(\omega_4) \times A(\omega - \omega_1 + \omega_2 - \omega_3 + \omega_4)d\omega_1 d\omega_2 d\omega_3 d\omega_4$$

$$(10.14)$$

where $A(\omega) = A(\omega, 0)$ —that is, the amplitude envelope of the optical pulses at the input of the fiber. Taking the Fourier transform of (10.7) and assuming $A(t, z)$ is of sinusoidal form we have

$$\frac{\partial A(\omega, z)}{\partial z} = G_1(\omega)A(\omega, z) + \int_{-\infty}^{\infty}\int_{-\infty}^{\infty} G_3(\omega_1, \omega_2, \omega - \omega_1 + \omega_2)A(\omega_1, z)A^*(\omega_2, z)$$

$$\times A(\omega - \omega_1 + \omega_2, z)d\omega_1 d\omega_2$$

$$(10.15)$$

where $G_1(\omega) = -\frac{\alpha_0}{2} + j\beta_1\omega + j\frac{\beta_2}{2}\omega^2 - j\frac{\beta_3}{6}\omega_3$ and $G_3(\omega_1, \omega_2, \omega_3) = j\gamma$. ω assume the values over the signal bandwidth and beyond in overlapping the signal spectrum of other optically modulated carriers while $\omega_1 \ldots \omega_3$ are all also taking values over a similar range as that of ω. For general expression, the limit of integration is indicated over the entire range to infinity. The first-order transfer function (10.7) is then given by (see details given in Grigoryan and Richter [9])

$$H_1(\omega, z) = e^{G_1(\omega)z} = e^{\left(-\frac{\alpha_0}{2} + j\beta_1\omega + j\frac{\beta_2}{2}\omega^2 - j\frac{\beta_3}{6}\omega^3\right)z}$$

$$(10.16)$$

This is in fact the linear transfer function of an optical fiber with the dispersion factors β_2 and β_2. Similarly for the third-order terms we have

$$\frac{\partial}{\partial z}\int_{-\infty}^{\infty}\int_{-\infty}^{\infty} H_3(\omega_1, \omega_2, \omega - \omega_1 + \omega_2, z) \times A(\omega_1)A^*(\omega_2)A(\omega - \omega_1 + \omega_2)d\omega_1 d\omega_2$$

$$= \int_{-\infty}^{\infty}\int_{-\infty}^{\infty} G_3(\omega_1, \omega_2, \omega - \omega_1 + \omega_2)H_1(\omega_1, z)A(\omega_1)H_2^*(\omega_2, z)$$

$$\times A(\omega_2)H_1(\omega - \omega_1 + \omega_2)A(\omega - \omega_1 + \omega_2)d\omega_1 d\omega_2$$

$$(10.17)$$

The third kernel transfer function can be obtained as

$$H_3(\omega_1, \omega_2, \omega_3, z) = G_3(\omega_1, \omega_2, \omega_3) \times \frac{e^{(G_1(\omega_1) + G_1^*(\omega_2) + G_1(\omega_3))z} - e^{G_1(\omega_1 - \omega_2 + \omega_3)z}}{G_1(\omega_1) + G_1^*(\omega_2) + G_1(\omega_3) - G_1(\omega_1 - \omega_2 + \omega_3)}$$

$$(10.18)$$

The lengthy fifth-order kernel, $H_5(\omega_1, \omega_2, \omega_3, \omega_4, \omega_5, z)$, can be obtained as listed in Grigoryan and Richter [9]. Other higher-order terms can be derived with ease if higher accuracy is required. However, in practice such a higher order would not exceed the fifth rank. We can understand that for a length of a uniform optical fiber the first- to nth-order frequency spectrum transfer can be evaluated indicating the linear to nonlinear effects of the optical signals transmitting through it. Indeed the third- and fifth-order kernel transfer functions based on the Volterra series indicate the optical field amplitude of the frequency components that contribute to the distortion of the propagated pulses. An inverse of these higher-order functions would give the signal distortion in the time domain. Thus, the VSTFs allow us to conduct distortion analysis of optical pulses and hence an evaluation of the bit-error rate (BER) of optical fiber communications systems. The superiority of such Volterra transfer function expressions allows us to evaluate each effect individually, especially the nonlinear effects so that we can design and manage the optical communications systems under linear or nonlinear operations. Currently this linear–nonlinear boundary of operations is critical for system implementation, especially for optical systems operating at 40 Gb/s where linear operation and carrier-suppressed return-to-zero format is employed. As a norm in the series expansion its convergence is necessary to reach the final solution. This convergence limits the validity of the Volterra model under the nonlinear effects in the optical transmission system.

10.3.2.2 Convergence Property of Volterra Series Transfer Function

As we can see from the previous section, the Volterra series transfer function takes the form of a power series, whose convergence can be examined with a number of well-established tests. The ratio test is chosen in this chapter to test the convergence of the VSTF as it would lead to the best estimation of the convergence of the series. Grigoryan and Richter [9] obtained the upper bounds for the inputs to each kernel of different orders can be expressed in terms of the integration of lower-order kernels. This expression can be simplified to, under the case of $n = 1$ or the convergence of the third-order kernel with respect to the linear kernel as

$$\frac{\left| \int_{-\infty}^{\infty} \int_{-\infty}^{\infty} H_3(\omega_1, \omega_2, \omega - \omega_1 - \omega_2) d\omega_1\, d\omega_2 \right|}{|H_1(\omega)|} \times |U_{max}|^2 < 1 \qquad (10.19)$$

$$\rightarrow \qquad |H_1(\omega)| > \left| \int_{-\infty}^{\infty} \int_{-\infty}^{\infty} H_3(\omega_1, \omega_2, \omega - \omega_1 - \omega_2) d\omega_1\, d\omega_2 \right| \times |U_{max}|^2 \qquad (10.20)$$

Equation (10.19) indicates the relationship between the linear transfer function or the effective bandwidth of an optical fiber operating in the linear dispersion region and the dispersion effect due to the self-phase modulation due to the intense optical pulse power contributing via the third-order kernel. In general, the radius of convergence as a function of the total input optical field amplitude can therefore be expressed as

$$
U_{max} = \left(\frac{1}{\delta_n}\right)^{\frac{1}{n-1}} = \inf \left\{ \left(\frac{|H_1(\omega)|}{\left| \int_{-\infty}^{\infty} \cdots \int_{-\infty}^{\infty} H_n(\omega_1, \ldots, \omega - \omega_1 - \ldots - \omega_{n-1}) d\omega_1 \ldots d\omega_{n-1} \right|} \right)^{\frac{1}{n-1}} \right\}
$$

(10.21)

Accordingly, the *peak power* of the input pulse so that the VSTF is convergent and hence computable is given by

$$
P_{peak} = U_{max}^2 \left(\frac{1}{\delta_n}\right)^{\frac{2}{n-1}} = \inf \left\{ \left(\frac{|H_1(\omega_1)|}{\left| \int_{-\infty}^{\infty} \cdots \int_{-\infty}^{\infty} H_n(\omega_1, \ldots, \omega - \omega_1 - \ldots - \omega_{n-1}) d\omega_1 \ldots d\omega_{n-1} \right|} \right)^{\frac{2}{n-1}} \right\}
$$

(10.22)

This is the most important result and can be used in the determination of the linear and nonlinear operation of optical fiber communications under high and intense optical input pulses, especially when numerous optical channels are propagating in a single fiber. Furthermore, the radius of convergence of the higher order of the VSTF indicates the manageable level of the input optical pulse power so that the linear dispersion effect can be compensated for by the nonlinear effects in the fiber. Otherwise the series would be divergent and the radiation or complete depletion of the optical pulses would be due to dispersion. The radius of convergence for a launched input power level into the input port of a fiber can be increased by reducing the length of a fiber segment. This is estimated using (10.22) and is depicted as a variation of the propagation length in Grigoryan and Richter [9] and in Figures 10.2 and 10.3. The convergence distance is in the order of 50 km for standard single-mode fiber with a launched input power up to 25 dBm.

10.3.2.3 Segmentation

As the radius of convergence (ROC) at a specific input power level would increase, better convergence can be achieved by segmentation of the propagation of the NLSE represented by the VSTF model. The drawback of dividing the fiber link into several segments is the increase in the computational cost, but it is affordable even for a standard desktop personal computer. The

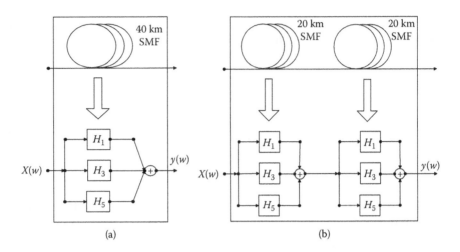

FIGURE 10.2
The Volterra series transfer function (VSTF) model for optic fiber used in communication systems: (a) single transfer function and (b) fiber section with two VSTF segments.

S-VSTF approach can therefore be regarded as a trade-off between the computational time and the accuracy of the solution obtained. A fiber segment can be represented by a VSTF model as depicted in Figure 10.2a and two segmented sections in Figure 10.2b.

For the implementation of the S-VSTF model, the fiber link is divided into two or more segments [5]. With the reduced transmission distance per each segment, the convergence of the S-VSTF model can be improved as compared to the ordinary VSTF model. The 40 km fiber link as shown in Figure 10.2a is segmented into two halves, each represented by a VSTF model as illustrated in Figure 10.2b. This S-VSTF model is equivalent to representing the entire fiber link as a nonlinear system consisting of two concatenated nonlinear components. The advantage of the S-VSTF approach is that each S-VSTF segment in the model is operating in a region that is farther away from the limit specified by the radius of convergence curve and hence is more convergent compared to the conventional VSTF. Accordingly the solution obtained would be more accurate [9].

Referring to the nonlinear algebraic rule for nonlinear systems as discussed in Atherton [20], the S-VSTF model can be mathematically represented as follows:

$$[\mathbf{H}] = [\mathbf{H}_1 + \mathbf{H}_3 + \mathbf{H}_5] \cdot [\mathbf{H}_1 + \mathbf{H}_3 + \mathbf{H}_5]$$
$$= \mathbf{H}_1 \cdot \mathbf{H}_1 + 2\mathbf{H}_1 \cdot \mathbf{H}_3 + 2\mathbf{H}_1 \cdot \mathbf{H}_5 + H.O.T. \tag{10.23}$$

where $\mathbf{H}$ is the overall transfer function; $\mathbf{H}_1$ is the linear transfer function; $\mathbf{H}_3$ and $\mathbf{H}_5$ are the third-order and fifth-order nonlinear transfer functions; and $H.O.T.$ is the higher-order nonlinear terms that are negligible under the weakly nonlinear approximation.

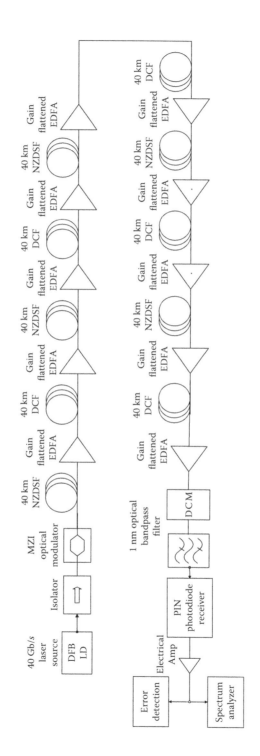

FIGURE 10.3
The 400 km dual-channel optic transmission system.

By expanding Equation (10.23), the operator in the frequency domain can be obtained as

$$y(\omega) = H_1(\omega)H_1(\omega)X(\omega)$$

$$+2\int_{-\infty}^{\infty}\int_{-\infty}^{\infty} H_1(\omega)H_3(\omega_1,\omega_2,\omega-\omega_1-\omega_2)$$

$$\times X(\omega_1)X(\omega_2)X(\omega-\omega_1-\omega_2)$$

$$+2\int_{-\infty}^{\infty}\int_{-\infty}^{\infty}\int_{-\infty}^{\infty}\int_{-\infty}^{\infty} H_1(\omega)H_5(\omega_1,\omega_2,\omega_3,\omega_4,\omega-\omega_1-\omega_2-\omega_3-\omega_4)$$

$$\times X(\omega_1)X(\omega_2)X(\omega_3)X(\omega_4)X(\omega-\omega_1-\omega_2-\omega_3-\omega_4)$$

$$+H.O.T. \tag{10.24}$$

This equation shows that the additional computational complex is relatively small for the S-VSTF model.

The S-VSTF approach can also be regarded as a trade-off between the additional computation time cost and the accuracy of the obtained solutions of the model. The optimal number of segments can be determined by trial and error or some other pseudo-analytic procedures that are subject to further research. Note that as the number of segments increases further and the smaller fiber length per segment, it is theoretically possible that the contribution from the fifth-order kernel H_5 of each segment becomes negligible and can be treated as part of the H.O.T. This can potentially reduce the computational time required because the calculation of the fifth-order nonlinear effect is rather time consuming. One such model is by an eight-segment S-VSTF with each segment of 5 km length.

Mathematically, the fiber link can be written in the following operator form:

$$[\mathbf{H}_1 + \mathbf{H}_3 + \mathbf{H}_5]^8 = [\mathbf{H}_1 \cdot \mathbf{H}_1 + 2\mathbf{H}_1 \cdot \mathbf{H}_3 + 2\mathbf{H}_1 \cdot \mathbf{H}_5 + H.O.T.]^4$$

$$= \mathbf{H}_1 \cdot \mathbf{H}_1 \cdot \mathbf{H}_1 \cdot \mathbf{H}_1 \cdot \mathbf{H}_1 \cdot \mathbf{H}_1 \cdot \mathbf{H}_1 \cdot \mathbf{H}_1$$

$$+8\mathbf{H}_1 \cdot \mathbf{H}_1 \cdot \mathbf{H}_1 \cdot \mathbf{H}_1 \cdot \mathbf{H}_1 \cdot \mathbf{H}_1 \cdot \mathbf{H}_1 \cdot \mathbf{H}_3 \tag{10.25}$$

$$+8\mathbf{H}_1 \cdot \mathbf{H}_1 \cdot \mathbf{H}_1 \cdot \mathbf{H}_1 \cdot \mathbf{H}_1 \cdot \mathbf{H}_1 \cdot \mathbf{H}_1 \cdot \mathbf{H}_5 + H.O.T.$$

In Equation (10.25) the fifth-order kernel operator $\mathbf{H}_5$ is still retained for the cases where the fifth-order nonlinear contribution is nontrivial. If the fifth-order nonlinear effects can be neglected, then (10.25) can be further simplified to

$$[\mathbf{H}_1 + \mathbf{H}_3 + H.O.T.]^8 = [\mathbf{H}_1 \cdot \mathbf{H}_1 + 2\mathbf{H}_1 \cdot \mathbf{H}_3 + H.O.T.]^4$$

$$= \mathbf{H}_1 \cdot \mathbf{H}_1 \cdot \mathbf{H}_1 \cdot \mathbf{H}_1 \cdot \mathbf{H}_1 \cdot \mathbf{H}_1 \cdot \mathbf{H}_1 \cdot \mathbf{H}_1$$

$$+8\mathbf{H}_1 \cdot \mathbf{H}_1 \cdot \mathbf{H}_1 \cdot \mathbf{H}_1 \cdot \mathbf{H}_1 \cdot \mathbf{H}_1 \cdot \mathbf{H}_1 \cdot \mathbf{H}_3 + H.O.T.$$

$$\tag{10.26}$$

Equation 10.26 can be transformed to the frequency domain leading to

$$Y(\omega) = H_1(\omega)H_1(\omega)H_1(\omega)H_1(\omega)H_1(\omega)H_1(\omega)H_1(\omega)H_1(\omega)X(\omega)$$

$$+8 \int_{-\infty}^{\infty} \int_{-\infty}^{\infty} H_1(\omega)H_1(\omega)H_1(\omega)H_1(\omega)H_1(\omega)H_1(\omega)H_1(\omega)H_1(\omega) \tag{10.27}$$

$$\times H_3(\omega_1, \omega_2, \omega - \omega_1 - \omega_2)X(\omega_1)X(\omega_2)X(\omega - \omega_1 - \omega_2)d\omega_1 d\omega_2$$

The computational time required could be lower for S-VSTF models with third-order transfer function than that for the fifth-order nonlinear contribution from $H_5(\omega_1, \omega_2, \omega_3, \omega_4, \omega_5)$ kernel. However, this is only valid when the length for the fiber segment is sufficiently short. Three sets of simulations are conducted, using the conventional VSTF model, two-segment S-VSTF and five-segment S-VSTF models. The result obtained from the SSF method is also used as the standard benchmarking of the accuracy of these three VSTF models. The result obtained from each model is then compared to that obtained by the SSFM. The following criterion for calculating normalized squared deviation (NSD) is used to quantify the discrepancy between various VSTF models and SSF methods.

$$NSD(z) = \frac{\int_{-\infty}^{\infty} |A_{SSF}(t,z) - A_{VSTF}(t,z)|^2 dt}{\int_{-\infty}^{\infty} |A_{SSF}(t,z)|^2 dt} \tag{10.28}$$

where A_{SSF} and A_{VSTF} are the envelopes of the optical signals obtained by the SSFM and VSTF methods.

10.4 Transmission of Multiplexed Wavelength Channels

10.4.1 Dual Channel

The interchannel cross-talk induced by XPM, which is affected by channel spacing and the bit pattern of signal pulses, also plays an important part in the design of the DWDM systems. In order to demonstrate the applicability of the VSTF/S-VSTF model for the analysis of DWDM systems, a simplified dual-channel optical transmission system is implemented for simulation. Both the SSF method and the VSTF model are used for the simulation and their results compared.

In Section 10.3.1, the design setup of the dual-channel transmission system of 400 km transmission distance is discussed in detail. In Section 10.4.1, two sets of simulations are conducted to test the effect on cross-talk interference with different bit patterns. In Section 10.3.1, the effect of channel spacing is analyzed with the cross-phase modulation effects.

10.4.1.1 Transmission Systems

The design of the 400 km dual-channel transmission system is similar to the 400 km single-channel system implemented in this Section 10.4, with some modification. The dispersion compensation fiber (DCF) and nonzero dispersion shifted fiber (NZDSF) are interleaved to provide the needed dispersion management. Further, dispersion compensation is done at the end of the link with respect to each channel. The entire link consists of 5 sections of NZDSF, 5 sections of DCF, and 10 erbium-doped fiber amplifiers (EDFAs). Two channels of data-stream are multiplexed before transmitting and are demultiplexed at the end of the link. The schematic diagram of the optically amplified fiber transmission system setup is shown in Figure 10.3. The dispersion and attenuation spectrum of the NZDSF and DCF used in the link are listed in Table 10.1. The interleaving of NZDSF and DCF along the transmission link creates a zigzag pattern for the dispersion parameter (GVD). The variation of the dispersion and attenuation along the link can be easily estimated.

As expected the total dispersion experienced by Channel 1 and Channel 2 are quite different at the end of the transmission link. Therefore, it is necessary to have further dispersion compensation adjustment at the end of the link with respect to the amount of dispersion experienced by the channel. On the other hand, the attenuations for the two channels are almost identical. In the next section, the mathematical models developed in Section 10.2 are used for simulations. The XPM effect, a major factor contributing to the interchannel cross-talk, is investigated by simulation.

TABLE 10.1

Fiber Parameters for Nonzero Dispersion Shifted Fiber (NZDSF) and Dispersion Compensating Fiber (DCF)

Fiber Type	NZDSF		DCF	
Operating Wavelength	1550 nm	1560 nm	1550 nm	1560 nm
Dispersion	3.0 ps/km-nm	3.4167 ps/km-nm	−9.0 ps/km-nm	−2.6225 ps/km-nm
Attenuation	0.24565 dB/km	0.24069 dB/km	0.28442 dB/km	0.26016 dB/km
Core index	1.444		1.444	
Index difference	0.56604%		0.9217%	
Core radius	2.3797 mm		1.8665 mm	
Core composition	Pure silica		Pure silica	

10.4.1.2 Cross-Phase Modulation Effect and Bit Pattern Dependence

The XPM effect is the major contributor to the interchannel interference. Both the SPM and XPM are closely related to the intensity of the pulses transmitted. The bit pattern and channel spacing (in frequency) also play an important role in the XPM effect. In this section, the dual-channel transmission system proposed in Section 10.3.1 is used to simulate the pulse propagation of a dual-channel system. Different bit patterns are assigned to Channel 1 and Channel 2 in the first simulation, and then the bit pattern is changed for Channel 2 in the second simulation to facilitate comparison between the XPM caused by different bit patterns.

Case 1—Overlapped Pulse Sequence
In this case study, the following bit patterns are assigned for Channel 1 and Channel 2, respectively: (a1 and b1) Bit Pattern for Channel 1: [1 0 1 1]; and (a2 and b2) Bit Pattern for Channel 2: [1 1 0 0]. The evolution of the pulse sequences of Channel 1 and Channel 2 is shown in Figures 10.4a1 and 10.4b1 and Figures 10.4a2 and 10.4b2, respectively.

The discrepancy between the results obtained from the SSF method and the VSTF method is calculated using the NSD criterion given in (10.28). The NSD plots for both Channel 1 and Channel 2 are shown in Figure 10.5.

It is shown in Figure 10.5 that the VSTF method is adequate in representing the dual-channel optical transmission system. To examine the distortion caused by the XPM effect, the signal spectra of both Channel 1 and Channel 2 are plotted. The pulse spectra before and after transmission are plotted in Figures 10.6a1 and 10.6a2 and Figures 10.6b1 and 10.6b2, respectively.

Comparing the evolution of the pulse spectra of Figure 10.8, the distortion in Channel 2 is quite visible near the center of the carrier frequency. This is later to be compared to the spectrum obtained with a different bit pattern in Case 2.

Case 2—Nonoverlapped Pulse Sequence
In this simulation, the following bit patterns are assigned for Channel 1 and Channel 2, respectively: (a) Bit Pattern for Channel 1: [1 0 1 1]; and (b) Bit Pattern for Channel 2: [0 1 0 0]. There is no overlapping of the one-bit, and this should effectively reduce the XPM when both channels are in synchronization with each other. Due to the walk-off effect, the original frames of the two channels shift out of each other after the walk-off distance while other frames shift in along the transmission link.

The discrepancy between the results obtained from the SSF method and the VSTF model is calculated using the NSD criterion given in Equation (10.28). The NSD plots for both Channel 1 and Channel 2 are shown in Figures 10.7a and 10.7b. The NSD seems improved slightly as compared to those obtained in Case 1. This is due to the lower XPM effect. To compare the XPM distortion,

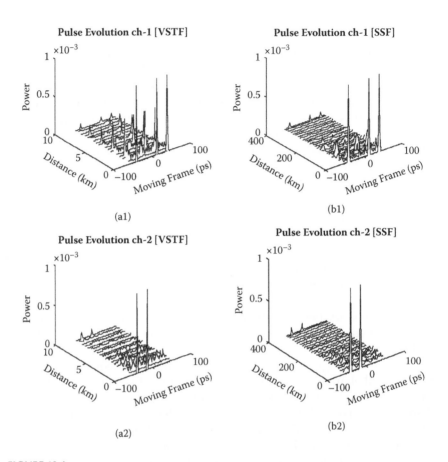

FIGURE 10.4
Pulse evolution along transmission line of Channel 1 (a1,b1); Channel 2 (a2,b2) using (a1,a2) split-step Fourier method and (b1,b2) Volterra series transfer function model.

the pulse spectrums before transmission and at the end of the link for Channel 1 and Channel 2, similar to Figure 10.6, are also obtained and analyzed for distortion. We can observe that the distortions experienced by both Channel 1 and Channel 2 are less. Thus, the bit pattern affects the contribution of XPM to the distortion of WDM transmission under the VSTF method.

10.4.1.3 Cross-Phase Modulation and Channel Spacing

It was demonstrated that the bit patterns of the transmitting channels could affect the XPM effect. In this section, the relationship between the XPM effect and channel spacing is investigated. Instead of using 1550 nm and spaced by

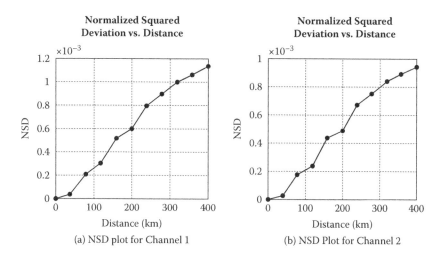

(a) NSD plot for Channel 1 (b) NSD Plot for Channel 2

FIGURE 10.5
Normalized squared deviation between split-step Fourier and Volterra series transfer function methods for both channels: (a) Channel 1, (b) Channel 2.

100 GHz in frequency as the carrier frequencies, this spacing is halved in this section with the new carrier frequencies at 1550 nm and spaced by 50 GHz. The same NZDSF and DCF fibers of Table 10.1 are used for the simulation setup. A new variation of the dispersion and attenuation along the link is expected. It can be seen that the difference in dispersion increases due to the larger channel spacing. The obvious effect of the increased dispersion difference is the larger group velocity difference and hence shorter walk-off distance. It is noted that the dispersion factor and dispersion slope of the fibers employed in the transmission links are included in the NLSE.

In this case study, the bit pattern used in Case 1 in Section 10.4.1.2 is used to facilitate the comparison with systems having different channel spacing. The following bit patterns are assigned for Channel 1 and Channel 2, respectively: (a) Bit Pattern for Channel 1: [1 0 1 1]; and (b) Bit Pattern for Channel 2: [1 1 0 0]. The pulse evolution for Channel 1 and Channel 2 are similarly obtained as given above. The discrepancy between the results obtained from the SSF and the VSTF methods is estimated using the NSD criterion. The NSD plots for both Channel 1 and Channel 2 are shown in Figure 10.8. The NSD obtained is similar in terms of magnitude as those obtained in Case 1. To compare the XPM distortion between the two simulations, the pulse spectrums before transmission and at the end of the link are, similar to Figure 10.6, also obtained and analyzed for distortion by both methods. The distortion experienced by Channel 2 is much less. Thus, it can be concluded that as the channel spacing increases, the contribution of XPM to the distortion of WDM transmission systems is less severe.

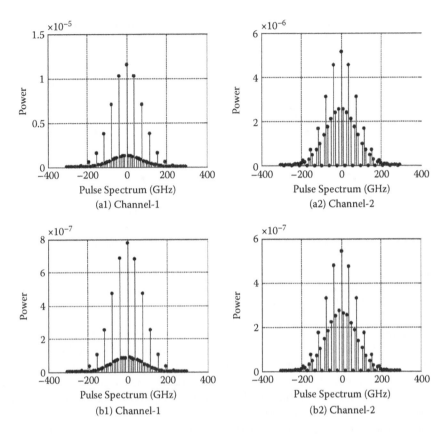

FIGURE 10.6
Normalized squared deviation power spectrum at the input launched end: (a1) Channel 1 and (a2) Channel 2, and at the output of pulse sequence under split-step Fourier and Volterra series transfer function methods for (b1) Channel 1 and (b2) Channel 2.

10.4.2 Triple Channel

The dual-channel optical transmission systems are simulated to investigate the effect of XPM and its relation with the bit pattern and channel spacing. In this section, the number of channels is three so that the effect of the overlapping of the bit pattern of either equal or uneven channel spacing between the channels can be examined. In this section the carrier frequencies of the three optical channels are selected to be equally spaced and nonequally spaced. The selected central wavelengths of the three channels are 1550 nm, 1555 nm, and 1560 nm. Each channel is 5 nm apart from the next channel. Closer spacing can be employed without minimum changes in the observed results. Then the carrier frequencies of the lowest and highest carriers are changed to 1549 nm and 1559 nm, thus channel spacing between Channel 1 and Channel 2, and 4 nm channel spacing between

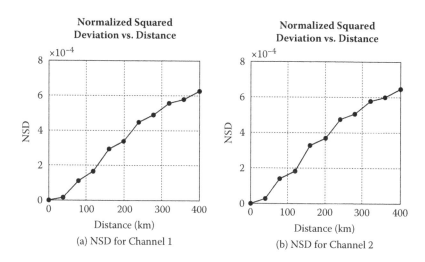

FIGURE 10.7
Normalized squared deviation between split-step Fourier and Volterra series transfer function methods for Channels 1 and 2: (a) and (b), respectively.

Channel 2 and Channel 3 is about 6 nm. The spacing is important for the case where the orthogonal frequency division multiplexed modulation schemes [21] are used in which the interactions between the subcarriers of different wavelength channels occur and can seriously affect the transmission quality.

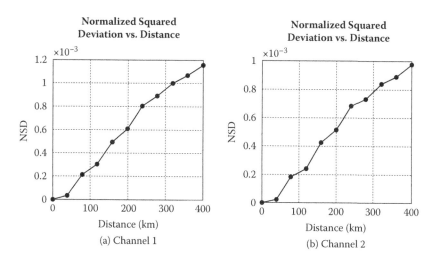

FIGURE 10.8
Normalized squared deviation between split-step Fourier and Volterra series transfer function methods for both channels: (a) Channel 1, (b) Channel 2.

10.4.2.1 Equal Channel Spacing

The transmission setup for this simulation is similar to the systems used in Section 10.3.2. The schematic diagram is shown in Figure 10.9. Optical amplifiers, such as EDFAs are incorporated either at 40 km or 80 km span length. Raman optical amplification can also be included by the Raman scattering terms in the NLSE. The carriers of the three channels are located at frequencies 1550 nm (central frequency), and channel spaced by 50 GHz on the left and right sides of this wavelength. A noise figure of 3 dB is used for all optical amplification blocks. The bit patterns used for Channel 1, Channel 2, and Channel 3 are as follows: (a) Bit Pattern for Channel 1: [1 0 1 1]; (b) Bit Pattern for Channel 2: [1 1 0 0]; and (c) Bit Pattern for Channel 3: [0 1 0 1]. The propagation evolution of the bit patterns and NSD are plotted using similar methods given in Section 10.4.1.2. It is noted also that a total number of multiplexed channels is 40, which are located in the C-band catered at 1550 nm and spaced by 50 GHz between adjacent channels.

The evolution of the pulse sequences of the channels over 400 km transmission link of the three wavelength channels can be obtained via the S-VSTF model and the SSF model. The discrepancy between the results obtained from the SSF method and the VSTF model is calculated using the NSD criterion given by (10.28). The NSD plots for the three channels are shown in Figure 10.10.

In order to quantify the nonlinear distortion experienced by each channel, the following normalized distortion criterion is proposed:

$$ND_{ch} = \sum_f \left[\frac{|A_{ch}(f,0)|}{\sum_f |A_{ch}(f,0)|} - \frac{|A_{ch}(f,z)|}{\sum_f |A_{ch}(f,z)|} \right]^2 \tag{10.29}$$

where *ch* denotes the channel of interest; $A(f, z)$ is the spectrum of the signal; f is the frequency variable; and z is the position.

Because the spectra are of a discrete form due to the periodic nature of the pulses, the summation is used in (10.29). The spectra at the beginning and the end of the fiber are normalized with respect to the total power, because only the shapes of the spectra are of interest for the nonlinear distortion analysis. Referring to Figure 10.11, we can observe that Channel 1 experiences less nonlinear distortion as compared to Channel 2 and Channel 3. This is due to the fact that the average power for Channel 1 is 50% higher as compared to the other two channels due to the assigned bit pattern. This shows that the XPM effects caused by other channels are more significant to the SPM effects caused by the channel itself.

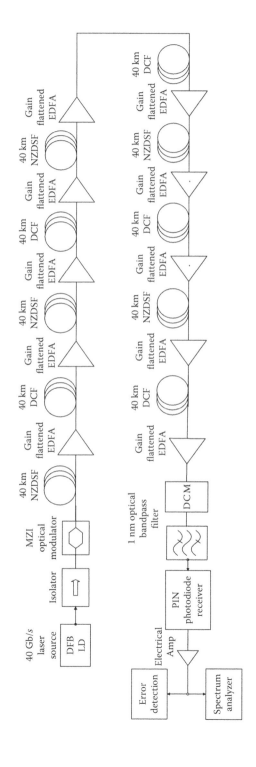

FIGURE 10.9
The 400 km multichannel optic transmission system.

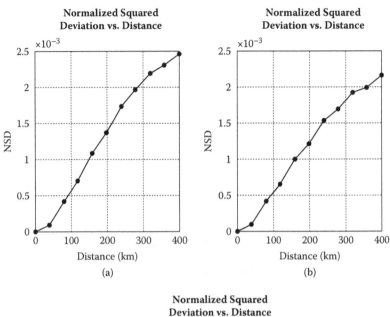

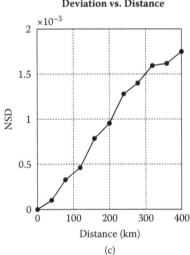

FIGURE 10.10
Normalized squared deviation plots of the three channels for the 400 km optic transmission system: (a) Channel 1, (b) Channel 2, and (c) Channel 3.

10.4.2.2 Unequal Channel Spacing

In Section 10.4.2.1, the optical transmission systems with three channels separated with equal channel spacing are simulated. In this section, unequal channel spacing is used to investigate the relation between XPM effects and channel

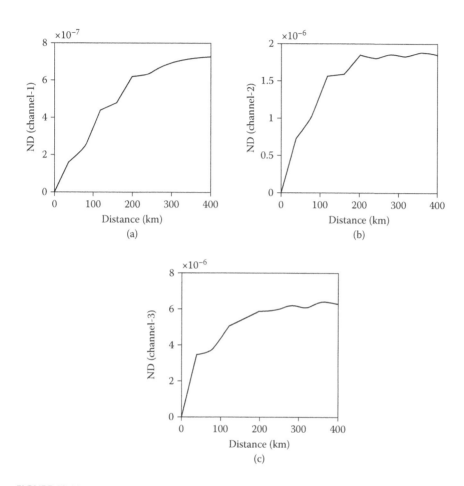

FIGURE 10.11
Nonlinear distortion of the three-channel transmission system: (a) Channel 1, (b) Channel 2, and (c) Channel 3.

spacing. The carriers of the three channels are located at wavelength 1555 nm (center channel) and spaced by 50 GHz on the left and 75 GHz the right frequency of the central wavelength. The GVD and attenuation of each channel can be evaluated from Table 10.1. The same bit pattern as used in Section 10.4.1.2 is employed in this simulation. The pulse evolution and NSD plots are depicted in Figure 10.12. The transmission was conducted over a 400 km link for three-multiplexed channels. Both VSTF and SSMF methods are used and compared. The evolution of the bit pattern over the propagation distance for the three channels under the VSTF model and SSFM can be obtained as illustrated in Figure 10.4. The discrepancy between the results obtained from the SSF method and the VSTF model is calculated using the NSD criterion given in (10.28). The NSD plots for the three channels are shown in Figure 10.12. To compare the nonlinear distortion caused by the XPM effects, the nonlinear

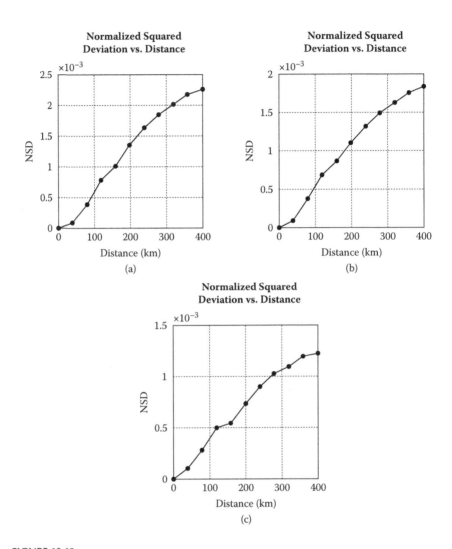

FIGURE 10.12
Normalized squared deviation evolution over the 400 km optic transmission system for three channels: (a) Channel 1, (b) Channel 2, and (c) Channel 3.

distortion computed using (10.28) and (10.29) is illustrated in Figure 10.13 for the three channels. Comparing the nonlinear distortion plots of Figure 10.12 and Figure 10.13, it can be observed that the nonlinear distortion is reduced by only 50% under unequal channel spacing between the three channels.

10.4.3 Remarks

In this chapter, the segmented VSTF model is developed for evaluating the transmission performance of wavelength multiplexed systems. The method is also compared with the most popular method of the SSFM in representing the

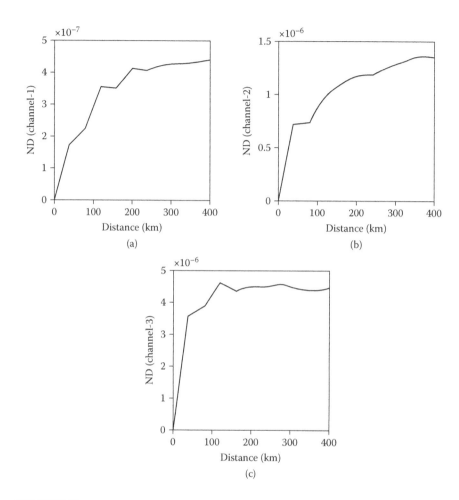

FIGURE 10.13
Nonlinear distortion of the three-channel transmission system: (a) Channel 1, (b) Channel 2, and (c) Channel 3.

propagation of the optical pulse sequence over single-mode dispersive fibers. The mathematical model derived in Section 10.3 can be applied to DWDM systems with any number of channels. For simplicity, a dual-channel WDM system is simulated to demonstrate that the VSTF model can be adequate in design and analysis of a WDM transmission system. It has also been shown that both the bit pattern and the spacing between wavelength channels can affect the XPM effect, the leading contributor to the interchannel cross-talk. In Section 10.4.2, two different channel spacing schemes are used for the simulation. It has been shown that by tuning the carrier frequencies by a small amount to detune the equal channel spacing, the XPM effects can be reduced substantially. This nonlinear distortion is also evaluated with the restrictions defined as the difference between the VSTF and SSFM techniques.

The segmentation of the fiber link is employed to ensure the convergence of the model as VSTF enabling the propagation of the optical pulse sequence over a very long length of fiber with minimum computing resources and time. However, the convergence of the VSTF influences the accuracy of the model. A dispersion criterion is given and coupled with the propagation of the pulses.

10.5 Inverse of Volterra Expansion and Nonlinearity Compensation in Electronic Domain

In this section, we describe compensation and equalization of nonlinear effects of optical signals transmitted over linear chromatic dispersion (CD) and nonlinear single-mode optical fiber. The mathematical representation of the equalization scheme is based on the inverse of the nonlinear transfer function represented by the Volterra series. The implementation of such a nonlinear equalization scheme is in the electronic domain. That is at the stage where the optical signals have been received and converted into the electrical domain as the voltage output of the electronic preamplifier, then digitized by an analog-to-digital converter (ADC) and processed in a digital signal processor.

The electronic nonlinearity compensation scheme based on the inverse of the Volterra expansion [30] is implemented in the electronic domain in a digital signal processor (DSP). It is first reported in Ip and Kahn [33] that 1.2 dB in quadrature (Q) improvement can be achieved with 256 Gb/s polarization division multiplexing (PDM)-16 quadrature amplitude modulation (QAM), and simultaneously reduce the compensation complexity by the reduction of the processing. To meet the ever-increasing demands of the data traffic, improvement in spectral efficiency is desired. Data signals modulate lightwaves via optical modulators using advanced modulation formats and multiplexing using subcarriers or polarization. Multilevel modulation formats such as 16QAM or 64QAM with higher spectrum efficiency are considered for realization of the future target rate of 400 Gb/s or 1Tb/s per channel. However, the multilevel modulation formats require that the received signal has a higher level of optical signal-to-noise ratio (OSNR) that significantly reduces the possible transmission distance. To achieve a higher received OSNR, suppression of the nonlinear penalty is inevitable to keep sufficient optical power from being launched into the transmission fiber. There are several approaches to suppress the nonlinearity such as dispersion management, employing new fibers with larger core diameter, and electronic nonlinear compensation in the digital domain. A typical structure of a digital coherent receiver for quadrature phase shift keying (QPSK) and polarization demultiplexing of optical channels is shown in Figure 10.14. First the optical channels are fed into a 90° hybrid coupler to mix with a local oscillator. Their

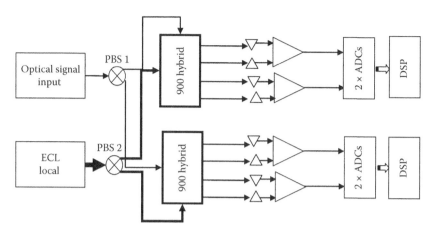

FIGURE 10.14

Typical structure of a digital coherent receiver incorporating analog-to-digital converters (ADCs) and digital processing units. One ADC is assigned per detected channel of the in-phase and quadrature quantities.

polarizations are split so that they can be aligned for maximum efficiency. The opto-electronic device, a balanced pair of photodiodes, converts the optical into electronic currents and then is amplified by an electronic wideband preamplifier. At this stage the signals are in an electronic analog domain. The signals are then conditioned in their analog form (e.g., by an automatic gain control) and converted into digital quantized levels and processed by the digital signal processing unit.

Digital signal processing (DSP) techniques make possible the compensation of large amounts of accumulated chromatic dispersion at the receiver. As a result, we can achieve the benefit of suppressing interchannel nonlinearities in the WDM system by removing inline optical dispersion compensation, and hence reducing inline optical amplifiers and thus increasing or extending the transmission distance. Under this scenario, intrachannel nonlinearity becomes the dominant impairment [22]. Fortunately due to its deterministic nature, intrachannel nonlinearity can be compensated. Several approaches have been proposed to compensate the intrachannel nonlinearity, such as the digital back-propagation algorithm [23], the adaptive nonlinear equalization [24], and the maximum likelihood sequence estimation (MLSE) [25,26]. All of the proposed methods suffer from the difficulty that the implementation complexity is too high, especially their demands on ultrafast memory storage. In this chapter, we propose a new electronic nonlinearity compensation scheme based on the inverse of the Volterra expansion. We show that 1.2 dB in Q improvement can be achieved with 256 Gb/s PDM-16QAM transmission over a fiber link of 1000 km without inline dispersion compensation at 3 dBm launch power. We also simplify the implementation complexity by reducing the nonlinear processing rate. Negligible performance degradation with the same *baud* rate of the modulated data sequence for the nonlinearity compensation can also be achieved.

10.5.1 Inverse of Volterra Transfer Function

The Manakov-polarization mode dispersion (PMD) equations that describe the evolution of optical electromagnetic field envelope in an optical fiber operating in the nonlinear SPM (self-phase modulation) region can be expressed as [27]

$$\frac{\partial A_x}{\partial z} - j\frac{\beta_2}{2}\frac{\partial^2 A_x}{\partial t^2} + \frac{\alpha}{2}A_x = -j\frac{8}{9}\gamma\left(|A_x|^2 + |A_y|^2\right)A_x \tag{10.30}$$

$$\frac{\partial A_y}{\partial z} - j\frac{\beta_2}{2}\frac{\partial^2 A_y}{\partial t^2} + \frac{\alpha}{2}A_y = -j\frac{8}{9}\gamma\left(|A_x|^2 + |A_y|^2\right)A_y \tag{10.31}$$

where A_x and A_y are the electric field envelopes of the optical signals measured relative to the axes of the linear polarized mode of the fiber, β_2 is the second-order dispersion parameter related to the group velocity dispersion (GVD) of the single-mode optical fiber, and α is the fiber attenuation coefficient. Both linear polarization-mode dispersion (PMD) and nonlinear polarization dispersion are not included for simplicity. The first- and third-order Volterra series transfer function for the solution of (10.30) and (10.31) can be written as [28]

$$H_1(\omega) = e^{-(\alpha + j\beta_2\omega^2)L/2} \tag{10.32}$$

$$H_3(\omega_1, \omega_2, \omega) = -\frac{8}{9}\frac{j\gamma}{4\pi^2}H_1(\omega) \times \frac{1 - e^{-(\alpha + j\beta_2(\omega_1 - \omega)(\omega_1 - \omega_2))L}}{\alpha + j\beta_2(\omega_1 - \omega)(\omega_1 - \omega_2)} \tag{10.33}$$

where $j\beta_2(\omega_1 - \omega)(\omega_1 - \omega_2)$ accounts for the impact of the waveform distortion due to the linear dispersion effects within a span. The complex term indicates the evolution of the phase of the carrier under the pulse envelope. The third-order kernel as a function of the nonlinear effects contains the main frequency component and the cross-coupling between different frequency components. Higher transfer functions H_5 can be used as well, but consuming a large chunk of memory may not be possible for ultra-high-speed ADC and DSP chips. Furthermore, the accuracy of the contribution of this order function would not be sufficiently high to warrant its inclusion. When the linear dispersion is extracted, the third-order transfer function can be further simplified as

$$H_3(\omega_1, \omega_2, \omega) \approx -\frac{8}{9}\frac{j\gamma}{4\pi^2} \times \frac{1 - e^{-\alpha L}}{\alpha}H_1(\omega) \tag{10.34}$$

For optically amplified N spans fiber link without using the dispersion compensating module (DCM), the whole fiber transfer functions are given by [29]

$$H_1^{(N)}(\omega) = e^{-j\omega^2 N\beta_2 L/2} \tag{10.35}$$

$$H_3^{(N)}(\omega_1, \omega_2, \omega_3) = -\frac{8}{9}\frac{j\gamma}{4\pi^2} H_1^{(N)}(\omega) \times \frac{1 - e^{-(\alpha + j\beta_2(\omega_1 - \omega)(\omega_1 - \omega_2))L}}{\alpha + j\beta_2(\omega_1 - \omega)(\omega_1 - \omega_2)}$$

$$\sum_{k=0}^{N-1} e^{-jk\beta_2 L(\omega_1 - \omega)(\omega_1 - \omega_2)} \tag{10.36}$$

where $\sum_{k=0}^{N-1} e^{-jk\beta_2 L(\omega_1 - \omega)(\omega_1 - \omega_2)}$ accounts for the waveform distortion at the input of each span. The third-order inverse kernel of this nonlinear system [30] can be obtained as

$$K_1^{(N)}(\omega) = e^{j\omega^2 N\beta_2 L/2} \tag{10.37}$$

$$K_3^{(N)}(\omega_1, \omega_2, \omega_3) = \frac{8}{9}\frac{j\gamma}{4\pi^2} K_1^{(N)}(\omega) \times \frac{1 - e^{-(\alpha + j\beta_2(\omega_1 - \omega)(\omega_1 - \omega_2))L}}{\alpha + j\beta_2(\omega_1 - \omega)(\omega_1 - \omega_2)}$$

$$\sum_{k=0}^{N-1} e^{j(N-k)\beta_2 L(\omega_1 - \omega)(\omega_1 - \omega_2)} \tag{10.38}$$

Taking the waveform distortion at the input of each span into consideration, and ignoring the waveform distortion within a span, (10.38) can be approximated by

$$K_3^{(N)}(\omega_1, \omega_2, \omega_3) \approx \frac{8}{9}\frac{j\gamma}{4\pi^2} \times \frac{1 - e^{-\alpha L}}{\alpha} K_1^{(N)}(\omega) \sum_{k=1}^{N} e^{jk\beta_2 L(\omega_1 - \omega)(\omega_1 - \omega_2)} \tag{10.39}$$

10.5.2 Electronic Compensation Structure

Equations (10.37) and (10.39) can be realized by the scheme shown in Figure 10.15. This structure or algorithm can be implemented in the digital domain in the DSP after the electronic preamplifier and the ADC as described in Figure 10.14. Here $c = \frac{8\gamma}{9} \times \frac{1-e^{-\alpha L}}{\alpha}$ is a constant, and $H_{CD} = e^{j\omega^2\beta_2 L/2}$ compensates the residue dispersion of each span. Figure 10.15a shows the general structure to realize (10.37) and (10.39). The compensation can be separated into a linear compensation part and a nonlinear compensation part. The linear compensation part is simply CD compensation. The nonlinear compensation part can be divided into N stages, where N is the span

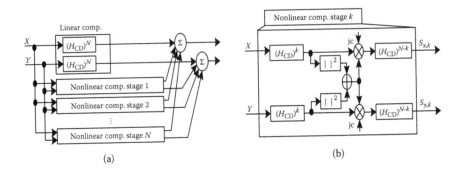

FIGURE 10.15
Electronic nonlinear compensation based on third-order inverse of Volterra expansion to be implemented in electronic digital signal processing: (a) block diagram of the proposed compensation scheme; (b) detailed realization of nonlinear compensation stage k. (Extracted from Ref [33]).

number. The detailed realization of each nonlinear inverse compensating stage is shown in Figure 10.15b, which is a realization of

$$K_{3,k}^{(N)}(\omega_1, \omega_2, \omega_3) \approx \frac{8}{9} \frac{j\gamma}{4\pi^2} \times \frac{1 - e^{-\alpha L}}{\alpha} K_1^{(N)}(\omega) e^{jk\beta_2 L(\omega_1 - \omega)(\omega_1 - \omega_2)} \tag{10.40}$$

$$S_{x,k}(\omega) = jc(H_{CD})^N \int\int_{-\infty}^{\infty} e^{jk\beta_2 L(\omega_1 - \omega)(\omega_1 - \omega_2)} \left[A_x(\omega_1) A_x^*(\omega_2) + A_y(\omega_1) A_y^*(\omega_2) \right] \tag{10.41}$$

$$\times \quad A_x(\omega - \omega_1 + \omega_2) d\omega_1 d\omega_2$$

$$S_{Y,k}(\omega) = jc(H_{CD})^N \int\int_{-\infty}^{\infty} e^{jk\beta_2 L(\omega_1 - \omega)(\omega_1 - \omega_2)} \left[A_x(\omega_1) A_x^*(\omega_2) + A_y(\omega_1) A_y^*(\omega_2) \right] \tag{10.42}$$

$$\times \quad A_Y(\omega - \omega_1 + \omega_2) d\omega_1 d\omega_2$$

$S_{x,k}(\omega)$ and $S_{y,k}(\omega)$ are derived by first passing the received signal of X and Y through $(H_{CD})^k$, and then implementing the nonlinear compensation of $jc(\|_X^2 + \|_Y^2).0_X$ and $jc(\|_X^2 + \|_Y^2).0_Y$ [31]. Finally the residual dispersion can be compensated for by passing through the linear inverse function $(H_{CD})^{N-k}$.

For linear compensation, the processing rate equal to doubling of the baud rate is common, but further reduction of the sampling rate can also be possible [31]. For the nonlinear compensation, it is possible by simulation

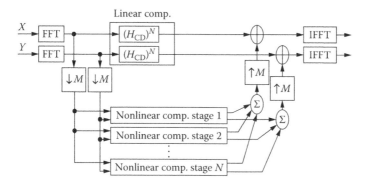

FIGURE 10.16

Simulation system setup and electronic nonlinearity compensation scheme at the digital coherent receiver with a nonlinear dispersion scheme in electronic domain employing the inverse Volterra series algorithm. (Extracted from Ref [33]).

that a single b*aud* rate result is comparable to that of the doubling *baud* rate, hence reduction of the implementation complexity. It is very critical as the DSP at ultrahigh sampling rate is very limited in the number of numerical operations.

The simulation platform is shown in Figure 10.16 [33]. The parameters of the transmission are also shown in the insert of Figure 10.16. The transmission scheme 256 Gb/s NRZ PDM-16QAM with periodically inserted pilots is generated at the transmitter and then transmitted through 13 spans of fiber link. Each span consists of a standard single-mode fiber (SSMF) with a CD coefficient of 16.8 ps/(nm.km), a Kerr nonlinearity coefficient of 0.0014 $m^{-1}W^{-1}$, a loss coefficient of 0.2 dB/km, and an EDFA with a noise figure of 5.5 dB. The span length is 80 km, and we assume 6 dB connector losses at the fiber span output, so the gain of each EDFA is 22 dB. There is also a preamplifier and a postamplifier at the transmitter and receiver to ensure a specified launch power and receiving power to satisfy the operational condition of the fiber nonlinear effects and the sensitivity of the coherent receiver. Random polarization rotation is also considered along the transmission line, but no polarization mode dispersion (PMD) is assumed. In order to have sufficient statistics, the same symbol sequence is transmitted 16 times with a different amplification stimulated emission (ASE) noise realization in the link and a total of 262,144 symbols are used for the estimation of the bit-error rate (BER). The transmitted signal is received with a polarization diversity coherent detector, then sampled at the rate of twice the baud rate and then processed in the digital electronic domain including the nonlinear inverse Volterra algorithm given herewith. The sampled signal first goes through the nonlinear compensator shown in Figure 10.16, where both CD and intrachannel nonlinearity are compensated. The quadruple butterfly-structured finite difference recursive filters and used for polarization demultiplexing and residual distortion compensation.

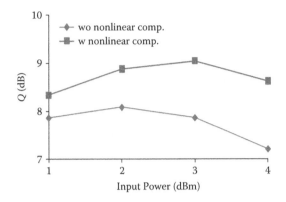

FIGURE 10.17
Performance of the proposed nonlinearity compensator. Quality factor Q versus input launched power with and without nonlinearity compensator. (From L. Liu, L. Li, Y. Huang, and Q. Xiong, Electronic Nonlinearity Compensation of 256 Gb/s PDM-16QAM Based on Inverse of Volterra Expansion, submitted to *IEEE Journal of Lightwave Technology*. With permission.)

Feed-forward carrier phase recovery is carried out before making a decision. With the help of periodically sent pilots, Gray mapping without differential coding can be used to minimize the complexity of the coder. The proposed nonlinear compensation scheme is shown in Figure 10.16. Both linear compensation and nonlinear compensation sections share the same first Fourier transform (FFT) and inverse (I) FFT to reduce the computing complexity. For the linear compensation, the processing rate is twice the *baud rate*. For the nonlinear compensation part, a reduced processing rate is possible to reduce the demand on the processing power of the DSP at extremely high speed.

Figure 10.17 shows the performance of the proposed electronic nonlinear compensator. Figure 10.17 shows that the proposed nonlinearity compensator improves Q by 1.2 dB at 3 dBm launch power. This improvement is significant in terms of sensitivity of the receiver. It was found that [31] there is only negligible performance degradation with a nonlinear processing rate of the baud rate.

10.5.3 Remarks

In this section, an electronic digital processor is proposed with the processing algorithm based on the analytical version of the inverse Volterra series for compensation of the nonlinear effects in single-mode optical fibers in transmission systems. That is the optical modulated signals under polarization multiplexed 16-QAM scheme are transmitted over noncompensating fiber links and received by the optical receiver and sampled by a real-time oscilloscope and stored in memory for off-line processing.

The inverse of the Volterra transfer function of the fiber is derived from the nonlinear transfer function of the Volterra series of single-mode optical transmission lines. We showed that 1.2 dB Q improvement can be achieved with 256 Gb/s PDM-16QAM transmission over a fiber link of 1000 km without inline dispersion compensation and at 3 dBm launch power. The implementation complexity can be simplified with the reduction of the nonlinear processing rate. Negligible performance degradation with the sampling rate equals that of the baud rate for the nonlinearity compensation is also observed. When the high sampling rate ADC and compatible DSP integrated circuit (IC) are available, this algorithm would offer significant advantages in the real-time compensation of nonlinear propagation effects in the fiber transmission line. This electronic processing technique currently emerges as an advanced technology for enhancing the receiver sensitivity in association with a coherent detection method. The inverse Volterra transfer function is a nonlinear processing technique. The convergence of the inverse function must be further investigated to ensure that a solution is obtained within a finite time window.

References

1. J. E. Simsarian, J. Gripp, A. H. Gnauck, G. Raybon, and P. J. Winzer, Fast-Tuning 224-Gb/s Intradyne Receiver for Optical Packet Networks, OFC 2010, Postdeadline paper PDPB5, San Diego, Optical Fiber Conference 2010, March 2010.
2. T. D. Vo, Hao Hu, M. Galili, E. Palushani, J. Xu, L. K. Oxenløwe, S. J. Madden, D. Y. Choi, D. A. P. Bulla, M. D. Pelusi, J. Schröder1, B. Luther-Davies, and B. J. Eggleton, Photonic Chip Based 1.28 Tbaud Transmitter Optimization and Receiver OTDM Demultiplexing, OFC2010 Postdeadline paper PDPC5, Optical Fiber Conference 2010, San Diego, 2010.
3. G. Agrawal, *Nonlinear Fiber Optics*, 3rd ed., Academic Press, New York, 2002.
4. L. N. Binh, *Digital Optical Communications*, CRC Press, Boca Raton, FL, 2009.
5. M. Suzuki, T. Tsuritani, and N. Edagawa, Multi-Terabit Long Haul DWDM Transmission with VSB-RZ Format, *IEEE LEOS Newsletter*, October 2001, pp. 26–28.
6. N. Takachio and H. Suzuki, Application of Raman-Distributed Amplification to WDM Transmission Systems Using 1.55-µm Dispersion-Shifted Fiber, *IEEE J. Lightwave Technol.*, 19(1), 60–69, January 2001.
7. V. Grigoryan and A. Richter, Efficient Approach for Modeling Collision-Induced Timing Jitter in WDM Return-to-Zero Dispersion-Managed Systems, *IEEE J. Lightwave Technol.*, 18(8), 1148–1154, eq.2, August 2000.
8. K. P. Ho, Statistical Properties of Stimulated Raman Crosstalk in WDM Systems, *IEEE J. Lightwave Technol.*, 18(7), 815–921, July 2000.

9. V. Grigoryan and A. Richter, Efficient Approach for Modeling Collision-Induced Timing Jitter in WDM Return-to-Zero Dispersion-Managed Systems, *IEEE J. Lightwave Technol.*, 18(8), 1148–1154, August 2000.

10. K. Chang and L. Binh, A Simulation Platform for Single and Multi-Channel Transmission Systems in Frequency Domain Using Volterra Series Transfer Functions, *Conference Proceedings, WARS02, Workshop on Applications of Radio Science*, D3, Sydney NSW, Australia, 2002.

11. L. N. Binh, Linear and Nonlinear Transfer Functions of Single Mode Fiber for Optical Transmission Systems, *J. Opt. Soc. Am. A*, 26(7), 1564–1575, 2009.

12. K. V. Peddanarappagari and M. Brandt-Pearce, Volterra Series Transfer Function of Single-Mode Fibers, *IEEE J. Lightwave Technol.*, 15(12), 2232–2241, 1997.

13. K. V. Peddanarappagari and M. Brandt-Pearce, Study of Fiber Nonlinearities in Communication Systems Using a Volterra Series Transfer Function Approach, *Proc. 13th Annu. Conf. Inform. Sci-Syst.*, 752–757, 1997.

14. K. V. Peddanarappagari and M. Brandt-Pearce, Volterra Series Approach for Optimizing Fiber-Optic Communications System Designs, *IEEE J. Lightwave Technol.*, 16(11), 2046–2055, 1998.

15. J. Tang, The Channel Capacity of a Multispan DWDM System Employing Dispersive Nonlinear Optical Fibers and an Ideal Coherent Optical Receiver, *IEEE J. Lightwave Technol.*, 20(7), 1095–1101, 2002.

16. J. Tang, A Comparison Study of the Shannon Channel Capacity of Various Nonlinear Optical Fibers, *IEEE J. Lightwave Technol.*, 24(5), 2070–2075, 2006.

17. J. Tang, The Shannon Channel Capacity of Dispersion-Free Nonlinear Optical Fiber Transmission, *IEEE J. Lightwave Technol.*, 19(8), 1104–1109, 2001.

18. B. Xu and M. Brandt-Pearce, Comparison of FWM- and XPM-Induced Crosstalk Using the Volterra Series Transfer Function Method, *IEEE J. Lightwave Technol.*, 21(1), 40–54, 2003.

19. L. N. Binh, On the Linear and Nonlinear Transfer Functions of Single Mode Fiber for Optical Transmission Systems, *J. Opt. Soc. Am. A*, 26(7), 1564–1575, 2009.

20. D. Atherton, *Nonlinear Control Engineering*, Van Nostrand Reinhold, New York, 1982.

21. W. Shieh and I. Djordjevic, *OFDM for Optical Communications*, Academic Press, New York, 2010.

22. E. Yamazaki et al., Mitigation of Nonlinearities in Optical Transmission Systems, OFC2011, OThF1, Optical Fiber Conference OFC 2001, Los Angeles, CA, March 6, 2001.

23. E. Ip and J. M. Kahn, Compensation of Dispersion and Nonlinear Impairments Using Digital Backpropagation, *J. Lightwave Technol.*, 26(20), 3416–3425, October 2008.

24. Y. Gao et al., Experimental Demonstration of Nonlinear Electrical Equalizer to Mitigate Intra-channel Nonlinearities in Coherent QPSK Systems, ECOC2009, Vienna, Austria, Paper 9.4.7.

25. N. Stojanovic et al., MLSE-Based Nonlinearity Mitigation for WDM 112 Gbit/s PDM-QPSK Transmission with Digital Coherent Receiver, OFC2011, paper OWW6, Optical Fiber Conference OFC 2011, Los Angeles, CA, March 6, 2001.

26. L. N. Binh, T. L. Huynh, K. K. Pang, and Thiru Sivahumara, MLSE Equalizers for Frequency Discrimination Receiver of MSK Optical Transmission Systems, *IEEE J. Lightwave Technol.*, 26(12), 1586–1595, June 15, 2008.

27. P. K. A. Wai et al., Analysis of Nonlinear Polarization-Mode Dispersion in Optical Fibers with Randomly Varying Birefringence, OFC'97, paper ThF4, 1997, Optical Fiber Conference OFC 97, Dallas, TX, USA.

28. K. V. Peddanarappagari and M. Brandt-Pearce, Volterra Series Transfer Function of Single-Mode Fibers, *J. Lightwave Technol.*, 15(12), 2232–2241, December 1997.

29. J. K. Fischer et al., Equivalent Single-Span Model for Dispersion-Managed Fiber-Optic Transmission Systems, *J. Lightwave Technol.*, 27(16), 3425–3432, August 2009.

30. M. Schetzen, Theory of *p*th-Order Inverse of Nonlinear Systems, *IEEE Trans. on Circuits and Syst.*, 23(5), 285–291, May 1976.

31. E. Ip and J. M. Kahn, Digital Equalization of Chromatic Dispersion and Polarization Mode Dispersion, *J. Lightwave Technol.*, 25(8), 2033–2043, August 2007.

32. T. Pfau et al., Hardware-Efficient Coherent Digital Receiver Concept with Feedforward Carrier Recovery for M-QAM Constellations, *J. Lightwave Technol.*, 27(8), 989–999, April 2009.

33. L. Liu, L. Li, Y. Huang, and Q. Xiong, Electronic Nonlinearity Compensation of 256 Gb/s PDM-16QAM Based on Inverse of Volterra Expansion, submitted to *IEEE Journal of Lightwave Technology*.

Appendix A: Derivation of the Generalized Nonlinear Schrödinger Equation

A.1 Wave Equation in Nonlinear Optics

In order to derive the wave equation for the propagation of light in a nonlinear optical medium, we begin from the Maxwell's equations for dielectric medium written as follows:

$$\nabla \times \vec{E} = -\frac{\partial \vec{B}}{\partial t} \tag{A.1}$$

$$\nabla \times \vec{H} = \frac{\partial \vec{D}}{\partial t} \tag{A.2}$$

$$\nabla \cdot \vec{D} = 0 \tag{A.3}$$

$$\nabla \cdot \vec{B} = 0 \tag{A.4}$$

where $\vec{E}$, $\vec{H}$ are the electric and magnetic fields, respectively. The electric and magnetic flux densities $\vec{D}$ and $\vec{B}$, respectively, are related to the electric and magnetic fields via

$$\vec{D} = \varepsilon_0 \vec{E} + \vec{P} \tag{A.5}$$

$$\vec{B} = \mu_0 \vec{H} \tag{A.6}$$

where ε_0 is the permittivity of free space, μ_0 is the permeability of free space, and $\vec{P}$ is the induced electric polarization.

By taking curl of (A.4) and using (A.3) we obtain

$$\nabla \times \nabla \times \vec{E} = -\mu_0 \nabla \times \frac{\partial \vec{H}}{\partial t} \tag{A.7}$$

Substitution of (A.2) into (A.7) yields the generic wave equation:

$$\nabla \times \nabla \times \vec{E} = -\frac{1}{c^2}\frac{\partial^2 \vec{E}}{\partial t^2} - \mu_0 \frac{\partial^2 \vec{P}}{\partial t^2} \tag{A.8}$$

where c is the speed of light in a vacuum and is given by $c = 1/\sqrt{\varepsilon_0 \mu_0}$, the induced polarization consists of linear and nonlinear components as

$$\vec{P} = \vec{P}_L + \vec{P}_{NL} \tag{A.9}$$

which are defined as

$$\vec{P}_L(\vec{r},t) = \varepsilon_0 \int_{-\infty}^{\infty} \chi^{(1)}(t-t') \cdot \vec{E}(\vec{r},t')dt' \tag{A.10}$$

$$\vec{P}_{NL}(\vec{r},t) = \varepsilon_0 \iint_{-\infty}^{\infty}\int \chi^{(3)}(t-t_1,t-t_2,t-t_3) \times \vec{E}(\vec{r},t_1)\vec{E}(\vec{r},t_2)\vec{E}(\vec{r},t_3) \tag{A.11}$$

where $\chi^{(1)}$ and $\chi^{(3)}$ are the first- and third-order susceptibility tensors.

A.2 Generalized Nonlinear Schrödinger Equation (NSE)

The starting point for the derivation of the NSE is the wave equation (A.8). In order to cover a larger number of nonlinear effects, the general form of nonlinear polarization in (A.11) must be used, and an approximation of the $\chi^{(3)}$ is given by

$$\chi^{(3)}(t-t_1,t-t_2,t-t_3) = \chi^{(3)}R(t-t_1)\delta(t-t_2)\delta(t-t_3) \tag{A.12}$$

where $R(t)$ is the nonlinear response function normalized in a manner similar to the delta function (i.e., $\int_{-\infty}^{\infty} R(t)dt = 1$). By introducing (A.12) into (A.11) with the slowly varying approximation, the nonlinear polarization in scalar form is given by

$$P_{NL}(\vec{r},t) = \frac{3}{4}\varepsilon_0\chi^{(3)}_{xxxx}E(\vec{r},t)\int_{-\infty}^{\infty} R(t-t_1)|E(\vec{r},t)|^2 dt_1 \tag{A.13}$$

The assumptions in (A.13) are also applied for simplification of the NSE derivation. It will be clearer to describe numerous effects in frequency domain with using the following Fourier transforms:

$$E(\vec{r},\omega) = \int_{-\infty}^{\infty} E(\vec{r},t)\exp[j(\omega-\omega_0)t]dt = FT\{E(\vec{r},t)\} \tag{A.14}$$

$$E(\vec{r},t) = \frac{1}{2\pi}\int_{-\infty}^{\infty} E(\vec{r},\omega)\exp[-j(\omega-\omega_0)t]d\omega = FT^{-1}\{E(\vec{r},\omega)\} \tag{A.15}$$

In the frequency domain, the convolution of Equation (A.15) becomes a simple multiplication and the time derivatives can be replaced by $\partial/\partial t \rightarrow -j\omega$ and $\partial^2/\partial t^2 \rightarrow -\omega^2$. Hence, a modified Helmholtz equation can be derived from (A.9) by using (A.10) and (A.13) and taking the Fourier transform to give

$$\nabla^2 E(\vec{r},\omega) + \varepsilon_L(\omega)\frac{\omega^2}{c^2}E(\vec{r},\omega) = -\mu_0\omega_0^2\left(1+\frac{\omega}{\omega_0}\right)P_{NL}(\vec{r},\omega) \tag{A.16}$$

where $P_{NL}(\omega)$ are Fourier transforms of P_{NL} in the time domain. Equation A.16 can be solved by using the method of separation of variables. The slowly varying part of the electric field $E(\vec{r},\omega)$ is approximated by

$$E(\vec{r},\omega) = F(x,y)A(z,\omega)\exp(j\beta_0 z) \tag{A.17}$$

where $A(z,\omega)$ is the slowly varying function of z, β_0 is the wave number, and $F(x,y)$ is the function of the transverse field distribution that is assumed to be independent of ω. Then substituting (A.17) into (A.16), the Helmholtz equation is split into two equations:

$$\frac{\partial^2 F}{\partial x^2} + \frac{\partial^2 F}{\partial y^2} + \left[\varepsilon_L(\omega)\frac{\omega^2}{c^2} - \beta^2\right]F = 0 \tag{A.18}$$

$$F(x,y)\left[2j\beta_0\frac{\partial A(z,\omega)}{\partial z} + \left(\beta^2-\beta_0^2\right)A(z,\omega)\right] = \mu_0\omega_0^2\left(1+\frac{\omega}{\omega_0}\right)P_{NL}(\vec{r},\omega) \tag{A.19}$$

Equation (A.18) is an eigenvalue equation that needs to be solved for the wave number β and the fiber modes. In (A.18), $\varepsilon_L(\omega)$ can be approximated by $\varepsilon_L \approx n^2 + 2n\Delta n$, where Δn is a small perturbation and can be determined from $\Delta n = j\alpha c/(2\omega)$.

In the case of a single mode, (B.19) can be solved using the first-order perturbation theory in which Δn does not affect $F(x,y)$, but only the eigenvalues. Hence, β in (A.19) becomes $\beta(\omega)+\Delta\beta$, where $\Delta\beta$ accounts for the effect of the perturbation term (referred to Δn) to change the propagation constant for the fundamental mode. Using (A.17) and (A.15), the electric field can be approximated by

$$E(\vec{r},t) = F(x,y)A(z,t)\exp(j\beta_0 z) \tag{A.20}$$

where $A(z,t)$ is the slowly varying complex envelope propagating along z in the optical fiber. From (A.19) after integrating over x and y, the following equation in the frequency domain is obtained:

$$\frac{\partial A(z,\omega)}{\partial z} = j\left(\beta(\omega)-\beta_0\right)A(z,\omega)-\frac{\alpha(\omega)}{2}A(z,\omega)+j\gamma(\omega)\left(1+\frac{\omega}{\omega_0}\right) \tag{A.21}$$

$$\times \int\int_{-\infty}^{\infty} R(\omega_1-\omega_2)A(z,\omega_1)A^*(z,\omega_2)A(z,\omega-\omega_1+\omega_2)d\omega_1\,d\omega_2$$

where $R(\omega)$ is the Fourier transform of $R(t)$, the nonlinear coefficient γ that has been introduced is given by

$$\gamma = \frac{n_2\omega_0}{cA_{eff}} = \frac{2\pi n_2}{\lambda A_{eff}} \tag{A.22}$$

and where A_{eff} is the effective area of the optical fiber and given by

$$A_{eff} = \frac{\left(\int_{-\infty}^{\infty}|F(x,y)|^2\,dx\,dy\right)^2}{\int\int_{-\infty}^{\infty}|F(x,y)|^4\,dx\,dy} \tag{A.23}$$

Equation (A.21) is the NSE that describes generally the pulse propagation in the frequency domain. It is useful to take into account the frequency dependence of the effects of the propagation constant β, the loss α, and the nonlinear coefficient γ by expanding them in the Taylor series as

$$\beta(\omega) = \beta_0 + \beta_1(\omega-\omega_0)+\tfrac{1}{2}\beta_2(\omega-\omega_0)^2+\tfrac{1}{6}\beta_2(\omega-\omega_0)^3+\cdots$$

$$= \sum_{n=0}^{\infty}\frac{\beta_n}{n!}(\omega-\omega_0)^n \tag{A.24}$$

$$\alpha(\omega) = \sum_{n=0}^{\infty} \frac{\alpha_n}{n!} (\omega - \omega_0)^n \tag{A.25}$$

$$\gamma(\omega) = \sum_{n=0}^{\infty} \frac{\gamma_n}{n!} (\omega - \omega_0)^n \tag{A.26}$$

However, the pulse spectrum in most cases of practical interest is narrow enough such that γ and α are constant over the pulse spectrum. Therefore, the NSE in the time domain can be obtained by using the inverse Fourier transform:

$$\frac{\partial A(z,t)}{\partial z} + \frac{\alpha}{2} A(z,t) - j \sum_{n=1}^{\infty} \frac{j^n \beta_n}{n!} \frac{\partial^n A(z,t)}{\partial t^n} = j\gamma \left(1 + \frac{j}{\omega_0} \frac{\partial}{\partial t} \right)$$

$$\times A(z,t) \int_{-\infty}^{\infty} R(t') \left| A(z,t-t') \right|^2 dt' \tag{A.27}$$

Equation (A.27) is the basic propagation equation, commonly known as the generalized nonlinear Schrödinger equation (NLSE) that is very useful for studying the evolution of the amplitude of the optical signal and the phase of the lightwave carrier under most effects of third-order nonlinearity in optical waveguides as well as optical fibers.

Appendix B: Calculation Procedures of Triple Correlation, Bispectrum, and Examples

Triple Correlation and Bispectrum Estimation

Definitions of triple correlation and bispectrum for continuous signal $x(t)$ are given in Chapter 6. However, calculation of both triple correlation and bispectrum is normally achieved in the discrete domain. Thus, the discrete triple correlation is estimated as follows:

$$C_3(\tau_1, \tau_2) = \sum_k x(kdt)x(kdt - \tau_1)x(kdt - \tau_2) \tag{B.1}$$

where $x(kdt)$ is the discrete version of $x(t)$, k is integer number, $dt = 1/f_s$ is the sampling period, f_s is the sampling frequency, and the delay variables are also discretized as $\tau_i = mdt$, $m = 0,1,2 \ldots$, N/2-1.

Similarly, the discrete bispectrum is estimated by the discrete Fourier transform of C_3:

$$B_i(f_1, f_2) = \sum_m \sum_n C_3(mdt, ndt)\exp(-2\pi j(f_1 mdt + f_2 ndt)) \tag{B.2}$$

where m,n are integers and the frequency variables $f_i = Kdf$, $K = 0,1,2\ldots$ N/2–1 and the frequency resolution $df = 2/(Ndt)$, N is the total number of samples in each computation window.

In this thesis, the following steps are used to estimate the bispectrum:

- A discrete process or signal is divided into M computation frames in which the number of samples N in each frame chosen is 1024. The sampling time dt is properly selected to ensure that the significant frequency components of $x(kdt)$ are in the range from $-B$ to B, where $B = 1/(2dt)$.

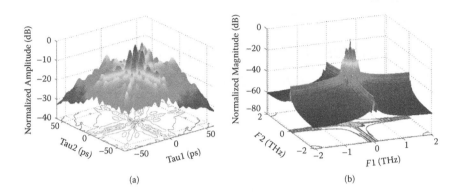

FIGURE B.1

(a) Triple correlation and (b) magnitude bispectrum (Bottoms: corresponding contour representations).

- The triple correlation of each data frame is computed by using (B.1). The result obtained is an array with the size 512 × 512. Each value in the array is represented for the amplitude of the triple correlation.
- The bispectrum of each frame is then calculated by using the discrete Fourier transform (B.2). Thus, the bispectrum is also an array with the same size. Finally, the bispectrum is averaged over M data frames via the following expression:

$$B(f_1, f_2) = \frac{1}{M} \sum_{i=1}^{M} B_i(f_1, f_2) \tag{B.3}$$

- Both the triple correlation and the bispectrum can be displayed in a three-dimensional graph as shown in Figure B.1 in which the magnitude is normalized in a logarithmic scale. However, the contour representation is selected to display effectively the variation in the bispectrum structure.

Properties of Bispectrum

Important properties of the bispectrum are briefly summarized as follows [1]:

- The bispectrum is generally complex. It contains both magnitude and phase information that is important for signal recovery as well as identifying nonlinear response and processes.

- The bispectrum has the lines of symmetry $f_1 = f_2$, $2f_1 = -f_2$ and $2f_2 = -f_1$ corresponding to permutation of the frequencies f_1, f_2.
- The bispectrum of a stationary, zero-mean Gaussian process is zero. Thus, a nonzero bispectrum indicates a non-Gaussian process.
- The bispectrum suppresses linear phase information or constant phase shift information.
- The bispectrum is flat for non-Gaussian white noise and zero for Gaussian white noise.

Bispectrum of Optical Pulse Propagation

In this section, propagation of optical pulses through optical fiber as an example is characterized and analyzed by the triple correlation and the bispectrum. Figures B.2 and B.3 show, respectively, the triple correlations

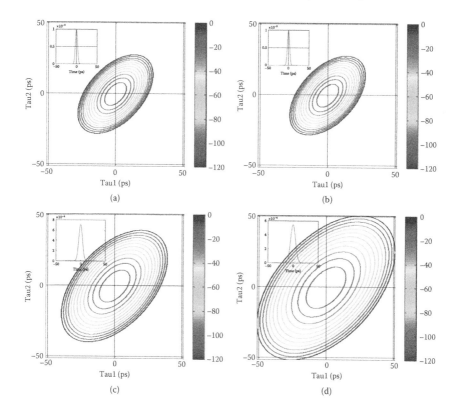

(a)

(b)

(c)

(d)

FIGURE B.2
Triple correlations of Gaussian pulse propagating in the fiber with $b_2 = -21.6 \ \text{ps}^2/\text{km}$ at different distances: (a) $z = 0$, (b) $z = 50 \ \text{m}$, (c) $z = 650 \ \text{m}$, (d) $z = 1 \ \text{km}$ (Insets: the waveforms in time domain).

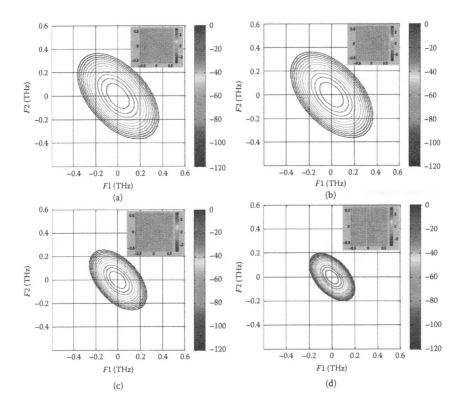

FIGURE B.3

Corresponding bispectra of Gaussian pulse propagating in the fiber at different distances: (a) $z = 0$, (b) $z = 50$ m, (c) $z = 650$ m, (d) $z = 1$ km (Insets: the corresponding phase bispectra).

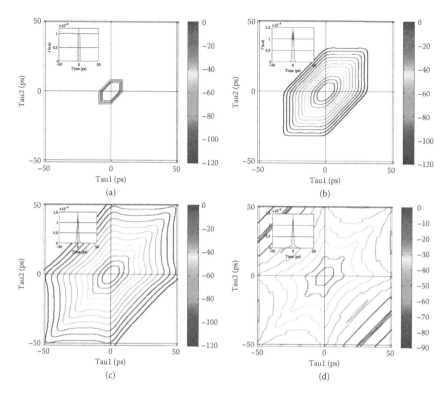

FIGURE B.4

Triple correlations of a super-Gaussian pulse propagating in the fiber with $b_2 = -21.6$ ps^2/km at different distances: (a) $z = 0$, (b) $z = 50$ m, (c) $z = 100$ m, (d) $z = 300$ m (Insets: the waveforms in time domain).

and the bispectra of the 6.25 ps Gaussian pulse propagating at different lengths of the optical fiber with the second-order group velocity dispersion (GVD) coefficient $b_2 = -21.6$ ps^2/km. Figures B.4 and B.5 show, respectively, the triple correlations and the bispectra of the 6.25 ps super-Gaussian pulse propagating at different lengths of the same fiber.

More important, the triple correlation can detect easily the asymmetrical distortion of the pulse that is impossible in an autocorrelation estimation. Figures B.6 and B.7 show, respectively, the triple correlations and the bispectra of the super-Gaussian pulse propagating through the fiber with the third-order dispersion coefficient $b_3 = 0.133$ ps^3/km.

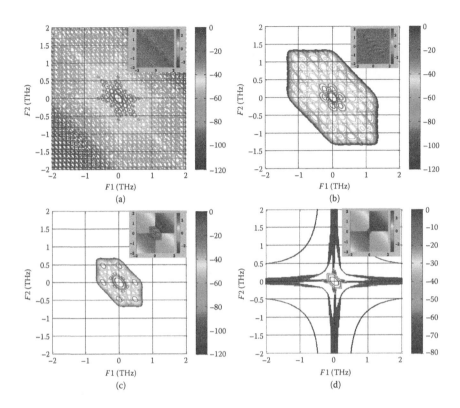

FIGURE B.5
Corresponding bispectra of the super-Gaussian pulse propagating in the fiber at different distances: (a) $z = 0$, (b) $z = 50$ m, (c) $z = 100$ m, (d) $z = 300$ m (Insets: the corresponding phase bispectra).

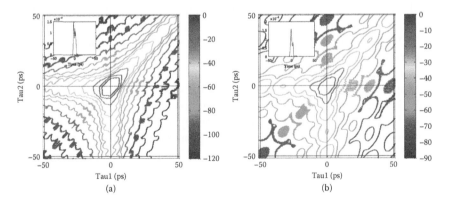

FIGURE B.6
Triple correlations of a super-Gaussian pulse propagating in the fiber with $b_2 = 0$, $b_3 = 0.133$ ps^3/km at different distances: (a) $z = 100$ m, (b) $z = 500$ m (Insets: the waveforms in time domain).

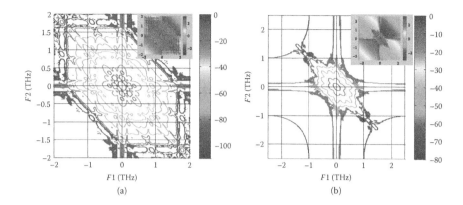

FIGURE B.7
Corresponding bispectra of the super-Gaussian pulse propagating in the fiber at different distances: (a) $z = 100$ m, (b) $z = 500$ m (Insets: the corresponding phase bispectra).

Reference

1. C. L. Nikias and M. R. Raghuveer, Bispectrum Estimation: A Digital Signal Processing Framework, *Proc. IEEE,* 75, 869–891, 1987.

Appendix C: Simulink® Models

C.1 MATLAB® and Simulink® Modeling Platforms

A modeling and simulation platform for optical fiber transmission systems has been developed using MATLAB and Simulink, an environment for simulation and model-based design [1]. There are some advantages of the Simulink modeling platform:

- Subsystem blocks for a complicated transmission system can be set up from the basic blocks available in the toolboxes and block-sets of Simulink. It is noted that there are no block-sets for optical communication in Simulink. Therefore, the main functional blocks of the optical communication system in the Simulink platform have been developed for years. Details of operational principles as well as examples of the optical components and transmission systems can be found in Binh [1].
- Signals can be easily monitored at any point along the simulation system by available scopes in Simulink block-sets.
- Numerical data can be stored for postprocessing in MATLAB to estimate the performance of the system.

An example of an optical fiber transmission system is shown in Figure C.1. Depending on different problems or targets, various Simulink models are set up for investigation in this thesis.

C.2 Wavelength Converter in Wavelength-Division Multiplexing System

See Figures C.2 and C.3.

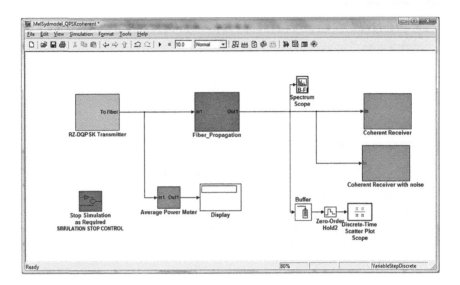

FIGURE C.1

An example of an optical fiber transmission system consisting of main blocks: optical transmitter, fiber transmission link, and optical receiver.

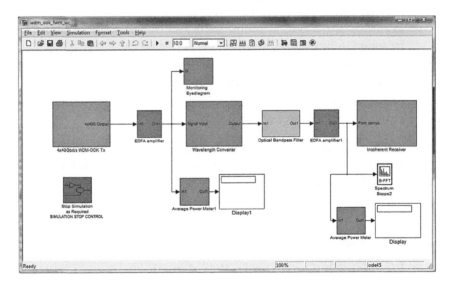

FIGURE C.2

The Simulink model of optical parametric amplifier used as a wavelength converter in the wavelength-division multiplexing system.

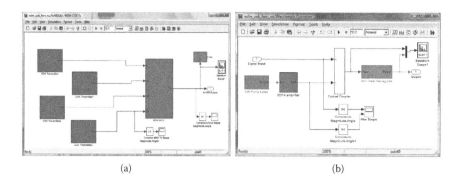

(a)   (b)

FIGURE C.3

(a) Simulink model of wavelength-division multiplexing (WDM) transmitter consisting of four optical transmitters at different wavelengths and a wavelength multiplexer. (b) Simulink setup of the parametric amplifier using the model of nonlinear waveguide used for wavelength conversion in a WDM system.

C.3 Nonlinear Phase Conjugation for Mid-link Spectral Inversion

See Figures C.4 through C.7.

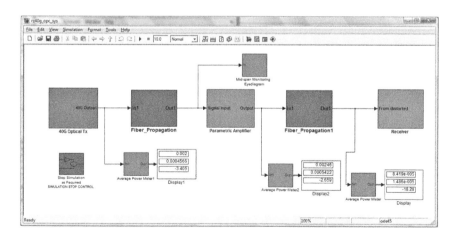

FIGURE C.4

Simulink setup of a long-haul 40 Gbit/s transmission system using nonlinear phase conjugation for distortion compensation.

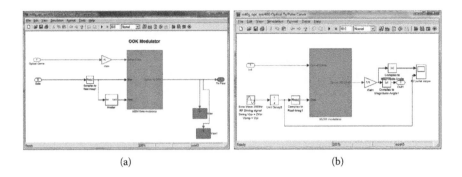

(a)   (b)

FIGURE C.5
(a) Simulink model of an intensity optical modulator driven by data. (b) Simulink model of an optical pulse carver driven by a sinusoidal signal for RZ pulse generation.

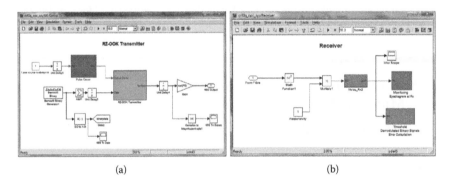

(a)   (b)

FIGURE C.6
(a) Simulink model of an optical transmitter for RZ-OOK modulation scheme. (b) Simulink model of an optical receiver for on-off keying signal.

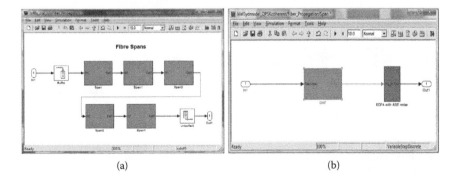

(a)   (b)

FIGURE C.7
(a) Simulink model of each fiber transmission section consisting of 5 spans. (b) Simulink model of each span consisting of one single-mode fiber and one erbium-doped fiber amplifier for loss compensation.

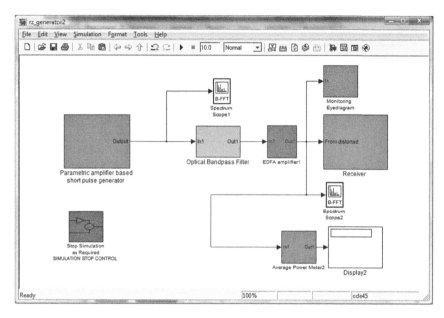

FIGURE C.8
Simulink setup of a short pulse generator at 40 GHz based on the parametric amplifier.

C.4 Pulse Generator

See Figures C.8 and C.9.

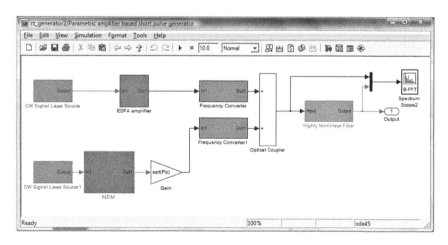

FIGURE C.9
Simulink blocks inside the 40 GHz short-pulse generator to demonstrate ultra-high-speed switching based on parametric amplification.

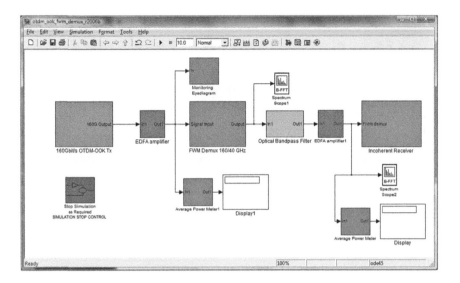

FIGURE C.10
Simulink setup of the 160 Gbit/s optical time-division multiplexing system using four-wave mixing process for demultiplexing.

C.5 Optical Time-Division Multiplexing (OTDM) Demultiplexer

For the OTDM system using an on-off keying (OOK) scheme, the Simulink models of the transmitter and receiver are similar to those shown in Figure C.6 with using a mode-locked fiber laser (MLFL) instead of a carver. Figure C.12 shows the Simulink models of the transmitter and receiver in the OTDM system using the differential quadrature phase shift keying (DQPSK) modulation scheme. (See Figures C.10 through C.13.)

C.6 Triple Correlation

Figure C.14 shows the Simulink model for the triple correlation based on four-wave mixing (FWM) in the nonlinear waveguide. The structural block consists of two variable delay lines to generate delayed versions of the original signal as shown in Figure C.15 and frequency converters to convert the signal into three different waves before combining at the optical coupler to launch into the nonlinear waveguide. (See Figures C.14 through C.18.)

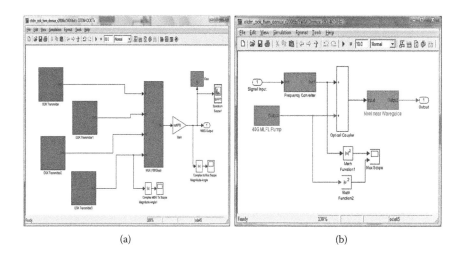

(a)   (b)

FIGURE C.11
(a) Simulink model of optical time-division multiplexing (OTDM) transmitter consisting of four optical transmitters and an OTDM multiplexer. (b) Simulink model of the four-wave mixing (FWM)-based demultiplexer.

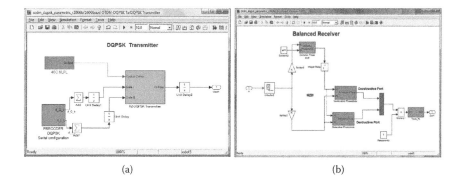

(a)   (b)

FIGURE C.12
(a) Simulink model of an optical transmitter for differential quadrature phase shift keying (DQPSK) modulation scheme. (b) Simulink model of an optical balanced receiver for DQPSK signal.

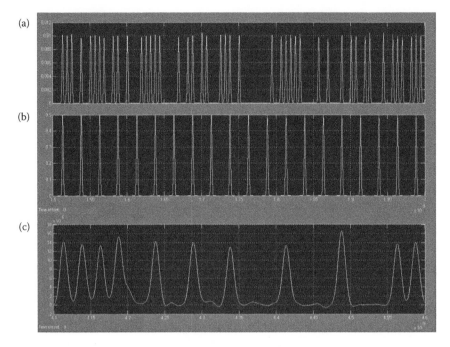

FIGURE C.13
Time traces of (a) the 160 Gbit/s optical time-division multiplexing signal, (b) the control signal, and (c) the 40 Gbit/s demultiplexed signal.

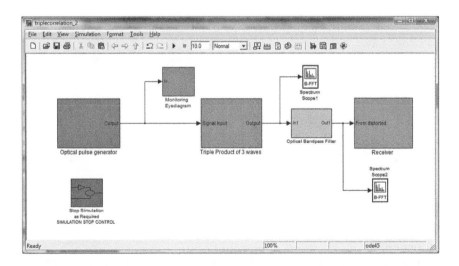

FIGURE C.14
Simulink setup for investigation of the triple correlation based on the four-wave mixing process.

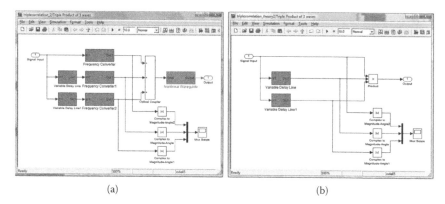

FIGURE C.15
(a) Simulink setup of the four-wave mixing (FWM)-based triple-product generation. (b) Simulink setup of the theory-based triple-product generation.

FIGURE C.16
The variation in time domain of the time delay variable, the original signal, and the delayed signal.

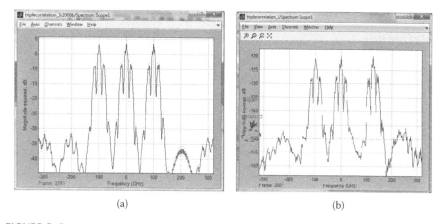

FIGURE C.17
(a) Spectrum with equal wavelength spacing at the output of the nonlinear waveguide. (b) Spectrum with unequal wavelength spacing at the output of the nonlinear waveguide.

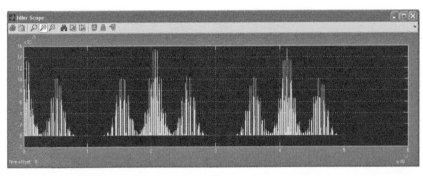

(a)

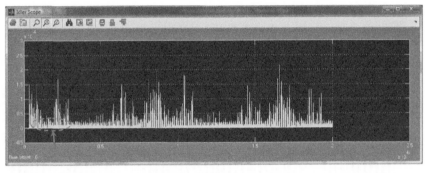

(b)

FIGURE C.18
Generated triple-product waves in time domain of the dual-pulse signal based on (a) theory
and (b) four-wave mixing in a nonlinear waveguide.

Reference

1. L. N. Binh, *Optical Fiber Communications Systems: Theory and Practice with
MATLAB® and Simulink® Models,* CRC Press, Boca Raton, FL, 2010.

Index